Small and Medium-Sized Enterprises and the Environment

Business Imperatives

Contributing Editor: Ruth Hillary

Small and Medium-Sized Enterprises and the Environment

BUSINESS IMPERATIVES

2000

To John Worontschak
and our lovely sons,
Max and Zak

Published by Greenleaf Publishing Limited
Aizlewood's Mill
Nursery Street
Sheffield S3 8GG
UK

Typeset by Greenleaf Publishing Limited and printed on environmentally friendly, acid-free paper from managed forests by Bookcraft, Midsomer Norton, UK.

British Library Cataloguing in Publication Data:

Small and medium-sized enterprises and the environment :
business imperatives
1.Small business - Environmental aspects 2.Social
responsibility of business
I.Hillary, Ruth
658'.022
ISBN 1874719225

Contents

Section 2: **Environmental Management in the Smaller Firm**

Section 3: **Practical Strategies for Reaching Small and Medium-Sized Enterprises**

Foreword

Chris Fay
Chairman, Advisory Committee on
Business and the Environment (ACBE), UK

Small and medium-sized enterprises—those classed as employing fewer than 250—have become one of the fastest-growing sections of the business community. Of the 3.7 million businesses in the UK, small businesses account for 99%; they account for 58% of all employment and for 38% of GDP. Small businesses are unquestionably a very important component of the UK economy. But they also account for their share of pollution, waste and other unsustainable practices.

Many small businesses do not perceive their own environmental impacts as significant when set against those of larger operators. But, collectively, they are. It is therefore important, both in terms of environmental protection and sustainable development more generally, that they too are encouraged and helped to improve their performance.

Raising their awareness and stimulating appropriate action is no easy task. By their nature, smaller businesses have limited capacity to deal with the many issues that confront them on a daily basis. In this context the priority will always be to address those concerns seen as having a more immediate bearing on their survival: their competitiveness, the skills and training of staff, productivity, interest rates, etc. However, as the environmental performance of business generally is increasingly being subjected to scrutiny by regulators, customers, employees, insurers, funders and the local community, doing nothing is no more an option for smaller businesses than it is for larger ones. What approach to take and how to set about it are now the critical questions.

There is a limit on how much can be achieved through traditional 'command-and-control' of regulatory systems or through fiscal measures. This needs to be complemented by voluntary market-based approaches. The trick is finding mechanisms that are appropriate for the size of the business. In the context of smaller businesses, size really does matter. The one-size-suits-all approach will not work. Methods of raising awareness and the types of mechanism or incentive to encourage and help them to improve their environmental performance must be proportionate to the size and complexity of the operation. Nor is this challenge peculiar to the UK.

In *SMEs and the Environment*, the complexities of how to engage small businesses in environmental stewardship are addressed by contributions from an international set of authors from academia, multilateral organisations and business. The book provides an excellent résumé of the issues and is peppered with practical ideas on how to achieve success.

Introduction

Ruth Hillary
Network for Environmental
Management and Auditing, UK

Small and medium-sized enterprises (SMEs) are the most important sector of a nation's economy. They provide and create jobs, especially during times of recession; they are a source of innovation and entrepreneurial spirit; they harness individual creative effort; and they create competition and are the seed bed for businesses of the future. In short, small and medium-sized firms are vitally important for a healthy dynamic market economy.

The SME sector is vast. In the UK alone 99.8% of all enterprises fall into this sector. In Europe around 90% of all enterprises are small or medium-sized. The percentages are similar in countries all over the world. And their numbers are set to increase. Advances in technology that allow more flexible production methods, the reorganisation, downsizing and outsourcing by large firms and the increase in franchising and self-employment will all result in greater numbers of small firms. But these crucial businesses are also plagued by problems: they are more likely to fail or experience stunted growth than large firms; they suffer more from financial problems, such as late payment of bills and access to loan finance; they can find it difficult to adapt to changing markets; and they lack the human and financial resources to tackle new pressures such as environmental regulation and stakeholders' concern about their environmental impacts.

The total environmental impact of SMEs is unknown. A figure of 70% is bandied around as SMEs' contribution to pollution levels—a figure that has taken on mythical status. Although unsubstantiated, it is quoted widely, as if it carries some authority. Generally, national economic statistics on SMEs do not tally with data collected on emissions, waste generation and effluents from firms, so it is doubtful whether smaller firms' contribution to pollution can be calculated at all. In fact, there is little hard data to determine the sector's contribution to pollution load. Collectively, their sheer numbers may mean their environmental impacts are substantial. While at a national level their combined impact is unknown, in pollution terms their significance at a local level can be important. Many smaller firms, especially those in developing countries, are characterised by their use of older technologies, their lack of awareness of legislation and of their own environmental impacts and their less structured management of such issues—all of which means that their impact on local ecosystems and communities can be potent.

The sector is under-researched. Little is known about its attitudes to and control of its environmental impacts. Its very vastness and importance for a healthy global economy (in the EU alone, the SME sector accounts for 70% of economic activity) means that its survival and growth is crucial. Its unquantified contribution to pollution load and unknown management of its environmental impacts means it has a critical role to play in contributing to sustainable development. This book goes some way to filling a yawning information gap by drawing experiences on SMEs from over 28 different business sectors in 18 different countries. The book is important in that it characterises the sector's diverse reactions to the environmental issues it faces. Its contributors understand the potential opportunities to be derived from engaging SMEs in the drive towards sustainable development.

What is an SME?

Any book on SMEs has to meet head on the use and abuse of the term 'small or medium-sized enterprise'. 'SME' is a catch-all term, bandied around whenever non-large firms are being discussed. At present, its popularity is high among policy-makers, support organisations, government and researchers—in fact, many groups except probably the most important group of all: SMEs themselves. Size matters, but daily businesses' interactions are normally couched in terms of customers and suppliers, profit margins and cash flow, growth and markets. Quite simply, SMEs regard themselves as businesses. Why, then, define an enterprise by size at all? What sort of categorisations are used and are they appropriate—especially since the sector is not a homogeneous group of companies?

Definitions for 'SME' are very blunt instruments when understanding the variety of businesses in the sector. There are broadly two categories of definition: operational definitions and theoretical definitions. Those of an operational nature are used for working purposes, e.g. to provide a cut-off level in the award of grants (see examples in Appendix 1, page 20), and those of a theoretical nature are employed to characterise the sector (see examples in Appendix 2, page 22). The failing of all of these definitions is that they cannot take into account the undeniable importance of the sector's diversity.

A degree of convergence in SME definitions has been brought about, in Europe at least, by the introduction in 1996 of the EU SME definition.[1] This uses a combination

1 The EU defines SMEs based on employee numbers, turnover or balance sheet total and ownership (CEC 1996b).
An *SME:*

of employee numbers, turnover or balance sheet total, and ownership to classify enterprises. While this is certainly a welcome move for the standardisation of data collection on enterprises (a major reason for defining SMEs), it does little to help us understand the diversity of the sector. In fact, harmonising definitions may obscure characteristics that more varied definitions try to draw out.

In this book the reader will come across a range of definitions for SMEs. This is totally appropriate because there is no single right definition of what an SME is and because the information in this book is drawn from such a wide range of sectors and countries. In employee numbers, for example, Danish SMEs have fewer than 500 staff (in the Pedersen study, Chapter 16), whereas the Mexican and Brazilian small and medium firms investigated by Dasgupta *et al.* in Chapter 23 have between 1 and 100 staff, and those in Australia in the Gerrans and Hutchinson chapter (Chapter 5) employee fewer than 200 people. In Bratasida's Indonesian case study (Chapter 24), SMEs are not defined by employee numbers at all but by assets and are classified as 'small' if they have less than $800,000-worth of assets. In contrast, Whalley (Chapter 9) suggests it is business culture not employee numbers or turnover that defines a small firm.

The variety of different ways of defining small and medium-sized organisations is relevant as we grapple with a useful way of grouping the organisations we wish to discuss. Studies that seek to investigate the sector as a whole and draw conclusions about it are comparing not just apples and pears but all the fruit in the fruit bowl. The SME definition changes depending on the sectorial or country context; a more sensitive method needs to be developed. It is recommended that, in future, the sector is considered in parts either as sub-groups by size—i.e. micro, small and medium—or by industrial sector.

About the book

Small and Medium-Sized Enterprises and the Environment: Business Imperatives begins with a section on the mind-sets of small and medium-sized firms. It details the attitudes, perceptions and behaviour of smaller firms to the environment and sustainability. A national survey of UK small and medium-sized companies by Smith *et al.* in Chapter 1 provides the opening snapshot of their attitudes. It highlights the sector's low level of awareness of environmental issues and the potential business benefits of positive

---

- Has fewer than 250 employees, and either
- An annual turnover not exceeding ECU 40 million, or
- An annual balance sheet total not exceeding ECU 27 million, and
- Is an independent enterprise, i.e. 25% or more of the capital or voting rights cannot be owned by a larger enterprise/s.

A *small enterprise:*

- Has fewer than 50 employees, and either
- An annual turnover not exceeding ECU 7 million, or
- An annual balance sheet total not exceeding ECU 5 million, and
- Is an independent enterprise, i.e. 25% or more of the capital or voting rights cannot be owned by a larger enterprise/s.

environmental management. Overridingly, small firms view their environmental impact in proportion to their size, i.e. small, and are woefully unaware of applicable environmental legislation.

Petts delves deeper into the attitudes of UK SMEs to environmental compliance in Chapter 2, revealing not just the attitudes of management but also non-management. She suggests smaller firms are 'vulnerably compliant'. Management in these firms believes that compliance is the right thing to achieve; but the workers have a more jaded view, believing that 'things are tidied up' before the inspector arrives. In fact, non-management thinks environmental legislation should be tougher. Both groups think self-regulation is an opportunity for the slacker firms to do nothing.

This view of self-regulation fostering and encouraging the 'cowboy' element in the sector is backed up by Tilley's research into the neglected area of small firms' environmental ethics (Chapter 3). She makes the point that small firms are not little versions of big firms and that their environmental attitudes and behaviour are distinctive and non-uniform. These companies are on the sidelines of the environmental debate, falling below or outside compliance requirements of environmental legislation, their environmental behaviour governed by regulation not an environmental ethos.

These first three chapters highlight a common characteristic of SMEs: i.e. their preference for environmental legislation to control environmental impacts and their sceptical view of self-regulation and the opportunities it might offer. Spanish SMEs prove to be little different to UK ones (Ludevid Anglada, Chapter 4). They perceive little opportunity from tackling environmental issues and have jaundiced views of fellow businesses, which they fear practise widespread 'environmental dumping' by moving to regions and countries with the lowest environmental standards. Environmental destruction is attributed to overpopulation and poverty and these Spanish companies believe developed countries are exonerated by their better management of the environment.

Sustainability and SMEs are considered in the last two chapters of Section 1. Gerrans and Hutchinson in Chapter 5 confirm that Australian small and medium-sized firms in the food production, manufacturing and services sectors are ignorant of their legislative requirements and oblivious of international environmental management standards. They are uninvolved in sustainability activities, remaining unconcerned with the most basic facets of environmental protection, let alone voluntary initiatives such as ISO 14001. In Chapter 6 Johannson develops the notion that SMEs' participation in ISO 14001 could be a practical step towards sustainable development. Citing Canadian smaller companies as an example, she highlights their general exclusion from the international standards development process and the consequent potential for the ISO 14000 series, and ISO 14001 in particular, to act as barriers to their trade.

Appraised of the unaware, unengaged and probably non-compliant SME, readers move onto the second section of this book where experts present a suite of environmental management tools firms can employ to manage their environmental impacts. Starkey opens the section by providing in Chapter 7 an overview of the voluntary management tools available to improve environmental performance, e.g. life-cycle analysis, environmental auditing, environmental management systems (EMSs), etc. He assesses the tools' relevance to SMEs, citing examples from Malaysia,

Germany, Denmark and the UK. He concludes that not all of the tools are appropriate to all SMEs, but that their careful selection and use is essential for improvement in the sector's environmental performance.

The next three chapters investigate the barriers and drivers for the uptake of such environmental management tools. Gerstenfeld and Roberts in Chapter 8 identify three drivers for environmental management in SMEs: legislation, business-to-business relationships and stakeholder pressures; however, they detail a longer list of seven barriers—such as the lack of training and awareness, expenses involved and ill-suited nature of standards for SMEs, etc.—which prevent change. Any solution to overcome these barriers needs to be inexpensive, co-operative, locally based, flexible, unique and accessible.

In Chapter 9 Whalley first challenges the notion of an all-encompassing definition for SME, suggesting 'independent-minded' or 'owner-managed' businesses better reflect the varied cultures found in the sector. Looking closely at owner-managed businesses, he suggests that, rather than the millions in the sector that could take up EMS standards, it is only those posing a substantial risk that are suitable candidates for such standards. In the UK, that means between 100,000 and 750,000 enterprises.

My own chapter (Chapter 10), presenting a major review of 33 studies on SMEs, indicates the lack of relevance of formal EMSs to many smaller organisations. It details the barriers, benefits and stakeholder drivers for SMEs' adoption of formal EMSs in the UK and the European Union (EU). While examples can be found of the extensive benefits accruing to SMEs adopting formal EMSs—ISO 14001 and the Eco-Management and Audit Scheme (EMAS)—the majority of the sector remains unaware of the initiatives, unconvinced of the benefits and beset with largely internal barriers—but some external barriers—to their adoption. Customers, the major stakeholders driving EMS adoption, are of little help, especially for micro firms. They seem disinterested or else satisfied with SMEs' current environmental performance.

Changes in SMEs' production and products to bring about sustainable development are the focus of Chapters 11 and 12. Hobbs suggests the most effective way of coping with unsustainable practices is to prevent pollution and waste occurring in the first place by using cleaner production strategies. He acknowledges that many cleaner production initiatives have been demand- as opposed to supply-led, inhibiting their uptake, but suggests roles for international agencies to alleviate this problem. He concludes that there is a long way to go in engaging SMEs in active sustainability actions. Holmberg *et al.* then present a qualitative model to enable smaller firms to build sustainable criteria into product design. This model purports to be more accessible and cost-efficient than life-cycle assessments (LCAs) and therefore could aid enterprises' improvements in product development.

Meredith argues in Chapter 13, the last of Section 2, that environmental legislation and its demands places SMEs in turbulent and unpredictable situations, which should provide the drivers for environmental innovations. The SME sector is the source of many innovators. Smaller firms have a distinct advantage over large firms when innovating because they are normally less bureaucratic, able to respond quickly to change and have efficient internal communication channels. However, Meredith also suggests three conditions crucial to environmental innovations: business competences, strategic orientation and network involvement, which are often missed by SMEs.

The assertion that small and medium-sized organisations are isolated from the mainstream movement in environmental management tools and thinking, but are nonetheless capable of using appropriately tailored techniques once engaged, provides the *raison d'être* for the next section of this book: 'Practical Strategies for Reaching SMEs'. The section begins with an insightful review of the most effective ways of communicating environmental best practice to SMEs (Clark, Chapter 14). International organisations are rarely good at communicating with SMEs because their remoteness makes them ineffective at reaching and stimulating the sector. She concludes this communication role is better left to trusted intermediary organisations such as trade associations.

In Chapter 15 Hunt supports the international analysis in Clark's chapter, with her analysis of how information reaches SMEs. She suggests networks that are trusted and/or have perceived relevance, e.g. supply chains, trade associations, etc., are key routes for information dissemination to SMEs, but that the idea that information will lead to rational behaviour fails to take account of the real experience of smaller enterprises. A powerful, established and effective network in Denmark is the local authority partnership with SMEs. Pedersen in Chapter 16 shows how small and medium-sized companies derived significant advantage by engaging in open dialogue with local authorities, which acted as a catalyst or coach to improve the environmental performance of these enterprises. Recognition was given to the conflict between the regulator and helper roles of local authorities; however, as guides they proved effective at bringing about environmental improvements in smaller enterprises.

Another partner that can coach and engage small and medium-sized companies in environmental improvements is the large company. Tunnessen in Chapter 17 shows how mentoring in the US has benefited not only the mentee, i.e. the SME, but also the mentor, i.e. the large firm. The non-large company benefits from free expertise while the large company can use mentoring along its supply chain as a route to go beyond compliance. It is suggested that mentoring provides a non-threatening, low-risk, low-cost and effective means of introducing and engaging SMEs in strategies to get better environmental results. And, as Powell points out in Chapter 18, supply chain management is a key business issue of the future. Customers will no longer be buying from a single company but be purchasing along an integrated supply chain. The incentive, therefore, for large companies to look at and drive environmental improvements along the supply chain is increasing. This means smaller firms, most frequently suppliers to large ones, will come under increasing pressure to modify their environmental behaviour.

The regulator has a crucial role to play in improving SMEs' environmental performance. However, for most small enterprises, the first time they come into contact with the regulator is when some environmental problem or non-compliance situation has arisen. In Chapter 19 Fanshawe argues that there is a role for the regulator in improving the environmental performance of non-large firms, in addition to the regulatory role. The focus on non-compliance by the regulators and support organisations has prevented effective communication with smaller companies in order to bring about real change. He proposes a model to engage SMEs in a long-term dialogue to achieve incremental improvements in their environmental performance.

Macro-level strategies are presented in the last two chapters in Section 3. Bichard in Chapter 20 takes an unusually critical look at the support services and advice offered to SMEs in the UK. He asserts that there is an over-supply of help with a vast array of organisations and a wide range in quality of advice—which results in confusion among companies. These shortfalls are powerful reasons for a radical review of support services. This experience contrasts sharply with the consensus-based community network model in the Netherlands. In Chapter 21 de Bruijn and Lulofs review the success of this policy network approach which uses intermediary organisations, such as trade associations, to build pacts with SMEs to improve their environmental performance. These voluntary networks are set against a backdrop of penalties for laggards and the threat of legislation if predefined sector environmental objectives are not achieved.

The final section of *Small and Medium-Sized Enterprises and the Environment: Business Imperatives* presents case studies from around the world. The first two are from small-scale plants in developing countries. In Chapter 22 Scott investigates small-scale industry in Zimbabwe and Bangladesh. He suggests small firms are significant in terms of pollution not on a national scale but on a local scale, where clusters of enterprises cause locally significant pollution. However, he found that, per unit output, small-scale enterprises do not necessarily pollute more than large industry. A ineffective regulatory framework plagues these two developing countries: it is poorly enforced and unable to detect most pollution incidents. These shortcomings, he argues, open up the opportunity to use self-regulatory mechanisms to improve small-scale industrial environmental performance.

Dasgupta *et al.* in Chapter 23 also seek to quantify pollution from small industry, in comparison to large industry, but in relation to its impact on local populations in Mexico and Brazil. The authors found that small plants were more pollution-intensive per employee than large plants. However, paradoxically, in Brazil it appears to be the wealthier regions and high-income populace that suffer the greater affects of air pollution because these richer areas hold the greater concentration of large industry.

A major economic crisis in Asia did not lead to the anticipated reduction in pollution from the closure of many industries unable to absorb the economic shock. Quite the reverse: smaller firms in Indonesia simply dispensed with end-of-pipe controls to cut costs. Hence, Bratasida in Chapter 24 shows the need for effective cleaner production and EMS approaches to engage SMEs in longer-lasting and more profound environmental improvements. In Chapter 25 Terui suggests an earlier crisis: the oil shocks in the 1970s started Japanese business on the route to energy efficiency and environmental awareness. The high numbers of Japanese businesses registered to ISO 14001, he suggests, is because EMSs are seen as meshing well with existing quality systems and because the rapid global uptake of the standard is seen as a competitive issue. SMEs are being strongly targeted with advice and support so that they, too, can prepare for ISO 14001 certification.

Innovative SMEs are the subject of the last pair of chapters in this book. In Chapter 26 Palmer looks at small and micro UK firms and their effort to improve environmental performance. This unique group of companies challenges the way SMEs are viewed. Many have strong social and ethical reasons for their actions and highlight personnel motivation and commitment as being the key factor in achieving environmental improvements. This finding is expanded on by Walley in Chapter 27, who highlights the importance of having an environmental 'champion' in a small company. In this case study company, the champion networked for support and 'greenjacked' the environmental initiative onto a mainstream business activity—*viz.* creating a paperless office—to give the whole process a fillip. The emphasis on training and communication was paramount to convey the new ethos of the firm.

General conclusions

In *Small and Medium-Sized Enterprises and the Environment: Business Imperatives* the reader is taken through a journey of the experiences of SMEs operating in a range of sectors in diverse locations. The reader is shown SMEs' attitudes and perceptions to the environment, the management tools they can adopt to manage their environmental performance and the strategies taken to engage such firms in environmental improvements. If generalisations can be made about such a vast and diverse sector, they are that the sector is:

- Largely ignorant of its environmental impacts and the legislation that governs it
- Oblivious of the importance of sustainability
- Cynical of the benefits of self-regulation and the management tools that could assist it in tackling its environmental performance
- Difficult to reach, mobilise or engage in any improvements to do with the environment

This dark and depressing picture of the sector as a whole contrasts with some of the shining examples that are presented in this book, which depict innovative and dynamic SME leaders on the environment; novel approaches to activate firms to tackle their environmental aspects; and real and tangible benefits from tackling environmental issues.

This stark contrast between the majority of inactive small and medium-sized firms and the very small minority of sector leaders on the environment poses a problem not only for policy-makers and legislators, but also for those that seek to support the sector. Current environmental policy seems not to touch the majority of SMEs, and support organisations fail to reach this group. However, for the benefit of the environment and the wider societal and policy need to engage all actors in the drive towards sustainability, this vast and important group needs to be reached. The focus of governments needs to shift away from large businesses, which can and do look after themselves, towards real and practical assistance of SMEs. In the UK, for example, the 6,600 large firms have disproportionately more influence than is warranted by their importance to the economy. They employ the lobbies, sit on the committees and are asked their opinions. However, the smaller firms do not have the strong associations to lobby effectively, do not have the personnel to get involved in networks that influence policy and, most importantly, are rarely asked for their opinions. This is particularly the case for the 2.33 million 'size class zero' businesses, i.e. sole traders and partnerships with no employees.

It is an error to neglect the SME sector. If sustainability is to become a meaningful objective for societies, and fully integrated into their structures, then small firms must be brought into the process. Whether or not SMEs pollute more or less than large firms—and there is little evidence to suggest either conclusion—becomes irrelevant if they are unengaged and inactive. Small and medium-sized organisations need to be part of the process not just because they employ so many people and have influence throughout the community but because their alienation and isolation from actions on the environment means that sustainability will never be achieved. SMEs comprise the fabric of all societies. Change their attitudes towards the environment and we have a chance to achieve sustainable development. Ignore them and we all suffer.

Appendix 1: Selected operational SME definitions

The following two tables present a summary of the range of definitions of SMEs used for operational functions by organisations with interests in such firms.

Function of definition	Maximum no. of employees	Turnover (ECU million)	Maximum balance sheet (ECU million)	Other criteria or information	Source
Statistics on SMEs					European Commission DG XXIII and Organisation for Economic Co-operation and Development (OECD)'s Working Body for SMEs.
Micro-sized enterprises	1–9				
Small-sized enterprises	10–19				
	20–49				
	50–99				
Medium-sized enterprises	100–199				
	200–249				
	250–499				
Large-sized enterprises	>500				
*State financial aid to SMEs**				Maximum 25% of capital to be owned by a big enterprise (with exceptions)	European Commission DG XXIII
Small-sized enterprises	50	5	2		
Medium-sized enterprises	250	20	10		
Annual Accounts†					European Commission DG XXIII
Small-sized enterprises	50	4 (5)	2 (2.5)		
Medium-sized enterprises	250	16 (20)	8 (10)		
Insurance (non-life)‡	250	12.8	6.2		European Commission DG XXIII
VAT exemption (VI Directive)		0.005			European Commission DG XXIII
European Investment Bank Global loans	500			Maximum one-third capital owned by a big enterprise (net amount real estate less than ECU 75 million)	European Investment Bank and European Commission DG XXIII
European Investment Bank (yet-to-be approved loans)	250	20	10	Maximum 25% capital owned by a big enterprise (with exceptions)	European Commission DG XXIII European Investment Bank
Research & Development Craft Feasibility primes	500	38		Maximum one-third capital owned by a big enterprise	European Commission DG XXIII

Notes

* It is enough that one of the two criteria (turnover or balance sheet) is verified besides the number of employees.

† It is enough that two of the three criteria (maximum number of employees, turnover, maximum balance sheet) are verified. Figures in brackets are concerned with the review proposed by the European Commission to the Council in 1993.

‡ It is enough that two of the three criteria are verified.

Function of definition	Maximum no. of employees	Turnover (ECU million)	Maximum balance sheet (ECU million)	Other criteria or information	Source
Government grants					Department of Trade and Industry (DTI)
Small-sized enterprises	100				DTI Small Firms Division
Medium-sized enterprises	250–500				
Membership policy	50			96% of their members employ fewer than 20 people	Small Business Federation
Lobbying for legislation and government tax policy					
Lobbying for:				No strict definition/no threshold because there is no difference in CBI membership. Definitions vary for different functions.	Confederation of British Industry (CBI)
Statutory sick pay	10				
Access to finance	500				
Administrative	500			Privately owned, owner-manager, not dominant in the marketplace 30% sole proprietor 30% partners 40% limited company Total membership = 22,029	The Forum of Private Business
Modal level of membership:					
33%	1–4				
32%	5–9				
18%	10–14				
3%	15–49				
1%	50–99				
	>99				
No set definition for SMEs				If loan amount is below US$15 million, it is passed onto a financial mediator who may have a definition for SMEs. (Investment breakdown: one-third EBRD, one-third company, one-third other investor)	European Bank for Reconstruction and Development (EBRD)
For some projects number of employees may be used.	250				
To define target population for projects				Some projects will define a target 'SME' population	International Finance Corporation (IFC)
Micro enterprise	<20				
SME	100–500				

Appendix 2: selected theoretical SME definitions

The following table presents a summary of selected theoretical definitions of SMEs.

Term/issue	Definition	
Small and medium-sized enterprises	Enterprises with fewer than 500 employees, whose capital is less than ECU 75 million and of which less than one-third may belong to a larger company (Gondrand 1992)	
General definition of small firms	A small firm is an independent business, managed in a personalised way by its owner or part-owners, and with a small market share (Bolton Report 1971)	
More specific definition	Manufacturing Retailing Wholesale trades: Construction: Mining/quarrying Motor trades Miscellaneous services Transport	200 employees or fewer Turnover of £50,000 per annum or less Turnover of £200,000 or less 25 employees or fewer 25 employees or fewer Turnover of £100,000 per annum or less Turnover of £50,000 per annum or less 5 vehicles or less
Update of turnover figures from the Bolton Report to 1983 figures	Update of turnover figures from the Bolton Report to 1983 figures (Sengenberger *et al.* 1990)	
	Retailing Wholesale trades Motor trade Miscellaneous services	£315,000 per annum or less £1,260,000 per annum or less £630,000 per annum or less £315,000 per annum or less
Best indicator of small enterprise	Major policy decisions are taken by one or two people who usually own, manage and risk their own money in the business (Clarke 1972)	
Companies Act 1981	Two (at least) of the following criteria must apply (Sengenberger *et al.* 1990):	
Small company	a) Turnover does not exceed £14 million b) Balance sheet total of assets does not exceed £0.7 million c) Average weekly number of employees does not exceed 50	
Medium company	a) Turnover does not exceed £57.7 million b) Balance sheet total of assets does not exceed £2.8 million c) Average weekly number of employees does not exceed 250	
General definition	A small firm is one that has only a small share of its market, is managed in a personalised way by its owner or part-owner and not through the medium of an elaborate management structure and which is not sufficiently large to have access to the capital market for the public issue or placing of securities. A branch of a large company cannot be a small firm, because, although it is small and may even be independent with regards to decision-making, it will still have access to capital and technical assistance from the parent company (Bannock 1981).	
Business format franchising—a form of small business	The franchise not only sells the franchiser's product or service but does so in accordance with precisely laid-down procedures. In return the franchiser provides the franchisee with assistance, e.g. training, marketing, management, research and development, in carrying on his/her business; however, like any other small business, the franchisee provides the capital for his/her business, but agrees to run the business in accordance with the franchiser's guidelines (Hough 1982).	
Problems with the Bolton Report definition of small firms	If a small firm, in statistical and organisational terms, was the sole supplier of particular goods to a large company, the small firm would have 100% of the market share but would still be considered by many as small. This may be because the market share of the firm has grown beyond the stage of smallness but whose ownership and management are still highly centralised (Chestermann 1982).	

I

ATTITUDES AND PERCEPTIONS OF SMALL FIRMS TO THE ENVIRONMENT AND SUSTAINABILITY

1
Small firms and the environment

Factors that influence small and medium-sized enterprises' environmental behaviour*

Ann Smith, Robert Kemp and Charles Duff

Since the Earth Summit held in Rio de Janeiro in 1992, the role of business and industry in achieving sustainable development has increasingly become the focus of attention of policy-makers throughout the world. Business and industry are seen to be the cause of, and an important part of the solution to, environmental problems. Indeed, Agenda 21, the action plan produced as a result of the Earth Summit, emphasises the importance of involving business and industry, workers and trade unions in sustainable development.

The EU has been an important driver for the promotion of sustainable development and the Single European Act and the Fifth Environmental Action Programme, *Towards Sustainability*, require environmental considerations to be incorporated into all EU policy. Recent EU initiatives such as eco-taxes, the EU eco-labelling scheme and the Eco-Management and Audit Scheme (EMAS) have complemented the more traditional 'command-and-control' approach of environmental legislation with economic drivers, voluntary and industry-led approaches.

* The methodology and results have been summarised from the report produced by MORI (1998) for Groundwork. The opinions expressed and the interpretation of the results are those of the authors and not necessarily those of MORI or Groundwork. The authors are grateful to the sponsors and supporters of the 1998 survey: Lloyds TSB Group plc, British Petroleum Company plc, EA, DTI, DETR, Association of Chartered Certified Accountants, Business in the Environment and Royal Sun Alliance.

Small firms make up a high proportion (99%) of businesses in the UK, they employ more than half the workforce (58%) and they make a significant contribution to the gross domestic product (Smith and Kemp 1998). Collectively, small and medium-sized enterprises (SMEs) are claimed to be the source of around 70% of environmental pollution (Hillary 1995). Although environmental legislation has been increasing year on year since the 1970s, it is only since the early 1990s that pressure to improve environmental performance has focused on SMEs. As a result, many initiatives have been introduced to assist small firms to improve their environmental performance and take up new environmental technologies (CEC 1989).

Much of the funding available through EU initiatives to improve the environmental performance of SMEs has been accessed by business-support organisations such as Business Links, Training and Enterprise Councils (TECs), Groundwork, business environmental associations and academic institutions. These organisations offer a variety of services to SMEs ranging from environmental awareness-raising seminars through more formal training to technical on-site support. The most successful projects have been waste minimisation schemes that demonstrate both cost savings and environmental improvement (CEST 1997, 1998). The Department of Trade and Industry (DTI) and the Department of the Environment, Transport and the Regions (DETR) have provided significant support for industry and business, seeking to improve their environmental performance through initiatives such as the Environmental Technology Best Practice Programme (ETBPP) and the Environment and Energy Awareness Programmes.

Although there have been a number of notable successes involving SMEs and environmental management (Smith and Kemp 1998; Sheldon 1998a; Robinson 1998), many of the EU and national funding programmes for SMEs are largely under-spent. For example, the Small Company Environmental and Energy Management Assistance Scheme (SCEEMAS), a UK government scheme that offered a grant of up to 40%–50% of the consultancy costs for SMEs to implement EMAS, has been withdrawn.

The difficulties in encouraging SMEs to take up free and highly subsidised opportunities of assistance to improve their environmental performance have been noted by a number of authors (BCC 1996a; Rowe and Hollingsworth 1996; Merritt 1998; Sheldon 1998b).

Based on the fundamental premise that improved environmental performance will lead to enhanced competitiveness, Groundwork commissioned Market Opinion Research International (MORI) to conduct a survey of small firms in the UK to identify the factors that influence their environmental behaviour (MORI 1998). The survey also sought to identify the organisations that have influence over small firms and to inform policy-makers and opinion-formers of the likely needs of their SME audiences. In other words, where might efforts best be directed to facilitate a greater uptake by small firms of activities to improve their environmental performance? This chapter provides an account of the research undertaken and considers the findings of the Groundwork survey—*Small Firms and the Environment 1998*—in the context of the so-called 'intractable SME sector' and environmental management.

Methodology

A questionnaire was designed jointly by MORI, Groundwork and the University of Hertfordshire to address the following key areas:

- The perceived role SMEs play in the national economy and the local environment
- Ratings of the environmental performance of respondents' own companies, and benefits of improved 'green' performance
- Which support bodies or initiatives have been or would be contacted?
- Awareness of environmental legislation
- Registration/certification currently held or planned
- Costs associated with environmental issues
- Which types of organisation or individual have enquired about the environmental performance of respondents' companies, and which might cause changes to practices?
- Preferred format for supply of environmental information

In order to check the flow, routing and wording of the questionnaire, a number of pilot interviews were conducted. No major changes to the questionnaire were required in light of the pilot study; however, a number of minor refinements were made.

Some questions were prompted in that they allowed the respondent to select an answer from a list of options provided. Other questions were unprompted and no 'hints' were provided. The results indicate where questions were prompted or unprompted.

Telephone interviews were conducted with 300 managers of SMEs (companies with fewer than 250 employees) between 17 February and 17 March 1998. All interviews were carried out by MORI In-Line Telephone Surveys Ltd, using computer-assisted telephone interviewing methodology. The interviews were conducted by telephone at the respondents' place of work and there was only one interview per company. The respondents were the designated senior managers within each company, specifically responsible for dealing with environmental issues.

The companies were sampled from The Business Database's corporate listings and, apart from industrial sector and company size, the selection of companies from the database was random (see Table 1). Non-commercial organisations were excluded.

The DTI Small Firms Statistical Unit (1993) stated that 94.4% of UK enterprises were 'very small' or 'micro' companies (1–9 employees), 5.1% were 'small' companies (10–99 employees) and 0.3% were 'medium' companies (100–249 employees). The sample profile did not match the national profile in that the proportion of 'micro' companies was under-represented and the small and medium-sized companies were over-represented (see Table 1).

	Number
Number of employees	
1–9	76
10–36	65
37–49	41
50–99	54
100–249	64
Turnover	
£0–£249,999	33
£250,000–£499,999	21
£500,000–£999,999	19
£1 million–$4.99 million	89
£5 million and over	73
Market sector	
Manufacturing	141
Construction	14
Service (including transport, wholesale/dealing/retail)	133
Region	
East Anglia	11
East Midlands	21
Greater London	27
North-East	18
North-West	35
Northern Ireland	14
Scotland	22
South-East	49
South-West	48
Wales	7
West Midlands	30
Yorkshire & Humberside	18

Table 1: SAMPLE PROFILE FOR THE 1998 SURVEY

All the results from the questionnaire are given in percentages. The figures are based on the full sample of 300 companies and are accurate to between ±3% and ±6%. Where the figures do not add up to 100%, this is due to computer rounding, multiple answers or the exclusion of 'don't know' or refusal answer categories. An asterisk (*) represents a figure less than 0.5% but more than zero.

Results

The questionnaire results were used to analyse the level of awareness of respondents, the activities undertaken by their companies to mitigate environmental impacts and the sources of advice and training consulted.

SMEs' perceived role in the national economy and the local environment

When asked what contribution SMEs in general make to the national economy, respondents perceived that their companies played an important role: very/fairly high, 77%; neither high nor low, 14%; and very/fairly low, 8%.

Respondents were asked (unprompted) to identify the two or three most important issues facing businesses in the UK today. Issues such as competitiveness, skills/training/staff shortages, interest rates, productivity and environmental issues ranked ahead of meeting statutory requirements, investment, exchange rate and European single currency issues. Other issues considered to be less important than the environment included credit control/being paid on time, maintenance of product quality, technology, customer satisfaction and staff morale (see Table 2).

When asked to rate the impact that their company had on the local environment, respondents considered their environmental impacts to be more positive than negative: very/fairly positive, 51%; neither positive nor negative, 39%; and very/fairly negative, 10%.

Ratings of environmental performance and benefits of improved 'green' performance

When asked to rate their company's environmental performance, respondents considered it to be good overall: very/fairly good, 74%; neither good nor poor, 21%; and very/fairly poor, 4%. Just under two-thirds of respondents' companies addressed environmental issues either through a formal environmental policy (23%) or as part of their business plan (38%). Just over a third (36%) of respondents' companies did not have a formal environmental policy or address environmental issues through a business plan.

In those companies with a formal environmental policy, responsibility for its implementation most commonly rested with the managing director (42%). Other SME managers likely to take responsibility for the environmental policy included health and safety managers (9%), site managers (8%), technical directors (8%), quality managers (4%), and production/operations managers (3%). Other managers, such as personnel and finance managers, were less likely to be responsible for environmental policy.

Respondents were asked (unprompted) to indicate the benefits gained by their company from improved environmental performance (see Table 3). The benefits included: better customer relations/image; helping the environment; cost savings in terms of materials used; cost savings in terms of waste reduction/disposal; improved

Issues	Frequency (%)
Competitiveness	30
Skills/training/manpower shortages	21
Interest rates	15
Productivity	15
Environmental issues	**13**
Meeting statutory requirements	10
Investment	8
Exchange rate/strength/cost of the pound/economic environment	7
Exports/lack of exports	6
European/single currency/Europe/EU/EEC	6
Role/support of government	5
Credit control/being paid on time	4
Maintenance of product quality/quality	2
Technology	2
Customer expectations/satisfaction	2
Staff morale/motivation	1
Lack of consumer demand/customer spending	1
Safety	1
Traffic/parking	1

Table 2: MOST IMPORTANT ISSUES FACING UK SMES

Benefits gained	Frequency (%)
Better customer relations/image	19
Helping the environment	**13**
Cost savings: materials used	9
Cost savings: waste reduction/disposal	9
Improved image in the community	7
Cost savings: energy	6
Higher employee morale	6
Commercial/competitive advantage	5
Attainment of environmental standards	**3**
Meeting/compliance with legislation/avoidance of penalties	3
Remaining on key customers' approved supplier lists	1
New markets and opportunities	1
Lower loan interest rates and insurance premiums	1

Table 3: BENEFITS GAINED FROM IMPROVED ENVIRONMENTAL PERFORMANCE

image in the community; cost savings in terms of energy; higher employee morale; and commercial/competitive advantage. Just over a third (36%) of respondents did not identify any of the 20 benefits given spontaneously by other respondents.[1]

Registration/certification to EMAS / ISO 14001 held or planned

Only 3% of respondents' companies were currently registered under EMAS and 7% were certified to ISO 14001; 13% of respondents did not know whether their company was registered under EMAS and 6% did not know whether their company was certified under ISO 14001. Of those respondents whose companies were not registered/certified and who did not know, 9% planned to gain registration to EMAS and 14% planned to gain certification under ISO 14001.

Awareness of environmental legislation/regulations

Respondents were asked (unprompted) to identify environmental legislation and regulations that they were aware of that directly affected their company (see Table 4). The legislation/regulations identified included the Special Waste Regulations, Control of Substances Hazardous to Health (COSHH) regulations and the 1997 Producer Responsibility Obligations (Packaging Waste) regulations 1997. Only 5% of respondents identified the waste 'duty of care' legislation and less than 5% identified the Environment Act 1995, landfill tax and the Health and Safety at Work etc. Act 1974. A quarter of respondents did not identify any of the 22 items of legislation and regulations given spontaneously by other respondents.[2]

Environmental legislation/regulation	Frequency (%)
Special waste regulations	17%
Control of Substances Hazardous to Health (COSHH)	17%
Packaging waste regulations	15%
Environmental Protection Act	12%
Controls on emissions of volatile organic compounds (VOCs)	6%
Water Industry Act	6%
Duty of Care regulations	5%
Integrated Pollution Control (IPC)	<5%
Environment Act 1995	<5%
Landfill Tax	<5%
Health & Safety Act	<5%

Table 4: ENVIRONMENTAL LEGISLATION/REGULATIONS IDENTIFIED BY RESPONDENTS

1 See Hillary's review of the full range of benefits cited by SMEs in Chapter 10.
2 See Petts's results on the awareness and attitudes to legislation of management and non-management in SMEs in Chapter 3.

In a separate question, 84% of respondents were aware, and 16% were not aware, of director liability whereby a company director personally may be held liable for pollution caused by their company.

Costs associated with environmental issues

Respondents were asked to estimate how much environmental compliance (meeting environmental regulations) was costing their company, how much their company spent per year on improving environmental performance beyond the requirements of compliance, and what annual cost savings their company could attribute to improved environmental performance.

The majority of respondents estimated that environmental compliance cost their company nothing and that their company made no cost savings (see Table 5). A quarter of respondents did not know the costs or cost savings related to compliance or improved environmental performance.

Costs/cost savings	Compliance Costs (%)	Costs beyond compliance (%)	Cost savings (%)
Nothing	16	46	61
£1–£100	6	4	1
£101–£500	9	3	2
£501–£1,000	9	3	2
£1,001–£2,000	6	1	1
£2,001–£5,000	10	4	2
£5,001–£10,000	8	3	1
£10,001–£25,000	6	3	1
£25,001–£50,000	2	2	*
£50,000–£100,000	1	1	1
£100,001–£250,000	1	1	*
£250,001–£500,000	0	*	*
£500,001 or more	0	1	1
Don't know	24	27	26

Table 5: ESTIMATED ANNUAL COSTS AND COST SAVINGS RELATED TO LEGISLATIVE/REGULATORY COMPLIANCE AND IMPROVED ENVIRONMENTAL PERFORMANCE

Types of organisation that have enquired about the environmental performance of SMEs

Respondents were asked (prompted) to identify organisations that had enquired about their company's environmental performance. These included local authorities, customers, insurers, Environment Agency (EA), suppliers of goods and services, trade associations, general public, environmental groups, investors/shareholders, lawyers, competitors and bankers. A quarter of respondents reported that none of these organisations had sought information about their company's environmental performance.

Respondents were then asked (prompted) which of the above organisations might persuade their companies to change their environmental practices. This resulted in a slightly different order: local authorities, customers, EA, insurers, general public, investors/shareholders, environmental groups, trade associations, suppliers of goods and services, lawyers, bankers and competitors. Fourteen per cent of respondents reported that none of these organisations might persuade their companies to change their environmental practices.[3]

Support bodies or initiatives that have been or would be contacted

Respondents were asked (unprompted) to identify environmental support bodies or initiatives with which they have had contact. These included local authorities, EA, DETR, Health and Safety Executive (HSE) and environmental consultancies (see Table 6). More than 25 other organisations were identified but were scored by less than 5% of respondents; 38% of respondents did not identify any of the 30-plus organisations identified by other respondents; 16% identified other organisations and 2% did not know.

Organisations	Enquiring about SMEs' environmental performance (%)	Able to persuade SME to change environmental practices (%)
Local authorities	41	60
Customers	40	59
Insurers	39	46
Environment Agency	25	54
Suppliers of goods and services	24	28
Trade associations	17	29
General public	14	38
Environmental groups	9	30
Investors/shareholders	8	32
Lawyers	6	28
Competitors	5	21
Bankers	4	23

Table 6: ORGANISATIONS SEEKING INFORMATION ON ENVIRONMENTAL PERFORMANCE OR ABLE TO INFLUENCE SMEs' ENVIRONMENTAL PRACTICES

3 See Hillary's review of stakeholder influence on SMEs' environmental performance in Chapter 10.

Respondents were then asked (unprompted) to identify environmental support bodies or initiatives that they would approach for information and training on business and environmental issues or their company's responsibilities. The responses were similar (see Table 7). However, 13% of respondents did not identify any of the organisations identified by the others, 17% identified other organisations and 16% did not know.

Organisation	Support body SME has had contact with	Support body SME would approach	Support body SME has direct contact with
Local authorities	16	15	68
Environment Agency	12	10	<5
DETR	7	9	<5
Health and Safety Executive	6	7	<5
Environmental consultancies	5	<5	<5
Business Links	<5	<5	43
Chamber of Commerce/TEC	<5	5	52
Trade association	<5	5	53
Local university/college	<5	<5	44
Government offices	<5	<5	33
Federation of Small Businesses	<5	<5	23
Business in the Community	<5	<5	21

Table 7: SMEs' RELATIONSHIPS WITH SUPPORT BODIES

Respondents were asked (prompted) which business support bodies their companies had direct contact with. These were local authorities, trade associations, chambers of commerce/TECs, local university/college, Business Links, Government Offices, the Federation of Small Businesses (FSB) and Business in the Community (BiC).

◻ *Preferred format for environmental information*

Respondents were asked (prompted) to indicate the formats for environmental help and training that they would find useful. Respondents gave a 'yes' response to the following: printed information—checklists/DIY guides (83%); free telephone help-line (78%); advice from regulators (74%); printed information—newsletters/updates on new technology (73%); on-site advice from local consultants/advisers (61%); local seminars/workshops (52%); videos (47%); Internet (44%); and national conferences (15%).

Discussion

The views of SME managers, along with information from other recent surveys (Merritt 1998; Holland and Gibbon 1997; Rowe and Hollingsworth 1996), reveal that small firms recognise that there are pressures to improve their environmental performance. Holland and Gibbon (1997) suggested that many small firms believe that their environmental impact is in proportion to their activities, i.e. small and minimal, and they found that SMEs responded quickly to environmental issues with relatively small environmental impacts but not to greater impacts.

Although SMEs make an important contribution to the economy both nationally and locally, they fall behind their larger counterparts in terms of environmental activity. However, SMEs are both concerned about the environment and are willing to address their environmental responsibilities (Merritt 1998; Smith and Kemp 1998). They are also aware of the benefits of improved environmental performance in terms of improved customer relations, cost savings and competitive advantage.

There is a substantial gap between environmental awareness of SMEs and the business benefits they can gain. Merritt (1998) found that SMEs had little knowledge in the field of environmental management and that they had not introduced formal practices to manage the environmental performance of their businesses. The present study (Smith and Kemp 1998) found that SMEs' awareness of environmental legislation directly affecting their companies was poor and there was a general perception that legislative compliance would be costly.

Efforts by government and business-support organisations to raise awareness of the cost savings and competitive advantage that result from improved environmental performance have had little impact on SME behaviour (Merritt 1998). Despite the difficulty in engaging and influencing SMEs, they *would* be persuaded to change their environmental behaviour by the regulators, customers and insurers (Smith and Kemp 1998). The supply chain was also considered to be a powerful tool in this respect.

The role of business support organisations in providing environmental management training and support was emphasised by Rowe and Hollingsworth (1996); however, SMEs made little use of these services, even when free or subsidised (Smith and Kemp 1998). SMEs generally agree that they require external assistance to meet their environmental responsibilities, but this assistance should be locally accessible and include best-practice case studies relevant to the size and sector of the company. More work needs to be done to demonstrate to SMEs the cost savings they can make, including the avoidance of potential costs represented by environmental liabilities.

2

Small firms' environmental ethics

How deep do they go?

Fiona Tilley

This chapter explores the relationship between ethics, small firms and the environment, using the empirical findings from a national study that investigated the perceived gap between environmental attitudes and behaviour of small firms in the UK.

Are environmental ethics relevant to small firms? At the heart of the environmental or sustainability challenge is the way people perceive, value and, therefore, behave towards each other and the non-human environment in all walks of life, including the workplace. Owner-managers and employees of small firms are no more exempt from environmental ethics than anyone else in society. Discussing small-firm environmental ethics in a theoretical and practical context is fundamentally important. This is because ethics, among other things, is the critical examination of what is morally right and wrong and the rules of social conduct. Henderson (1982) writes 'ethics in the broadest sense provides the basic conditions of acceptance for any activity'. As a subject ethics is more than the description of moral behaviour or a prescription of what is right or wrong. Ethics is about subjecting the description of people's attitudes and behaviour alongside that of the rules and principles on which moral behaviour is supposedly founded, to rigorous critical examination (Chryssides and Kaler 1993).

The chapter is divided into four parts, beginning with an outline of the link between ethics, business and the environment. This is followed by a rationale for investigating small firms in the context of environmental ethics and an overview of the emerging literature in this field. Evidence is then extracted from the analysis of 60 in-depth interviews with small UK firms in order to identify their dominant environmental ethic. Finally, consideration is given to the implications and issues raised by a 'shallow ecology' approach to small-firm environmental ethics.[1]

The link between ethics, business and the environment

Environmental ethics is, like business ethics, a new field of inquiry within the wider discipline of ethics. Dobson (1995) describes the development of environmental ethics from two perspectives (see Table 1). The first reflects the dominant paradigm that governs the conventional discourse in ethics. Theorists in this camp are seeking to define a code of conduct of 'shallow ecology' for environmental ethics. This can be interpreted as the 'business-as-usual' approach. In contrast the second perspective reflects the work of 'deep ecology' theorists,[2] who have sought to highlight the inadequacies of conventional ethical discourse (Naess 1986). They advocate that an ecologically sound ethics will only emerge from a new paradigm—a new perception of sustainability. The attempt to interpret deep ecology and its bio-ethics or biological egalitarianism using the conventional ethical discourse has proved to be somewhat of a conundrum for environmental ethicists. Fox (1990) argues in favour of a new ecological consciousness, or state of being, from which an ecological ethics can emerge to govern the way people value and behave toward the environment.

Environmental ethics	
Code of conduct	***State of being***
Dominant world-view	New world-view
Shallow ecology	Deep ecology
Conventional ethical discourse	New ethical discourse
Hierarchical	Systemic
Individualistic/atomistic	Holistic
Anthropocentric	Ecocentric

Table 1: ENVIRONMENTAL ETHICS: SUMMARY CHARACTERISTICS OF THE TWO MAIN APPROACHES

1 'Shallow ecology' refers to an anthropocentric world-view. A machine is often used as a metaphor for nature. This hierarchical approach is reductionist in thinking and places the human race at the top of the value chain; as such, nature is an instrument of human endeavour.

2 'Deep ecology' refers to a holistic world-view. The metaphor for nature is that of an organism. This systemic approach recognises that humanity is only part of the whole global system and therefore must respect and live in balance with the intrinsic biological patterns and needs of nature.

It is the conventional ethical discourse, that is the code of conduct of shallow ecology, which dominates much of business thinking today. The ethics of deep ecology still has some way to go before it can be said to have transformed mainstream business thinking.

Small firms, ethics and the environment

Small-firm ethics and corporate social responsibility emerged in the business literature in the US during the 1970s and 1980s. To begin with, this literature was mainly concerned with moral issues of employee behaviour and other areas of social responsibility. The late 1990s have seen some early explorations into the environmental ethics of small firms. However, like many aspects of business research, the predominant focus of the business ethics literature has been on the activities of larger organisations. Thompson and Smith (1991) suggest this is because:

- Small firms are perceived as lacking sufficient resources.
- Research methodologies created for large companies are not readily adapted to small firms.
- More information is accessible to research large firms.
- Large companies have a higher public profile which generates more interest in the theories and research about these organisations and their corporate social responsibility.

Interestingly, similar reasons are given to explain the relatively low research interest in the relationship between small firms and the environment (Tilley 1999). Some may argue that small firms are a less worthy, needy or relevant research subject compared to larger firms. There are three arguments to counter this viewpoint. The first is based on the premise that small firms are significant to the UK in economic and environmental terms. Defining small firms has often been an arbitrary exercise. In the context of this chapter the definition used is that which classifies firms employing less than 50 employees as small. According to the latest government statistics, small firms constitute approximately 99% of all businesses in the UK (DTI 1998). In 1996, of the 3.724 million firms in the UK, 3.693 million employed less than 50 employees. Even if you discount the self-employed from this number this still leaves 1.176 million firms contributing to the UK economy, which account for 32.3% of total employment.

Nor should the potential environmental impact of the small-firms sector be underestimated. Even though there is little quantitative data available that measures the environmental impact of individual small firms, it is estimated that the cumulative environmental impact of the sector as a whole could be quite considerable. As measures of environmental impact and sustainability indicators become more prevalent it should be possible to calculate with greater accuracy the precise environmental impact of the small-firms sector. Until that time, however, estimates must be relied on. It has been suggested that small firms cumulatively could contribute as much as 70% of all industrial pollution (Hillary 1995). Hillary points out that this is an unsubstantiated figure and should be treated with some caution. Nevertheless, it is possible to conclude that small firms can no longer be viewed, individually or collectively, as an insignificant component either with regard to the economy or their impact on the environment.

The second argument is concerned with addressing the quality and quantity issues associated with a comparatively under-researched subject, such as small firms, in academic and other business literature. Such research has, in recent times, come in for some criticism for its lack of theoretical rigour and conceptual development (Goss 1991). For example, there has been much debate as to what constitutes a small firm. The environment and ethics literature has, in the main, neglected small firms, instead focusing its attention on the activities of large corporations (Smith 1993; Thompson and Smith 1991). In addition, the emerging literature on small firms and the environment has also been accused of insufficient analytical inquiry, relying too heavily on anecdote (Geiser and Crul 1996). There is, therefore, a need for more research on the environmental ethics of small firms in order to contribute to this important yet neglected subject.

The final argument to support the importance of small firms' environmental ethics research is based on the premise that theory generated for large firms cannot necessarily be applied to small firms. It has been noted that small firms often differ from larger firms in their management style, organisational structure and in the characteristics of the owner-managers (Dandridge 1979). Small firms are, by comparison to large, often resource-poor, and thus may have problems accessing finance, labour and finding the necessary time to manage environmental matters (Welsh and White 1981). Small firms, therefore, need their own unique ethical understanding of the difficult environmental problems they face.

The indications to date suggest there is a need for further research to consider the relative importance of the small-firms sector in terms of its contribution to the economy and its possible cumulative impact on the environment.

The studies that have been completed have drawn attention to the barriers between small firms and ethics. Vyakarnam *et al.* (1997) warned that the rise in environmental and social issues on the business agenda of larger firms has not been matched in the small-firm sector. In a survey of small firms to investigate the perceived gap between business goals, professed values and actual behaviour, Russell (1993: 3) found that 'ethical codes alone are insufficient to change either attitudes or behaviour because they have been notoriously difficult to implement'. Russell suggests the four most common attitudinal barriers to ethics among small firms are:

- Ethics and business don't mix.
- It doesn't pay to be ethical.
- If it's legal, it's ethical.
- Compared to others, this company is ethical.

Despite these barriers of perception, what empirical evidence is there to show that small firms are widening their outlook, from that of commercial success, to take account of their social and environmental responsibilities?

The dominant theme in the small-firm ethics literature has been social responsibility. For example, studies have investigated the ethical relationship between small firms and their customers (Humphreys *et al.* 1993), and the difference between small-firm ethics and the ethics of the owner-manager (Vyakarnam *et al.* 1997; Quinn 1997).

Although this research agenda is making a modest contribution to the business ethics literature, there have been very few studies reporting on the environmental dimension to small-firm ethics, although two recent studies have reported the low acceptance of the environment as a business issue among small firms (Tilley 1999; Rutherfoord and Spence 1998). These acknowledge that the reasons for this behaviour are complex, but Joyce *et al.* (1996) suggest that the continued separation of issues of social responsibility from business performance may contribute to a small-firm business culture that has difficulty integrating non-economic responsibilities into core business values.

Small-firm environmental ethics is an under-researched area of study. In view of the widening societal concern for environmental issues (Dunlap 1997), the impact of sustainability on the activities of public-sector organisations (Grubb *et al.* 1993) and the increasing environmental regulation likely to affect the behaviour of business (James 1998), the environmental ethics of small firms is an important issue that needs to be explored in greater depth. A better understanding of small-firm environmental ethics may help to explain the problems many such businesses encounter when embarking on activities to improve their environmental performance.

> Despite the proliferation of industry initiatives on the environment and environmental laws and regulations, accompanied by a wealth of research projects and publications during the 1980s and 1990s, research indicates that management, and SMEs in particular, have been slow to progress from a reactive to a proactive response to environmental pressures (Hutchinson and Hutchinson 1997: 305).

Organisations seeking to reduce the environmental impacts of the small-firms sector have often found it difficult to engage owner-managers into taking positive action. The remainder of this chapter will begin to explore the state of small-firm environmental ethics.

Small-firm environmental ethics: some empirical findings

The findings presented below have been extracted from a much bigger investigation which explored the perceived gap between the environmental attitudes and behaviour of small firms in the mechanical engineering and business services sectors in Leeds, West Yorkshire (Tilley 1998). The two industrial sectors were chosen because of their importance to the economy of Leeds (Leeds Development Agency 1992). The purpose of the research was to uncover greater understanding and meaning of the underlying processes (the 'why' and 'how' questions) rather than a description of the problem (the 'what' and 'where' questions). For this reason, a qualitative approach was taken, keeping the sample size low in order to attain information-richness. In selecting the small firms the following parameters were applied: the firm must employ less than 55 people; must not be a subsidiary; and must be owner–managed in a personalised way. In qualitative research the sample size is governed by the principle of information saturation, at the point when no new data is being collected. This point was reached after 60 semi-structured interviews with owner-managers. The interviews, 29 with mechanical engineering firms and 31 with business services firms took place between October 1994 and February 1995. The research findings are grounded in a qualitative analysis of the data from the interview transcripts.

Environmental regulation was a major discussion theme during each interview and it is this data that is presented below. Environmental regulation refers to the means by which the environmental behaviour of business is controlled. At one end of the spectrum is **state regulation** (such as case/statute law and other forms of market intervention) and at the other end is **self-regulation** (such as voluntary codes of practice).[3] There is much debate about the relationship between business ethics and the law and codes of practice. It would be wrong to assume that the author is implying that business ethics can only be interpreted as compliance with the law. Nevertheless, it can be argued that 'the domain of ethics includes the legal domain, but extends beyond it to include the ethical standards and issues the law does not address' (Trevino and Nelson 1995: 15). From the perspective of business environmental ethics, the merit of state regulation compared with self-regulation is a contested issue. State regulation and self-regulation should not be viewed as mutually exclusive, as it is possible to argue from an ethical standpoint that there is a value and a need for both forms (Schokkaert and Eyckmans 1994).

The interview data on environmental regulation is used in this chapter to draw out the ethical concerns of the small-firm owner-managers and to identify the dominant environmental ethic held by the small firms in the study. The analysis from the owner-managers' responses reveals their thoughts on the moral rights and wrongs of their environmental behaviour and the methods they use to control their environmental practices. Although these responses do not provide a complete picture of the corporate environmental ethics exhibited by small firms, they do begin to make an important contribution to our understanding in an emerging area of inquiry.

Self-regulation

Is the small-firms sector capable of self-regulating its environmental responsibilities? Self-regulation ought to appeal to the business community because it allows autonomy to interpret and regulate acceptable standards of environmental behaviour. Experiments in the use of voluntary schemes to green business practices and cut pollution damage have also become attractive at a political level because, it is argued, they are popular among business, have successfully saved companies money and lead

3 See Petts's discussion in Chapter 3.

to greater resource efficiencies (Cairncross 1995). Environmental groups predominantly lobby on a moral platform demanding increases in environmental regulation, whereas business representatives often take a more pragmatic position demanding liberalised trade agreements which, they believe, will foster greater competitiveness and lead to economic growth, thus generating the wealth needed to pay for the cost of environmental clean-up (Cairncross 1995).

Nevertheless, as a means of controlling small firms' environmental practice, self-regulation was not well supported among the participants in this study.[4] A considerable proportion of the sample remarked outright that self-regulation would not work, although there was some recognition that voluntary agreements may form part of the regulatory mix, simply because of the onerous resource demands needed to administer and enforce state regulation. A solicitor in a firm of general practitioners explained:

> [We] have got to rely on self-regulation to some extent because of the financing of it. I'm sure that the government doesn't have the funds to regulate it themselves. Which is why we have to rely on business . . . I think it is certainly a good start but I think it maybe does need to be monitored, perhaps by government as well. They can't rely solely on businesses to do it themselves. At the end of the day businesses are in business for the money. If making a hell of a lot of money is the option as opposed to looking at an environmental issue, I reckon nine out of ten businesses will ignore the environmental issue. I think they would do something about it as second choice.

The reasons given by the owner-managers as to why self-regulation could not be relied on as a means of controlling the environmental performance of small firms were varied. In general, they recognise that they are not best placed to identify what actions they need to take in order to manage their environmental responsibilities in an acceptable manner. Although owner-managers may have expert knowledge of their industry, this does not necessarily mean they also have the requisite environmental expertise. Low standards of ecoliteracy[5] are common among small-firm owner-managers and employees which, in turn, reduces their awareness of the environmental issues they should, or could, be responding to within their businesses. Consequently, despite the best intentions, the typical small firm in this sample displayed limited internal motivation to take steps to reduce its environmental impacts. The proprietor of a company that supplies equipment to businesses operating in the petroleum industry explained:

> Self-regulation wouldn't work in our industry. Although we do certain self-regulation things, a lot of companies, because they are small to medium-sized, don't move to carry out certain matters until it is imposed on them because of cost. Plenty of people say they would like to do this, but we really can't afford it.

---

4 See Chapter 3.

5 In this context ecoliteracy is defined as the understanding of the principles of ecology and the environment and an ability to use these principles to create sustainable business organisations.

As such, the environment remains relatively low down the business agenda of small firms. These circumstances are further compounded by a governing economic system which, as the owner-managers described, rewards self-interest over collective interest. The environment is still perceived by many to be an economic cost, a burden that restricts their competitiveness. For the typical small firm there is still a great deal of tension between what is economically appropriate behaviour and environmentally acceptable behaviour. It is not that the owner-managers in this study did not value the environment, but, in any given situation, economic priorities will come out on top if a choice has to be made. Consequently, the economic system and business climate is operating as a dominant resistant force preventing many of the small firms in this study from voluntarily taking steps to behave with greater environmental responsibility than their competitors.

Self-regulation, in the minds of many of the small-firm owner-managers, fosters and encourages 'cowboy' activities—the illicit operation of what economists term 'free-riders' (Hardin 1983). Owner-managers claim they could not trust a 'certain element' to uphold the principles laid down by self-regulation and, as such, self-regulation offers them little protection from cowboy operators who, by flouting voluntary agreements, could gain a competitive advantage over the more responsible firms acting in good faith. Clearly, there appears to be a problem of trust within the business community. The managing director of a company manufacturing mechanical handling equipment remarked:

> I don't think it [self-regulation] will work one little bit. It is not workable. I don't trust people that far, particularly financial people. The responsible companies are not necessarily the large ones. I think there is an awful lot of irresponsible action going on.

To overcome the problem of self-interest, the small firms pointed to the role of government which supposedly represents all sections of society, not just business. The managing director of a company manufacturing tanker vehicles for the waste-handling industry stated:

> Self-regulation clearly does not work. In an ideal world okay, but it lets the government off the hook. The government should be an agency that takes a long-term view of the environment for the whole country.

Commoner (1990) supports this opinion, suggesting that pro-environmental behaviour, such as investing in cleaner technology, can conflict with short-term profit-maximising goals; therefore social responsibility must be implemented at the political level. This is further supported by research on the UK retail sector that concluded:

> Where environmental criteria match economic criteria so that pro-environmental change yields positive benefits, environmental ethics are incorporated into retailers' ethics (Eden 1993: 105).

State regulation[6]

The small firms in this study demonstrated a preference for external regulation, as opposed to self-regulation, as a means of controlling the environmental behaviour of businesses. It could be said that this contradicts survey findings that rank government regulation and taxes as more important than any other single problem facing small firms (Bannock and Peacock 1989). However, the opinions of the small firms in this study need to be tempered against the fact that many of the owner-managers know little about their own regulatory obligations, nor do they believe their activities are worthy of regulatory control.[7] This attitude inevitably fuels low standards of compliance. This problem is further compounded by a perception that enforcement is low, as too are the penalties imposed on companies prosecuted under environmental law, causing the senior partner of a chartered accountancy practice to comment:

> The cowboys that produce things more cheaply by cutting corners, by leaving debris, will prosper against those who have a public conscience. There have to be penalties to put cowboys out. Once you have the penalties you have changed people's approach. It then becomes not just a virtue but a business advantage to be environmentally sound.

The small firms recognise that passing more, and stricter, legislation is not a panacea. The managing director of a company manufacturing effluent and water treatment equipment observed:

> We see legislation affecting business significantly in the water sector, but the water companies have not been spending money. The legislation is not regulated strictly enough. People pay lip service. You can't prevent this attitude as it is built into the bureaucratic system.

Despite these problems, the small firms still expect state regulation to play an important role in governing their environmental activities. The reasons why they favour government to be responsible and enforce regulation are, in large part, to counter the problems associated with self-regulation. Analysis of the interview data indicates that small firms look to institutions, particularly government departments and agencies, to provide clear environmental guidelines and set standards that they and other businesses should be expected to follow. In many respects the owner-managers look to the government to provide leadership on issues of international importance, such as the environment. The managing director of a company designing computer systems claimed:

> I don't genuinely believe it [the environment] can be left as a discretionary issue for businesses to address. You have to have fairly strong leadership backed up by legislation. There is legislation but it is so difficult to get it in to process.

---

6 Regulation includes legislation and other market interventions to control environmental behaviour. In this study most of the small firms limited their comments to the role of legislation.

7 See the 1998 survey results in Chapter 1.

Due to the nature of environmental issues, small firms consider it to be the government's responsibility to act as a legitimate and unequivocal source of guidance and support, the reasons being that government can take a longer-term view than business and represent the wider ethical interests of all members of society. Small firms also hold the view that legislation provides a level playing field, establishes a minimum acceptable standard of behaviour and provides a source of external pressure that has, arguably, been lacking to date. The proprietor of a newspaper-cutting company remarked:

> We need the government to take a stronger lead in guiding environmental standards. Businesses need clear unambiguous guidelines which are specific and actionable.

In summary, the typical small firms in this study do not support the principle of self-regulation as the primary mechanism governing their environmental behaviour. From an ethical perspective, the reasons given were the lack of equity associated with voluntary codes of practice or agreements and the problem of free-riders. In addition, if environmental ethics is to be administered by self-regulation, a higher degree of expertise of the issues and ecoliteracy skills is required among small firms. For the above reasons it is understandable that, currently, a largely state-imposed regulatory framework is seen to be a fairer, more ethical system of control. It is perceived to be the government's responsibility to communicate environmental values, to establish a code of environmental conduct and to provide a benchmark of acceptable environmental standards for the business sector. The government is expected to take a leadership role concerning the environment, rather than individual businesses.

The results of the study indicate that the typical small firm lacks a clear appreciation of sustainability and environmentalism and consequently finds it difficult to relate its business practices with specific environmental issues presented in the media. Understandably, if a small firm cannot identify specific environmental problems within its own business, it is unlikely to consider a need to implement any solutions offered to it. Even if small firms do hold strong ethical principles in relation to the environment, the typical companies in this study prefer state regulation to embody their environmental ethics and control their behaviour.

Discussion

Analysis of the interview data indicates that most of the small firms surveyed are operating within an ethical system derived from the conventional ethical discourse. In this respect, the small firms have largely framed the environmental challenge from a 'shallow ecology' perspective in that they are seeking a code of conduct to govern their environmental behaviour, rather than seeking a new perception of the role and structure of business organisations in society today.

The other salient point to draw from the analysis is the lack of trust the small firms have with regard to each others' ability to behave in a responsible manner without the external threat of regulation. The owner-managers' moral behaviour is restricted by their low awareness of environmental issues combined with a low level of ecoliteracy. This forces them to seek and rely on external experts to guide their decision-making processes and to provide them with appropriate solutions: a situation that both disempowers the small firm and further empowers the dominant paradigm that resides in the institutions of power (government organisations, financial institutions and international agencies).

If small firms are to change their environmental attitudes and behaviour to become more environmentally responsible, they need to become more explicitly aware of their business culture, values and ethics. The owner-managers interviewed in this study acknowledge the important influence that their ethics and moral conscience can play in motivating and shaping their environmental behaviour and business practices.

Welford (1995) suggests that the vehicle that will steer business towards a sustainable model of business enterprise is a 'cultural change programme' that enables small firms to re-evaluate their own value system. Integrating sustainable development into a small firm's decision-making processes will, claims Welford, only be achieved by refocusing perceptions, which is a gradual long-term process. Hoffman (1993: 10) and Welford (1995) separately build the argument that, if businesses do not adopt and incorporate new environmental values into their own corporate value systems, their culture will be out of balance and incongruent with that of their stakeholders and other members of society. In the future, it is expected that environmentalism will become of increasing importance to individuals and society and that 'companies will find it increasingly difficult to resist bringing their corporate values in line with those of the individual and society' (Hoffman 1993). However, Dunlap (1997) states that the public suffers difficulties in converting its pro-environmental attitudes into similar pro-environmental behaviour. In this regard small firms do not differ that much from the general public.

The value system of most small firms is driven by the goal of economic prosperity. It has been claimed that the environment is, possibly, now accepted as a first-order value, resulting in a new relationship between environmental values and other first-order values, namely, social justice (equity), economic prosperity, national security and democracy (Paehlke 1995). Small firms do not operate in isolation; they are influenced and affected by the value systems of the individuals and organisations in their supply chain, their immediate stakeholder network and more distant societal networks.[8] The problem is that most of the small firms interviewed in this study have not yet been sufficiently motivated, or are still unwilling, to integrate their own personal environmental values, or the values of others, into their businesses. The longer this continues, arguably, the more out of touch small firms will become with the ethics of the rest of society.

Another explanation for the failure of small firms to integrate their own individual environmental values with the environmental values of society is their inability to identify a clear environmental vision emerging from the collection of different, often competing, environmental value systems. Environmental problems do not always have

8 See Hunt's discussion on networks in Chapter 15.

simple, clear-cut undisputed solutions. This predicament does not absolve small firms from the responsibility of tackling these difficult value-based environmental issues. Paehlke (1995) suggests that the authoritative allocation of values is the primary function of politics. The question is: are our political representatives circumnavigating the ethical, value-based component of environmental decision-making by pursuing environmental solutions that are based on a narrow technological and economic orientation?

Environmental policy in the UK has, arguably, been overly dependent on scientific knowledge, hence the tendency for technical solutions (DoE 1990). It is questionable to what extent the environmental ethics of 'deep ecology' has shaped solutions being offered to small firms. This implies that environmental initiatives targeted at small firms have tended to take the 'code of conduct' approach to environmental ethics by framing solutions within the existing paradigm. Consequently, the focus has been placed on the benefits to people and business, usually through bottom-line economic savings gained from environmental improvements, suggesting that the environmental ethics of the small-firms sector do not go that deep.[9]

There is a conflict between shallow and deep ecology as to what the prevailing environmental ethic ought to be. Those environmentalists that follow a deep ecology approach to business ethics are more likely to see the value system as the root cause of the 'ecological dilemma' with the solution requiring social or political restructuring rather than an economic or technological fix. Theoretical research into the 'greening of business' from a philosophical rationale has to recognise the tension created among the competing environmental schools of thought (Gandy 1996). Environmentalists from the deep ecology school believe ethical change has to take place in the public domain, whereas environmentalists from a more shallow perspective believe that ethical change can be administered by the individual in the private domain.

There are few fully developed codes of practice that can offer the basis of an environmental ethic for business. To date, there has been little attempt to link the ethical theory of deep ecology to practice. Lessem (1991) acknowledges this position, but suggests that there is a guiding philosophy developed by Schumacher (1973) in his publication *Small is Beautiful*. Schumacher proposed the bringing together of opposites, to find the 'middle way' between economics and ecology. Other attempts to join up ethical theory and practice have taken a more shallow ecology approach. Berry (1990, 1993) has attempted to develop a code of practice based on the concepts of stewardship and sustainability. He cites the work of the International Chamber of Commerce (1991), the Institute of Business Ethics (Burke and Hill 1990) and the Coalition for Environmentally Responsible Economics (CERES), publishers of the Valdez Principles, as examples of how business responds to the need for an environmental ethic to guide its practices. Until the theoretical debate is opened up for wider discussion, it is more than likely that small firms will continue to seek, and be influenced by, environmental ethics within the conventional discourse, rather than challenging these principles with deep ecology thinking.

---

9 Admittedly, there are individual exceptions to be found.

Conclusion

Whether small-firm environmental ethics should go any deeper is not a question to be answered by the author, but what can be said is that the dominant environmental ethic is being formed by the conventional ethical discourse. This discourse operates within the dominant paradigm, which in the Western world is represented *inter alia* by the neoclassical economic system. It is this economic system that acts as a dominant barrier to change small firms' environmental behaviour and, therefore, ethics. Although small firms are responsible for their own behaviour, justified by their own values and ethics, this is inextricably linked to the structures and institutions prevalent in society.

The typical small firm in this study has yet to become fully engaged in the environmental debate. Without institutional reform and a restructuring of the economic system in the public domain, it is unlikely that there will be widespread deepening of environmental ethics among small firms. Furthermore, there is only so much individual owner-managers and small-firm employees can do. If Friedman is to be believed, business must play within the rules of the game. The rules in this context are established by environmental regulations that are being framed using a shallow ecology ethic. In the UK, however, small firms too often fall below or outside the compliance requirements of environmental regulation. Consequently, they are not required to take responsibility for the environment to the same extent as larger businesses.

Hoffman (1991) suggests that the shallow ecology strategy promotes ethical and environmentally responsible behaviour to business on the basis that it is 'good business' to do so. This is interpreted as a 'win–win' situation that increases profits by saving costs or improving the efficiency of the business. This message may effectively attract small firms to take the first step to enhance their environmental performance and behave more responsibly, but what happens once the low-hanging fruit has been plucked from the tree of eco-efficiency? Once small firms are confronted with environmental problems that do not save them money, and may indeed cost them substantial sums with little or no financial return, the environmental ethic comes into conflict with 'good business'. The small firm is no longer in a 'win–win' situation within the shallow ecology environmental ethic.

Over-simplifying environmental ethics to the small-firm sector in this way has its attractions because it appeals directly to the profit motive of the owner-manager, but it also has dangers and limitations. This is why it is so important to build a new small-firm environmental ethic that will be able to guide their environmental behaviour in all circumstances, including those where the financial or economic costs are outweighed by the environmental benefits. Fox (1996), therefore, bestows a cautionary warning to those who work outside the field of environmental philosophy, but who use its tools:

> . . . treat these tools—these limited, often flawed, and, yet, nevertheless indispensable tools!—with a certain degree of caution. No one has the answer yet—even though we needed it yesterday—and to act as if one of these

> approaches is the answer is to forget that 'for every complex problem there is a solution that's simple, neat—and wrong'.

The discipline of small-firm environmental ethics is in an embryonic state. Much work still has to be done to develop better ethical tools and to integrate new theories with practice in small firms. These are early days; the challenge ahead is an important one if better and more appropriate solutions are to be found to enable small firms to become fully fledged participants in the development of sustainable societies.

3
Small and medium-sized enterprises and environmental compliance

Attitudes among management and non-management

Judith Petts

This chapter explores the regulatory context in which small and medium-sized enterprises (SMEs) operate, particularly the attitudes of individuals within SMEs to environmental compliance. The business and the environment literature has tended in the past to promote regulatory compliance as the baseline of corporate responses to the environment—without compliance improved environmental performance will not be possible. This argument has begun to be questioned, however, not least in relation to SMEs that may not have the resources to ensure compliance. Most of the literature has focused on managers within businesses when exploring the perceived importance of regulatory compliance, paying less attention to non-management, presumably because they have not been perceived to be the motivators of corporate responses. However, the increasing attention to issues of organisational learning and employee involvement must lead us to question of what are the attitudes of *all* employees.

This chapter focuses on the discussion of the results of empirical work which explored the attitudes of management and non-management within SMEs in England and Wales to the environment and environmental compliance (Petts *et al.* 1998a; Petts 1999; Petts *et al.* 1999). By means of introduction, the next section discusses some of the literature and issues that surround environmental compliance motivation and the styles and types of regulation to which SMEs are subject. This is followed by discussion of the results of the study in relation to attitudes to the importance of compliance, attitudes towards to the regulator, perceived differences between health and safety and environmental legislation, and perceptions on the importance of compliance as a basis for environmental performance. It concludes with consideration of the policy, regulatory performance and business implications.

Importance of regulations and compliance

Compliance as the motivator of greening?

Typologies of corporate greening (e.g. Welford and Gouldson 1993; Welford 1995) have suggested that the regulatory domain that affects businesses, specifically the driver of compliance with legislation, has to form a significant baseline for development of culturally and internally driven change. On the 'ROAST' scale of environmental performance (Dodge 1998), this baseline is encapsulated in the strategic corporate response that reacts to the forces of change and is reflective of a compliant and observant organisation. Surveys of businesses would seem to partially support these typologies with companies suggesting that compliance is a key motivating factor behind their environmental consciousness (e.g. IoD 1994; BCC 1994; Johnston and Stokes 1995: Hillary 1995; Baylis *et al.* 1997; Ross and Rowan-Robinson 1997; KPMG 1997b).

Unfortunately, research that is focused only on a large questionnaire, while providing us with some degree of statistical comfort about 'what' people think, does not provide for a full understanding of 'why'. The type of question asked will direct a certain reply such that the nuances that underpin the answers are lost. A general questionnaire that asks a manager to identify factors influencing a company's environmental performance could be predicted to identify legislation as a driving force. Certainly it could be predicted that in any survey the majority will identify the environment as an important issue, simply because respondents will consider it socially unacceptable to suggest anything to the contrary. Regulation will always get 'grudging acceptance' because most people want a 'liveable planet' (Porter and van der Linde 1995). In a similar vein, if legislation is not ticked as being important it may be thought to suggest that the company is non-compliant. Compliance is likely to be viewed as promoting the legitimacy of business. As legislation in general is viewed as a burden on business, not least among SMEs (House of Commons Select Committee 1998), environmental response should seek to ensure compliance and so lessen, or at least stabilise, the burden.

There is no doubt that models of corporate greening are conceptually attractive. However, their simplicity, not least in terms of their consideration of the processes by which change may take place within companies, has been questioned (e.g. Hass 1996; Eden 1996; Neale 1997). Activities are more likely to go on at several levels within companies. A single company might be compliant, reactive, anticipatory and proactive on different environmental issues, depending on the centrality of these to its own business, its public profile, legislative interest and the ease with which technological responses can be made (Eden 1996; Petts *et al.* 1999). While most companies take a compliant stance, for SMEs anything more than compliance is 'unrealistic' due to a lack of resources and ability to react rapidly and flexibly to pressures (Eden 1996: 67).

Compliance with what?

Command-and-control regulation represents a system of direct control over activities and organisations that has a legal basis (Jacobs 1991; Ball and Bell 1995; Gouldson and Murphy 1998). Traditionally viewed as the mechanism by which government protects public interests that could be compromised by the free market, command-and-control is the primary means by which environmental protection policies are achieved. Command-and-control is characterised by reactive enforcement. This emphasises the use of quantitative standards such as consents for discharge of effluent to water and limits on emissions to air, i.e. end-of-pipe control. The principles of operation to meet required standards are often encompassed in some type of permit. A monitoring regime then provides for ongoing contact between regulator and regulated, albeit of varying degrees from regular (even several times a week as for some landfills subject to the Waste Management Licensing Regulations 1994 in England and Wales) to very occasional, as is frequently the case with consents for discharge to water. Here, compliance has to be demonstrated, but there is less direct and regular intervention by the regulator.

Under command-and-control regimes the dependence, ultimately, on often time-consuming and expensive enforcement presents a problem. Enforcement action by the regulatory body is more likely to be prompted by a third party lodging a complaint or reporting an accident than to result from routine inspection by a regulator (Hutter 1986). For example, in England and Wales it is estimated that over 80% of all environmental enforcement actions stem directly from third-party complaints and/or information.[1] Furthermore, in Britain until very recently there has been considerable concern that the final penalties awarded by the courts are too small to have any deterrent effect (ENDS 1997). The understanding of the courts of the implications of environmental non-compliance has been questioned.

Where the public distrust industry to act in interests other than their own profit margins, reactive enforcement may provide for confidence in the regulatory process and perceived effectiveness in dealing with environmental protection policies (as long as the regulator is not perceived to be supporting or 'in league with' industry [Petts 1998]). For the regulated, the reactive approach based in formal procedures should encourage consistency of treatment—the 'level playing field'.

In contrast, self-regulation, while still promoted as 'in the public interest', allows the regulated to manage the regulatory process (Baggott 1989). Self-regulation is actually a very broad concept covering everything from voluntary action that is unenforced by law and unpersuaded by financial incentives (Jacobs 1991: 134), such as the Eco-Management and Audit Scheme (EMAS) (CEC 1993), through regulation via industry alliances (e.g. the chemical industry's Responsible Care programme), to self-

1 Environment Agency, Central Legal Department, personal communication, 1997.

regulation through a framework of rules (such as Britain's 'Duty of Care' for waste management; s. 34 of the Environmental Protection Act 1990). In regulatory regimes that encourage self-regulation, enforcement utilises inspection to encourage industry to take appropriate actions using qualitative principles such as 'best available techniques'. This arm's-length style relies more on co-operation and discussion than prescription. The Confederation of British Industry (CBI) has suggested that voluntary action should always be the first recourse of government because it is the most likely way of securing competitive advantage and maximising environmental action (CBI 1994). Recently, the CBI (1998) has called for a shift away from prescriptive environmental regulation to a more goal-oriented risk-based approach to control, drawing parallels with health and safety legislation.

Free-market doctrine has led to a search to remove or at least lessen bureaucracy. Integrated Pollution Control (IPC) (Part 1 of the Environmental Protection Act 1990) was 'sold' as a changed approach to dealing with environmental problems through the integration of the consideration of effects on the three environmental media and a move from a primary focus on end-of-pipe control to encouragement of prevention of problems through waste minimisation and technology enhancement. It was a new policy approach to dealing with the environment, albeit first promulgated in the late 1970s by the Royal Commission on Environmental Pollution (1976). IPC simplifies the administrative and regulatory system, responding in part to the deregulation demands of industry, but more directly dealing with complaints about complex regulation which left companies confused and uncertain (Haigh and Irwin 1990).

While self-regulation might be viewed as a means of bolstering the deficiencies of command-and-control regulation, it also potentially has more positive or supportive functions in terms of extending the role of regulation to influence corporate behaviour to respond to social goals, such as sustainable development. However, it is easy to argue that in an ideal world a balance should be achieved between proactive and reactive regulation, and it is apparent that application of environmental management initiatives appears to be strongest in those developed economies that already have strong command-and-control regulatory systems (Gouldson and Murphy 1998: 22). However, key questions remain as to effects on businesses in terms particularly of promoting environmental responsiveness.

Environment versus health and safety

Following the CBI's (1998) concerns, it is relevant to question whether any distinctions can be drawn between environmental compliance compared to legislation dealing with health and safety. The environmental and safety dimensions of industrial activity and management have some similarities. Both can significantly affect the ability of a company to perform well and both have potentially adverse cost implications if management fails, but have been perceived (at least in the past) as relatively small cost components, largely because they have not been measured effectively.

There has been evidence of an increasing management linkage between health and safety and the environment, particularly within larger companies that have established SHE (safety, health and environment) teams. In general, health and safety is more likely to have been incorporated formally into management structures than is the case with the environment. This is because the primary health and safety legislation in Britain (Health and Safety at Work etc. Act 1974) has one distinguishing feature compared to environmental legislation—it places a direct responsibility on the individual employee as well as the employer. Implementation by companies requires that all individuals are trained and are aware of their responsibilities. Health and safety legislation has, arguably, had a longer time-span of concerted impact on companies than the piecemeal, media-specific environmental legislation and hence could be expected to have a higher profile within companies.

Environmental legislation that impacts on SMEs

Clearly, the relationship of individuals within SMEs with the regulator must be founded on the nature of the regulation to which their companies are exposed. Herein also lies an important difference with large companies. The primary British legislation that supports the proactive regulatory style is IPC. IPC integrates consideration of emissions to air, water and land, its main objective being to minimise and, as a last option, render harmless emissions of prescribed substances. Operators of prescribed processes have to apply for an authorisation demonstrating that they understand the potential impact of the operations, have selected the Best Practicable Environmental Option which will minimise the impact of disposal and are applying BATNEEC (Best Available Techniques Not Entailing Excessive Cost) to operate the processes to meet the objectives of the legislation.

IPC focuses on the most potentially polluting industries and only currently relates to a few thousand processes, although this number was extended with implementation of Integrated Pollution Prevention and Control (IPPC) from October 1999 (CEC 1996a). However, only about 5% or less of SMEs are subject to Part A of IPC (in the absence of full national statistics, this is an estimate based on regional analysis in the research discussed here and from other studies, e.g. Baylis *et al.* 1997). Similar analysis suggests that perhaps about 15% of SMEs might be involved in Part B processes subject to local authority control (as opposed to Environment Agency regulation in England and Wales for Part A processes). While these Part B processes are subject to the preventative and proactive regulatory style inherent in consideration of the BATNEEC, this only relates to air pollution control and does not take the fully integrated view of air, water and land as under Part A.

A larger proportion of SMEs are subject to reactive traditional end-of-pipe controls or permit-based regulation by means of trade effluent consents for discharges to sewer and discharge consents for effluent discharged direct to watercourses, or waste management licences. However, more importantly, the majority of SMEs are only affected by incident-based (rather than permit-based) regulation so that companies only come into contact with an environmental regulator if a general pollution law has been contravened and has been traced back to them (Baylis *et al.* 1997). Like large companies, SMEs are subject to the 'duty of care' in relation to waste management, although previous surveys have identified a low awareness level in relation to this self-regulatory approach (Hillary 1995; Baylis *et al.* 1997).[2] Overall, it was hypothesised that the environmental regulator is having little real impact on the environmental responsiveness of SMEs.

SME attitudes to compliance

The following discussion draws on the views of over 1,000 individuals in SMEs (defined as companies with 10–249 employees, i.e. excluding the micro companies), both managers and non-management, who contributed to a multi-method research project.[3] The project combined a questionnaire with interviews and focus groups so as to understand better not only *what* people think but *why*.

The most important element of the research was the inclusion of non-management. Most previous SME work has focused on owner-managers, and there has been a strong message that, as in large firms, managers in many small firms possess positive environmental attitudes (e.g. Hutchinson and Chaston 1993). However, non-management will be vital to the translation of management commitment to the environment into real initiatives and if environmental concerns are to permeate the whole organisation (Hunt and Auster 1990; Smith 1993; Zeffane *et al.* 1995). Central to Agenda 21 are calls for workers and their unions to be more involved in the industrial change that will be necessary to achieve sustainable development. Yet previous work has done little to ascertain what non-management thinks and it is easy to argue that companies, particularly SMEs, have done little.

The importance of compliance

The research confirmed that both management and non-management believe that compliance is important. As predicted, compliance was considered to be morally right, with the strength of agreement between management and non-management on this matter being particularly important. The 'importance of compliance' was the only one of four factors identified by factor analysis of responses to 42 statements in the questionnaire which identified no differences between the views of managers and workers (Petts 1999).

Breaking the law was not viewed as acceptable, despite the costs or difficulties of compliance. Furthermore, the majority view was that compliance was not simply a means of ensuring a company does not end up in court. Rather, compliance was a means of protecting the environment. However, many people in discussion, particularly non-management, had pessimistic views as to whether companies, as

2 See the 1998 survey results in Chapter 1.

3 The survey was funded by the UK ESRC Global Environmental Change research programme (award reference no. L320253210) and conducted over the period October 1995–December 1997. The full survey methodology is described in Petts 1999 and Petts *et al.* 1999. The author acknowledges the contribution of Andrew Herd, Simon Gerrard and Chris Horne to the research.

opposed to individuals, have a positive attitude to compliance: 'there are always companies who will cut corners'; 'a firm will not comply unless it is told to do so'; 'the motivation to comply is based in what the company can get in return'; 'compliance is a marketplace consideration', and 'most companies will do the minimum which is necessary'. Some non-management employees expressed scepticism about their own managers, with suggestions in particular that they 'tidy up before the inspector is due but otherwise don't bother'. Fifty-one per cent of non-management agreed that 'industry's profit is made at the expense of the environment' whereas only 26% of management agreed with this statement.

Managers tended to stress the importance of their own stance in ensuring compliance, e.g. 'compliance will only happen with individual [director-level] backing' and saw compliance as an 'honest' approach to business, the problems of non-compliance (for example, interruption to business, bad press, penalties, prosecution) being 'too important for a manager to ignore'. Such statements tend to reinforce the business imperative as opposed to the individual preference, although in terms of outcome the two are not greatly dissimilar.

Overwhelmingly, 'compliance' was equated with compliance with end-of-pipe-focused legislation, the latter focus not being surprising in that such a small percentage of SMEs are subject to IPC. It was only in discussion with managers that concepts of voluntary actions and of industry good practice were equated with compliance. Because non-management seemed to see the law as essential in making their management comply (Petts *et al.* 1998a), it is perhaps not surprising that they do not have a concept of the value of self-regulation. Of course, it is unlikely that management have discussed such concepts directly with their workforce, or that the latter are exposed to the political discussions of self-regulation that are apparent in trade and industry organisations. This seems an important point to make. Any message that self-regulation can be a relevant way to protect the environment is likely to be a difficult one to convey to the workforce, not least when as individuals, just like the lay public, they view tough legislation as essential in keeping recalcitrant management in place.

Understanding legislation

Although the majority of individuals in discussion defined compliance as 'complying with current legislation', the specifics of exactly which legislation and its elements (regulation, codes of practice, licences, etc.) were articulated extremely poorly (Petts *et al.* 1999), particularly among non-management. The lack of individual responsibility, the fact that it is not essential to train staff in relation to environmental legislation and regulatory requirements all seem to have a limiting effect on awareness, certainly compared with health and safety legislation.

Focus group discussions revealed evidence that the purpose of some regulation (in terms of environmental protection) was not immediately apparent (e.g. the registration of waste carriers) and hence potentially open to abuse because of a perceived lack of importance. When discussing health and safety regulation, many people identified a higher profile in terms of compliance within their companies, not simply because of regulatory requirements, but because non-compliance could have immediately obvious and serious effects (i.e. death) in the workplace.

The research found that harm to the environment from non-compliance was separated from causation in people's minds. Thus, if the cause was accidental rather than deliberate this was regarded less seriously (Petts *et al.* 1999). Supporting other work (Baylis *et al.* 1997) there was not a strong correlation between views on the importance of compliance and whether an individual was from a company subject directly to a permit (Petts 1999). Being subject to a permit only seemed to have effect on management systems and structures.

Interestingly, there was still a majority of respondents in non-permit-holding companies who believed that their company (whether manufacturing or non-manufacturing) could pollute the environment. In general, managers, and those in medium as opposed to small firms, had a greater perception of their company's ability to pollute the environment. This may indicate a potential difficulty in convincing some smaller companies that they can have an environmental impact. However, caution is required in this interpretation as a sectoral analysis was not possible, which may show a more complex correlation between size of firm and potentially polluting activities. It also re-emphasises the difficulty in understanding how people interpret 'the environment' and 'environmental impact' and how they relate these to regulation.

When commencing both interviews with individuals and the focus groups, an open question was used 'What do you perceive as "the environment"?'. Non-management in particular tended to adopt a Darwinian concept of the environment 'being everything around me' and hence drew little distinction between their workplace and the outdoor environment. Indeed the condition of their workplace was often identified as an indicator of their management's commitment to the external environment. This perceived link is important to note because it stresses the potential for health and safety and environmental management to be linked.

In general, the environment was perceived to be a broad issue encompassing global, national and local issues, and pollution as well as amenity issues. For example, in just one focus group of eight non-management, the following were suggested as environmental concerns: 'concepts of pollution'; 'keeping Britain tidy'; 'the visual impact of our towns'; 'the impact of products and the consumer society'; 'the loss of green spaces and the spread of towns'; 'global concerns such as the hole in the ozone layer'; 'fumes in town centres caused by traffic'; 'increased poor health such as asthma'; 'loss of wildlife species forever'; and 'noise from neighbours'. The environment seemed to have both personal and social components; it was viewed as under stress rather than being in 'good health', although there was a mixed response in the questionnaire as to whether the state of the environment had improved or deteriorated over the last decade. Interestingly, those in rural areas were more likely to think that it had deteriorated (one of the few differences between rural and urban areas identified). Managers from companies subject to a permit were more likely to think that it had improved, perhaps because they had more involvement in controlling and also understanding environmental impacts. In the discussion sessions, older people, who could remember the smogs of the 1950s in urban areas for example, seemed to be more optimistic about the current state of the environment, and those who had been fishing for many years would talk about the water being cleaner now, but still lamented poor catches!

The study suggested a need to clarify what is perceived to be important to people in terms of environmental impact and hence where environmental policy should focus. This should also help to define criteria of environmental performance that have meaning to individuals. It is debatable whether criteria focused only on reducing emissions and waste, or on improving energy efficiency for example, which are often important indicators used in company environmental reports, address all the issues of concern to individuals. It is most important to identify any mismatch between rankings of serious impacts between lay people and experts/policy-makers (Petts 1999).

It is unlikely that the importance of compliance with environmental regulations will be accepted without question by non-management. There is a need to make the direct link to environmental harm as a result of non-compliance. The study leaves little doubt about the importance of protecting the environment as perceived by the majority of the workforce.

Attitudes towards the regulator

While compliance with regulation was regarded as important, the effectiveness of the regulatory system in terms of impact on companies—i.e. making them comply and helping to provide a baseline for self-regulation—was regarded as weak, with too small penalties being the most significant deficiency identified (Petts 1999; Petts *et al.* 1999). However, 91% of respondents to the questionnaire believed that regulation alone could not protect the environment, not only because no regulatory system can provide for full environmental protection, having to focus for efficiency on key elements of the environment, but also because any regulatory system will always be plagued by inadequate resources. This perception of weakness seemed to correlate also with individuals' views on the nature of pollution problems, not least in relation to amenity issues, some of which are not directly regulated.

Effectiveness was measured clearly by individuals in terms of inspection frequencies, enforcement and penalties, i.e. a reactive implementation style. The focus of consideration of effectiveness was on the outcome rather than the efficiency of the process. In the focus groups non-management frequently referred to the need to 'put managers in jail' if they were to be made to take compliance seriously, although overall only a slight majority (51%) agreed that this would be effective. Non-management tended to talk about 'too few inspectors to cover the whole country'; 'too few inspections'; and 'too small fines'. Management tended to talk about the complexity of legislation and the fact that there was too much legislation for them to be sure that they were complying. The latter is a significant issue in relation to raising awareness, explaining and justifying regulations to companies.

There was some ambivalence among individuals as to whether enforcement in companies such as their own was tough enough, which may reflect an inherent tendency to view what happens in one's own organisation as different to what happens in others. Supportive of the work in South Wales (Baylis *et al.* 1997), enforcement and application of regulation was not perceived to be a level playing field, either geographically, across sectors, or on the basis of the potential to pollute (Petts *et al.* 1999). A credible enforcement threat was perceived to be required to motivate compliance. However, while this opinion was based in an ethical view that compliance is right, it also reflected self-interest, i.e. maintaining the level playing field by making sure that all competitors are subject to the same stringent controls.

Health, safety and environment

Environmental compliance was not perceived to be more important than health and safety compliance. It was perceived clearly as being relatively new—or, rather, people had only become aware of its importance in their own workplaces relatively recently, i.e. since the beginning of the 1990s. One manager spoke of management and non-management in his company (an engineering firm) being on a par at this time—aware that the environmental legislation exists but still unsure as to how to implement a management structure that would ensure that the requirements are taken on board fully. Often there seemed to be a divide between the understanding of environmental issues and making a connection to the relevance of that issue in their own work environment. While over 60% of respondents to the questionnaire reported that their management had disciplinary procedures for failure to follow procedures relevant to health and safety issues, only 42% reported similar procedures in relation to environmental controls.

It is interesting that health and safety legislation in Britain does incorporate elements of self-regulation, for example through the Management of Health and Safety at Work Regulations 1994, which require a company to identify risks arising from its activities and to put in place systems to minimise these. Therefore, it is not fully prescriptive, one of the characteristics that non-management seemed to suggest was important for environmental legislation. Neither was the enforcement of health and safety legislation necessarily perceived to be effective. For example, in two focus groups there was much discussion of the Health and Safety Executive (HSE) not 'having the staff to enforce' and a view that the HSE therefore concentrated its efforts 'on showcase events where death was involved' so 'picking a big one and making an example'. Nevertheless, health and safety legislation was perceived as having a greater impact in the workplace than environmental regulations, due to a developed culture of response and the visibility and immediacy of the events arising from failure of control.

The health and safety literature often talks about the need to 'win hearts and minds' (e.g. Cox and Cox 1996) if the culture of safety is to be ingrained at all levels of an organisation. The study discussed here suggests a need to understand how the same can be achieved in relation to the environment, not least when environmental legislation itself seems only to be impacting directly on one or two individuals in SMEs and when the rationale behind the legislation and the benefits in performance that it may bring lack clarity to many.

Does regulation drive environmental performance?

Overwhelmingly, the study found little evidence that regulation and compliance was driving environmental performance, i.e. providing the baseline. Rather, it seemed that the majority of companies were probably 'vulnerably compliant' (Petts *et al.* 1999), i.e. they were aware and supportive of the importance of compliance, but they had not developed the management systems that could ensure that they were always compliant.

In the 12 companies that were studied in more detail because they were working to put in place an environmental management system, there was evidence that compliance and performance were not perceived to be linked. The need to meet certain regulatory requirements was not driving their work, rather the potential to save money and remain competitive, for example, through waste minimisation and energy saving activities (Petts *et al.* 1998a). However, there was evidence that, for individual managers who were responsible for the introduction of new systems, personal support for the environment provided an important foundation for their actions.

In the automotive supply chain (six of the companies in the proactive group) there was evidence of pressures to improve environmental performance and to ensure a baseline of environmental compliance. Performance was seen as a means of remaining competitive and of retaining an image appropriate to the needs of the large car manufacturers, although some managers expressed scepticism that they would be thrown out of the chain if they did not meet all targets. In other supply chains there was perceived to be virtually no influence from the large manufacturers and retailers—not least the clothing manufacturing sector.[4] Further work would be needed to confirm whether supply chains experience less pressure. For the regulator these might provide a good focus for activity, not least in the encouragement of self-regulatory activities.

The mismatch between perceived environmental impacts and the focus of the regulatory system—for example concerns about vehicle emissions and the need to manage transport that is not directly regulated by environmental legislation—undoubtedly contributes to the mismatch between compliance and performance. There could be companies who are performing well in terms of recycling, saving energy, contributing to the environmental improvement of the area in which they are based, reducing their waste arisings, etc., but who are still non-compliant at times in relation to permits that they hold. This, of course is the reason that in Britain, at least, the environmental regulators have so far not been willing to reduce inspection frequencies where a company has a certified environmental management system.

Key questions remain as to how we measure environmental performance and what are the criteria—ecological, social, economic—that should be applied, and whether environmental legislation brings environmental benefits as measured against these performance criteria.

4 The disinterest of customers in SME environmental performance is discussed by Hillary in Chapter 10.

Conclusions

The study discussed in this chapter has provided valuable insight into the responses of individuals in SMEs to environmental compliance. More work is needed to understand sector-specific differences and responses. There was little evidence from the research that there are differences between sizes of companies. Individuals believe that environmental compliance is important, a moral issue that is essential if industry is to legitimate its activities. However, the traditional command-and-control system is perceived to be weak, not least by non-management. Environmental legislation is not thought to have made the impact, as yet, that health and safety legislation has achieved. Part of the problem undoubtedly lies in the fact that the environmental impacts of industry are less visible to individuals within the workplace. Individuals' responsibilities are lacking in law and have, as yet, not been communicated effectively within the majority of SMEs.

Tough prescriptive regulation is perceived to be essential by non-management. Therefore, it will be necessary for both regulators and management to explain and convey the benefits of self-regulation if this is to be effectively adopted within companies.

Compliance with regulation is not in itself driving environmental performance in the minority of SMEs who are taking action. Indeed, looking at the activities in which more proactive companies are engaged, it is clear that there is a need to derive effective and acceptable criteria for measuring the benefits of environmental action. This will be essential if companies are to attempt to meet sustainability criteria. Social and economic criteria may be as important to individuals as physical or ecological impacts. Further work is needed to understand criteria that are relevant to individuals.

There appears to be scope for linking health and safety and environmental management systems and for closer working relationships between the two sets of regulators in Britain. The environment gets individual support; however, its visibility and direct relevance in the workplace need to be conveyed and promoted by regulation in a similar way to that of health and safety regulation.

Finally, the environmental regulator must justify new regulations and carefully explain their environmental benefits. The justification of new regulations has traditionally focused on a cost analysis, in terms of proving minimal additional cost to industry. This needs to be supported by its environmental justification.

4

Small and medium-sized enterprises' perceptions of the environment

A study from Spain

Manuel Ludevid Anglada

Before introducing any policy aimed at encouraging environmental management, it is essential to know how the business community perceives environmental problems, their causes and possible solutions. This is particularly important in the case of small and medium-sized enterprises (SMEs)[1] whose perceptions may differ considerably from those of large multinational companies.

This was the rationale for a study, started in 1996, of 20 Spanish-owned SMEs located in Catalonia.[2] The study was commissioned by the Fundació Jaume Bofill and supported by the Catalan regional government, the Generalitat de Catalunya. The aim was to reconstruct the mental 'discourse' of local SMEs in order to design

1 European Union definition of SME used, i.e. *small*: less than 50 employees, independent, annual turnover not exceeding ECU 7 million or annual balance sheet total not exceeding ECU 5 million; *medium*: less than 250 employees, annual turnover not exceeding ECU 40 million or annual balance sheet not exceeding ECU 27 million.

2 Catalonia is an autonomous region located in the north-east of Spain. It is bordered by France and the Mediterranean coast. Its six million inhabitants account for 16% of the Spanish population and it produces 19% of Spain's gross domestic product. Barcelona is the region's capital. The industrialisation of Spain began in Catalonia, whose industrial structure continues to be highly diversified, active and export-oriented. The Catalan government has made use of its autonomous powers to enact Spain's most advanced environmental laws and has applied European Union standards even before they became Spanish law. It is, therefore, the most interesting part of Spain in which to analyse corporate environmental policies.

appropriate public policies to encourage environmentally friendly management practices.[3]

This chapter summarises the conclusions of the study. It identifies, categorises and analyses SMEs perceptions about the existence of environmental problems, the characteristics of these problems, the geographic areas where they occur, the agents responsible for them and the impact of the new challenge of environmental protection on firms, the economy and society. The findings are illustrated with excerpts from the statements made by SME representatives during the in-depth interviews which are the basis of the study. The chapter concludes with recommendations aimed at encouraging firms to manage their environmental impacts.

Perception: a complex process

Our research is based on the human cognitive process. This is a complex process, which can be divided into four principal levels or stages (see Fig. 1) (Ludevid 1995):

- **Level of knowledge.** This is the information, of any type and quality, available on a particular subject. An example would be a television programme on global warming and the greenhouse effect.
- **Perception.** This is the individual's mental image or understanding of the problem. For example, having seen the television programme, the individual clearly understands what the global warming process involves and its possible consequences. This is known as the *cognitive process*.
- **Valuation.** Here the individual incorporates the problem, analysing it in accordance with his or her particular scale of values. Individual attitudes, values and preferences are derived from this 'incorporation'. To continue our example: the individual is aware that global warming is a serious problem, that something must be done to prevent it from getting worse, and that the government must intervene. This is known as the *affective process*.
- **Behaviour.** In the final phases of the process individuals compare their stated values with actual options: this is what we call *behaviour*. In our example, the individual decides to act in accordance with his or her stated values, joining an ecological association and changing buying habits.

Each one of these levels has its own difficulties and problems, and people do not automatically move from one stage to another. Many voice opinions in favour of environmental protection, but few behave in consonance with their beliefs.[4]

---

3 We chose in-depth interviews as our methodology because we were more interested in classifying opinions than in quantifying actions. Our aim was to reconstruct 'typical' opinions of managers of industrial SMEs in Catalonia. The Fundació Jaume Bofill, which specialises in social science research, is located at: C/. Provença, 324, pral. 08037 Barcelona, Spain. Tel: +34 9 3 458 87 00, +34 9 3 458 87 09; Fax: +34 9 3 458 87 08.

4 See the discussions on attitudes to environmental compliance in Chapter 3 and to environmental ethics in Chapter 2.

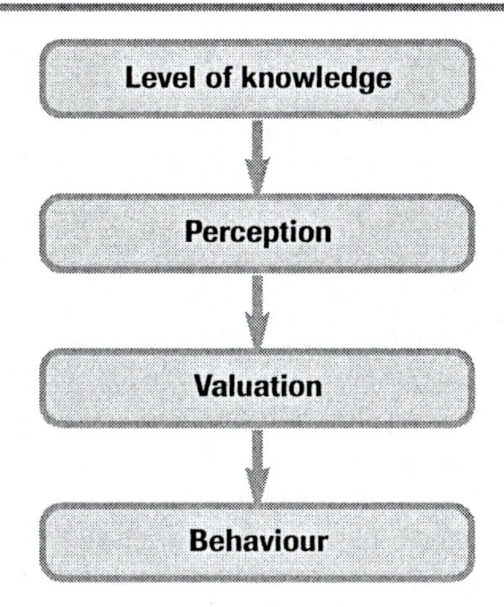

Figure 1: LEVELS OF HUMAN COGNITIVE PROCESS

The study findings discussed here shed some light on this complex learning process. They enable us to identify the views that prevail among Catalonia's SMEs and present a typology of their main positions.

Study methodology

The study is based on 20 in-depth interviews with the managing directors of 20 companies operating in ten different industries in Catalonia.[5] The industries were selected from those whose operations have the greatest impact on the environment.[6]

Two people from each industrial sector were interviewed: one was the managing director of a medium-sized company and the other his counterpart in a small company. We attempted to cover a wide geographical area (within Catalonia). As mentioned at the outset, all the companies studied were largely Spanish-owned. We deliberately did not study foreign-owned multinationals, whose approach to environmental issues is often motivated by concerns that are very different from those of local SMEs.[7]

5 The ten industrial sectors were: petrochemicals; chemical-pharmaceuticals; food and beverages; leather; energy and mining; concrete and building materials; paper; automobile; metal works; textiles.

6 Dr Enric Pol, professor of environmental psychology at the Universitat de Barcelona's School of Psychology, was commissioned by Manuel Ludevid to carry out the fieldwork for this study. The fieldwork was co-ordinated by Professor Tomeu Vidal, also of the School of Psychology. Their interviews served as the basis for this analysis.

7 Indeed, large multinational companies often adopt environmentally friendly management processes in response to head office policy, regardless of the situation in the country where they are located. For example, the first companies in Catalonia to be awarded ISO 14001 certificates were a number of multinationals acting in accordance with guidelines for their entire corporate group.

Some of the findings of the study are presented below and illustrated with comments from the interviewees.

Level of information

Following the steps in the human cognitive process, we first needed to determine the **level of knowledge**, i.e. the quantity and quality of available information.

Our first conclusion is that the managing directors interviewed had little information about environmental problems and that the information they did have was based mainly on reports appearing in the mass media, especially television.

Local entrepreneurs in Catalonia are not particularly knowledgeable about scientific matters and the population in general has little information or training in the field of environmental protection. Interest in scientific matters is still very limited in Spain and is, in fact, lower than the country's economic development would lead one to believe. Unlike in northern Europe or the United States, the mass media in Catalonia is only just beginning to report on scientific matters. Moreover, television has somewhat more impact here than in more highly developed countries.

One of the small business managers interviewed put it like this:

> We get our information from the media: the press, communications media in general. Only specialists read the *Butlletí Oficial de l'Estat*.[8]

In addition to these general sources of information, interviewees also receive an increasing amount of specialised information from employers' and industry associations. Government information—particularly from the Catalan regional government—ranks third as a source of information, but at a considerable distance behind the first two sources.

Acknowledging the problem

In terms of **perception**, the first thing we wanted to know was whether SME managers actually believe there is an environmental problem caused by human activities or if they think the whole issue has simply been invented by militant environmentalists and a sensationalist press. Assuming that they believe a problem exists, we also wanted to know how serious they think it is.

Almost all the people interviewed acknowledge that the world has an environmental problem, which is partly caused by human actions. Many of the interviewees feel that the problem is serious and several think that it will get progressively worse.

> Of course the problem exists. If it is not solved it will naturally be a problem. In the mid-term it will get more serious. In the long term, it will be extremely serious, and there will be no going back. We'll have to start taking steps. In the short term it's not a problem.

---

8 The *Butlletí Oficial de l'Estat* (BOE) is Spain's official gazette, which publishes new laws and regulations enacted by the country's executive and legislative branches.

Some people report that they have changed their opinions on the subject:

> Yes, there is an environmental problem. Badly run industries cause environmental problems. My opinion about this has changed at lot. Before, I thought it was a bunch of nonsense, but now I think there really is a problem.

Although the problem is fairly generally acknowledged, a few of the people interviewed feel that it is both fashionable and exaggerated:

> The problem exists. The thing is that certain groups have blown it up out of all proportion.

> It's fashionable now. The serious environmental damage was done a long time ago. Nowadays companies are much less aggressive towards the environment, but there's more talk about it.

Some of those interviewed say they are bewildered and confused, and even feel anguished and insecure when attempting to grasp and define the problem. They were very outspoken about the problems caused by the lack of clear and unambiguous environmental standards and point out that current scientific uncertainty makes people feel insecure in terms of what is or is not legal:

> The problem exists, but it has different facets . . . Everybody talks about it, but we still don't even know what a real ecological product is.

> I believe the problem exists. People are not very aware of it; generally speaking, people don't know much about it. It's a problem that might be serious, but we don't know how serious because people are so unaware.

> No government can guarantee that what you are doing right now might not be bad in the future.

This uncertainty makes industrialists feel somewhat defenceless and more reticent to invest in environmental protection measures that are not guaranteed to be 'correct'.

◻ *Identifying the problem*

SME managers find it considerably more difficult to identify and accurately describe the characteristics of environmental problems. This requires a great deal of perception.

The replies to our questions ranged from very vague or general ideas to extremely specific definitions of problems that directly affect the interviewees. Almost no one expressed the idea of an integrated system in which every action affects some part of the system. They frequently revealed more detailed knowledge of secondary factors (for example, dumping practices or government negligence) than of the essentials (for example, the most polluting production processes), and placed more emphasis on sporadic environmental problems than on structural ones.

These findings are understandable considering the complexity of many environmental phenomena and the low likelihood (or impossibility) of directly experiencing them. For example, it is impossible to 'experience' global warming. This explains why business people (and 'non-experts' in general) have a tendency to pay more attention to local matters that either directly affect them or which they can experience and verify. However, these local problems often play a very minor role in overall environmental problems.

The people interviewed also tend to talk more about causal agents (for example, who is responsible for a specific environmental problem) than about the phenomenon itself. They start assigning blame even before describing what is happening.

Thus, for example, SME managers often link environmental problems with very specific aspects of daily business operations:

> Some things aren't recycled enough. There are no landfills or waste recycling plants.
>
> The amount of waste in the air, the rivers and all over the place.
>
> It has many different facets: dumping waste in the rivers, oceans, in the air, sound pollution, etc.
>
> The tremendous use of materials that cannot be recycled. Increased industrialisation and industrial processes create waste faster than we can handle it today. Industry has grown at a breakneck pace.

At the opposite end of the scale, some interviewees referred to issues that are extremely global and far removed from their particular companies:

> The way industrial transformation processes and population growth are managed causes environmental problems. The more densely populated an area the more aggressive the attack on the environment. Anything we do wrong, even farming, can cause pollution.
>
> We've got forests, but they are getting more and more polluted. Some species are disappearing. All this is simply the logical evolution of industrial development. But if pollution keeps increasing we'll ruin our rivers and forests.
>
> It's very widespread. It isn't just industry. There is also household waste, packaging materials. It's hard to sum it up in a couple of words. But the main thing is to maintain the quality we have, not let things get worse—there's no question about it. Protect the ozone layer, outlaw noxious gases, reduce CFC, cut out as many of those things as possible. There's also the matter of clean-up campaigns and getting people to stop dumping dangerous products in the rivers. Dumping this kind of waste is going to increasingly pollute our rivers and oceans, both of which are part of our water system. Just a few drops of mercury, the whole business of heavy metal, can poison thousands of litres of water. I mean, it's a really broad field.

As indicated from the above quotations, the people interviewed did not have an integrated perception that relates concrete examples with more global or general environmental problems. It is as though our interviewees perceive the problem in disconnected 'flashes', either because they have directly experienced it in their own companies or because of media reports.

Geographic location

When SME managers were asked where the most important ecological problems occur, their replies revealed that the most important environmental problems are perceived as embracing the entire planet but often caused in some *distant* part of the world. The closer to home the problem, the less important it seems. There is a widespread belief that the poorer countries play a key role in environmental destruction:

> Although the reaction of the industrialised countries is very important, Europe and North America do run campaigns to promote environmental protection and better management of natural spaces. But the problem is really serious in the countries that are just starting to industrialise. The developed countries have the industry they need, but in the southern countries, which have the greatest natural assets, they don't have the technology they need to meet the environmental challenge. The balance has to be redressed.

> The underdeveloped countries have to survive. I don't know what the solutions are. The most serious problem is deforestation in South America and Africa. In Western countries we have technology enough to avoid destroying our forests and rivers.

The managers interviewed almost always consider that the environmental problem depends on the zones involved, once again reinforcing the idea that they do not conceive the problem as part of an interlinking system of phenomena. They almost always see urban and industrial zones as the areas with the most environmental problems.

> How serious it is depends on the zone. The problem is worse in cities than in rural areas.

> It's serious in urban areas and industrial zones, and less so in other places.

To sum up, the SME managers interviewed do not see environmental problems as affecting all parts of the world, all countries and all businesses equally. Instead, they tend to consider these problems as both local and isolated.

Direct responsibilities? Business and society

Opinions on who is to blame for environmental degradation are very interesting because they reflect both the interviewees' *perceptions* of the problem and their *valuation* of it.

Our first research conclusion is that the SME managers interviewed see a clear link between business activity and environmental destruction. Nevertheless, the majority feel that these are specific, *sporadic* problems that affect only a few companies, either as the result of poor management or because they operate in industries where the risk of pollution is particularly high. It was frequently stated that pollution is an inevitable cost of progress and increased social well-being. Among the opinions expressed were the following:

> When *poorly run*, the business world is very environmentally destructive.

> There is a fairly direct link between environmental problems and the business world. I think industrial operations are the first thing that need to be regulated. If we want to preserve the quality of life we have to support business activity, but it is also the thing that can do the most harm.

> Industry has the greatest responsibility; it has the technologies and the means with which to provide developing countries with the necessary solutions. They are macroeconomic solutions.

A second, but equally important conclusion—and one that was strongly emphasised by the SME managers interviewed—is that industry *alone* is not responsible for environmental destruction. Society as a whole and every individual member within it are responsible:

> Industry produces all sorts of waste, but it supplies the consumer society. Which came first, the chicken or the egg? We consumers are responsible for a lot of environmental waste.

> There is no single source of the problem. It is an inherent part of the way we are. The source is partly people, partly companies, etc. We attack the environment in a lot of ways. Our industries attack the environment and steps must be taken, but actually everybody attacks it . . . Business and individuals are equally to blame.

> All human activity influences it. It depends on density. The more human activity there is in a small space, the more problems there are or will be.

> The first cause is industry. But industrial waste emission here is totally controlled. Where does that waste go next? Everyone is somehow to blame. It's a chain.

Our third conclusion is that SME managers tend to blame the government, accusing it of not enforcing environmental regulations strictly enough. There is a tendency to shift responsibility, considering that environmental protection is more a matter for the government than for individual companies:

> It's easy to accuse business; companies are more visible. But politicians play a very important role.

> The government has to be the standard bearer. It has to do things and set an example. Then things would really work.

> The fundamental problem is that they make laws and then they don't enforce them strictly. They catch one company and really clamp down on it, but they don't control the others as tightly.

Moreover, SME managers often feel that government agencies are more sensitive to environmentalists than to the needs of the business community, and accuse them of designing non-realistic and overly theoretical projects:

> Everybody talks a lot about the environment, but let's not be ridiculous about it; we don't have to enclose our forests in glass cases. We have to make sure that forests and recreation are compatible.

At the root of these opinions is the belief that environmental protection is a matter of defending the 'common good' and that this is primarily a government responsibility. The majority of the managers surveyed feel that the government should signal the paths to follow for both the business community and society in general.

Deep-rooted causes

Not all the opinions expressed dealt with direct responsibilities, observable in the day-to-day behaviour of companies and society. The managers interviewed also voice opinions about the 'deep-rooted' causes underlying environmentally destructive behaviour.

The SME managers interviewed tend to feel that these deep-rooted causes of environmental problems are linked, on the one hand to the 'inevitable' cost exacted by high standards of living in the rich countries and, on the other (and very particularly), to the existence of serious problems of world poverty which are synonymous with environmental degradation.

One of the managers interviewed summed up the situation as follows:

> Up to now we have been improving living standards and increasingly industrialising, without giving any thought to the cost involved, i.e. the trade-off. It's like what happened in London with coal heating. This is the downside of efforts to improve living standards.

Furthermore, the managers interviewed tend to think that environmental destruction is mainly caused by overpopulation and poverty in much of the world. They also believe that wealthy countries are less environmentally destructive than their poorer counterparts. A possible corollary to this viewpoint is that the problem could be solved by stopping population growth and 'enriching the poor'. Several opinions in this vein were cited in the section on the geographic location of environmental hazards. Here is another example:

> The rich countries are not to blame because they have already enacted measures, they have made changes that take the environment into account. The dangerous time is before people are aware of the issue. That's when developing countries start industrialising. They need to industrialise. They're hungry! They can't bother with the niceties of the developed countries: they exploit their forests, their sources of energy, all their raw materials and they destroy the environment.

These opinions convey a disturbing message: they tend to exonerate the wealthy countries, i.e. the interviewees' country, as well as their companies.

A cost and a threat to companies

Practically all the SME managers interviewed perceive environmental protection as a *cost* to their companies and a *threat* to their competitiveness:

> Environmental policies are shutting down businesses because the cost involved is so high. This is also going to happen to big companies with outdated technology.

> All this has a high cost. It's a really big investment. Meeting the requirements of the 13 or 14 environmental laws that regulate our operations means higher costs and not everyone is prepared to incur these.

> If we had money, we would activate things in order to legalise our situation, but as it is we never catch up due to administrative problems. After we did this, we'd modernise our machinery.

> Environmental policies always have a negative effect on businesses because they require major investments. They are positive for consumers. But companies have to invest funds they often do not have. That's why the effect is negative. Still, that doesn't mean we don't have to do anything; it doesn't mean that these measures don't have to be taken.

The majority of the SME managers interviewed *do not* consider environmental management to be a business opportunity and feel that it does not help increase sales:

> We spend a lot of money on it. In the short term it increases the cost of our products. Later on, you try to recoup your investment. Everybody does it. You increase your prices. If society changes and starts demanding these products, we'll give them what they want, but clean manufacturing processes don't mean that you sell more. With all these labelling requirements, inspections, etc. consumers are overwhelmed; they don't even know what they're getting.

The majority of the people interviewed say that small businesses cannot afford costs like this and, unless they are given enough time to gradually change, they will be faced with the choice of quitting or going underground:

> We're talking about a lot of money: waste purifiers, neutralisers. It's a lot of money. If SMEs have to invest in things like that, the majority of them are going to fold.

> This issue is driving a lot of SMEs into the shadow economy: it's the only way they can survive.

> Small businesses have enough problems just surviving. The environmental issue is an added burden.

In a similar vein, our interviewees expressed a special concern about possible environmental 'dumping' from other parts of the country[9] and from poorer nations:

> Logically enough, the business world tries to make its profits at the lowest possible cost. Pollution control measures are an expense. If your competitors don't introduce control measures and you do, you increase your costs and you are going to avoid increasing your costs if you can possibly help it.

9 Industrialists operating in Catalonia frequently object to the fact that environmental legislation and inspection in other parts of Spain are more tolerant of businesses than their Catalan counterparts. They also complain about 'negative tax discrimination' because companies located in Catalonia have to pay environmental taxes (on liquid waste, for example) that do not exist in other parts of Spain.

> The Eastern European countries don't care about the environment; they only care about keeping down their prices. The environmental issue could end up creating a new border. The Germans realise that. I don't mean that we've got to open the borders. I mean that it could be a wall for keeping out imports.
>
> Moreover, we've been hard hit by European competition. Some countries really respect the environment, others not so much and still others don't respect it at all, like Italy, where people do whatever they feel like because nobody pays any attention to their laws.

Very few of the SME managers interviewed consider that environmental protection investments and costs translate into returns or gains in terms of image or market share. The following quotation was an exception to the rule:

> It makes production a little more expensive, but not exaggeratedly so. It may make the cost of every product unit five pesetas higher and that's not a lot. All the controls we perform make the product more expensive, but that isn't so important. It's like allocating 0.7% to the developing countries. It's more or less the same thing. And, in exchange, your image is really good.

Some interviewees mention savings, but point out that they turn into net costs as the company becomes increasingly 'green'.[10]

> We've cut down our expenses and are aiming to make it to legal requirements on the pollution chart because otherwise they might even shut down the company. Up to now, we've saved on what we've done, but from now on we'll have more expenses. Everything we've done so far enabled us to cut costs, but the new laws mean we will have to invest a lot of money without getting anything extra in return.

Opinions voiced by the few SME managers who feel that environmental protection measures have advantages for their companies indicate that the idea of safety or lower risk might do more to inspire companies to adopt environmental management practices than the idea of obtaining direct profits:

> Good management improves your company's image. That's very important because if your image is well founded it gives you a certain clout, It means you can ask for grants. It can even be a guarantee if there is ever an accident. I mean, earning money isn't everything.

The conclusion, then, is: SME managers regard the challenge of environmental protection as a cost and a threat, aggravated by the problem of environmental 'dumping' by other companies and regions. The principal arguments in favour of environmental management are increased safety and an improved image, rather than tangible economic returns.

---

10 Tilley (Chapter 2) raises the issue of SMEs picking the low-hanging fruit and then halting environmental improvements when easy 'win–win' pickings are exhausted.

The change process: sporadic and specific

Given the perceptions and valuations described, it is not surprising that the SME managers interviewed indicated that, although their companies are changing, these changes tend to be sporadic and specific, aimed more at solving concrete problems than completely changing the overall management system.

The following statements are a good example of this:

> There's not much we can do. We use recycled materials as much as possible.
>
> We recycle the water used in our production process. We also treat the water we draw from wells.
>
> We've installed a purifier for the water used to clean out furnaces and water heaters.
>
> More than changing, we're adapting.

These specific actions go hand in hand with ideas about the collective action that should be undertaken: 'dumping' should not be tolerated and SMEs should help one another:

> Collaboration between companies is also important . . . bringing them together in a sort of confederation, with the government helping out so that more companies can use by-products as their raw materials . . .

A few SME managers warned that this change process could make companies even more technologically dependent on the most highly developed countries:

> Environmental protection is very important in terms of competition, both as an advantage and a drawback. For example, a lot of countries are already very highly developed and they think everybody should eventually reach this level. We still have to do a lot of developing before we catch up with them. If the Germans think we have to be as developed as they are we will have to buy technology from them, because we don't have the technology we need.

Environmental policies are almost always informal. No one is specifically responsible for environmental issues and there is no written statement of intent:

> We don't have it down in writing, but we do it anyway. We have to do whatever we're told. Maybe we ought to draw up some kind of guidelines. We have somebody in charge, but he does other things (he's the plant manager, just under the factory's general manager) as well as handling these projects.

To sum up, environmental reasons have *not* led to any major changes in the overall management system of the majority of SMEs. A few industries have made some very specific changes. A small number of companies say they plan to make changes in the future, but the great majority have not budgeted for these changes.

Generally speaking, change is a gradual process and begins with actions that will most clearly and immediately benefit the company in economic or financial terms and have the most obvious and identifiable impact on the environment, among them saving energy or water. The process concludes with actions that are less noticeable and/or which may only bring economic benefits in the longer term.

Conclusions for environmental policy-makers

Our study led to the following conclusions which could be helpful when making public policy decisions aimed at encouraging environmental management in SMEs:

1. Interviewed companies see environmental problems as isolated incidents, not as interrelated events. Most companies see them as sporadically occurring specific problems for which there are specific, sporadically applicable solutions.

2. Generally speaking, SME interviewees believe that the business community is often blamed for environmental problems when in fact everyone is responsible. Some of the managers interviewed complain that initiatives by public agencies are aimed more at companies and pay little attention to society in general.

3. Many SMEs believe that it is up to governments and public agencies to seek solutions to environmental problems, particularly by defining standards and building infrastructures. Only a minority feel that corporate response to the challenge should be proactive.

4. The majority of the SME managers interviewed consider responses to environmental challenges as an additional cost and an extra obstacle to their competitiveness. Almost none feel that environmentally friendly practices lead to increased earnings or competitiveness, at least not in the short term.

5. By far the most important reasons for adopting environmental protection measures are legal and government pressure; for reasons of safety and risk reduction. Image was also mentioned, although to a lesser extent. Marketing reasons (environmental awareness increases sales) were scarcely mentioned at all.

6. SMEs interviewed complain that regulations are too strict and they are not given sufficient time to adapt to them. They want regulations that are applicable, flexible and easy to update. They say that current regulations are not properly enforced. They also complain that there is insufficient control to ensure that the regulations are applied equally in all companies. They feel that more consistent measures are required and complain about widespread environmental 'dumping'. They feel that there are fewer stimuli for companies willing to become environmentally friendly than for those who violate all manner of regulations and standards. They stress that there are more environmental taxes in Catalonia than in other parts of Spain or elsewhere in the world and maintain that this further undermines their companies' competitiveness.

Based on the above, the following recommendations could be helpful when designing policy initiatives aimed at encouraging environmentally aware management practices in SMEs:

1. Present an integrated and systematic vision of corporate environmental problems, emphasising the interplay between the different aspects of the environment.
2. Stress that environmental problems affect all businesses, all sectors of the economy and all production processes alike. Industry, agriculture and the service sector are all responsible, not to mention transport and construction.
3. Explain, through some practical examples, how an SME can obtain savings and economic gains thanks to better environmental management.
4. Make people aware that environmental challenges affect the entire company and not just the production or operations departments. They affect all functional areas and all phases of a product's life-cycle.
5. It should likewise be stressed that environmental problems affect pre-production processes (suppliers), as well as the post-production period when the product passes through the distributor to the final user(s). Manufacturers' responsibility must extend to these other parts of the life-cycle, including disposal.
6. SMEs should be encouraged to create associations in order to jointly resolve certain environmental problems.
7. Public agencies must make an effort to ensure that laws are strictly enforced and applied equally to everyone, doing their utmost to avoid flagrant cases of environmental 'dumping' that rob public policies of their credibility.
8. Lastly, SMEs expect government aid in order to adapt their processes and products to the new situation. They want financial aid, assistance with training and information, and more time to adapt to environmental laws.

5
Sustainable development and small to medium-sized enterprises

A long way to go

Paul Gerrans and Bill Hutchinson

Small and medium-sized enterprises (SMEs) are playing an ever-increasing role in the Australian economy. While definitions of SMEs vary, by any measure they make up a significant component of the economy. It has been suggested that, based on 1994–95 figures, small businesses account for approximately 97% of all Australian companies and half of Australia's private-sector employment. Further, small business alone accounts for 32% of goods and services and 'is the engine room of the Australian economy' (Department of Workplace Relations and Small Business 1998). In Western Australia, small business currently provides more private-sector employment than big business, at 51% of the workforce (Small Business Development Corporation 1999). Welford (1992: 201) suggests that 70% of global national product is derived from SMEs. Not only do these figures indicate the impact of SMEs on the economy, they also suggest a significant impact on the environment in terms of input and output requirements.

The impact of business and the economy on the environment, at the micro and macro levels respectively, and the existence of and potential for further impacts has been well documented over recent decades. Two separate revolutions in attitudes towards the environment have been identified by Pearce and Warford (1993). Initially, the focus was on environmental quality versus economic growth, whereas the second, and continuing, revolution of the late 1980s brought into focus the idea of sustainable development.

The concept of sustainable development was popularised and advanced by the World Commission on Environment and Development (WCED; the Brundtland Commission) in 1987 and has become a dominant stated objective of both public- and private-sector organisations. While there has been widespread endorsement of the concept, there does appear to be a divergence in its translation to an actionable checklist for behaviour.

The Brundtland Commission defined sustainable development as '. . . development that meets the needs of the present without compromising the ability of future generations to meet their own needs' (WCED 1987). Further elaborations emphasised the concept of needs, in particular those of the world's poor to whom priority should be given and the limitations placed by technology and various social organisations in meeting those needs or, more specifically, on the environment's ability to do so. Following the adoption of such objectives by most nations, pressure mounted for coherent and actionable environmental management practices consistent with them.

Sustainable development and emerging environmental management systems

The push toward adoption of explicit environmental management systems (EMSs) has seen the development of a variety of standards. The International Organization for Standardization (ISO) is working on the ISO 14000 series of international EMS standards. The EMS standard ISO 14001 does not itself set standards of performance *per se*. It is more focused on setting up an EMS that would be consistent with, or help a company move towards, sustainable development.

The argument for ISO 14001 as a tool for achieving sustainable development is predicated on the idea that better environmental performance can only occur where there is an awareness of all the environmental impacts of a company's activities which, hopefully, the EMS will provide.

If, for the moment, we accept the merits of the particular standards,[1] why should business adopt them? Brown and van der Wiele (1996) provide a typology of approaches to implementing quality standards in the ISO 9000 series and Total Quality Management (TQM) which provides a useful framework for analysing ISO 14001 and sustainable development. They identify **minimalists**, **converts** and **committed** organisation types. 'Minimalists' are, in a sense, forced to comply with one of the ISO 9000 standards, but are unlikely to move to a TQM approach as a result of certification to the standard. 'Converts' are also forced to be certified to a quality standard, but in the process discover beneficial outcomes of the process and thus move onto a quality path. The 'committed' group view international quality standards as a means of improving their business consistent with their TQM approach.

According to Richard Clements (1997), a member of the technical committee who drafted the ISO 14000 series, a 'free lunch' seems possible:

---

1 See *Institute of Environmental Management Journal* 1996 for a discussion of these.

> I shouldn't have to justify the use of an environmental standard. Registering your company as complying with an international standard on the environment means a huge marketing advantage. Customers like to see this, investors like it, and the local community likes it too.

No matter what the type of organisation or activity, the adoption of an EMS standard such as ISO 14001 will produce benefits in the form of profits, community satisfaction and sustainable development. It is as if Adam Smith's invisible hand is pushing all interests towards sustainable development.

Though not in relation to a particular standard, similar sentiments have been expressed by the Australian Institute of Company Directors (1992) which urged its members to voluntarily take up programmes as:

> . . . responsible environmental management can be achieved and can make an operation more efficient. As well as this, members of the public, including customers, will respond positively to organisations acting in an environmentally responsible and responsive way.

Proponents also identify possible supply-chain pressures on organisations. For example IBM recently became the first US company to issue a formal letter to its suppliers encouraging them to implement an EMS based on ISO 14001 (GlobeNet 1998). It is, ultimately, profits that will push companies towards sustainable development.

Is this the case for SMEs in Australia? Are the views and perceptions of EMS proponents also held by SMEs? This chapter reports the results of a survey of awareness of existing environmental standards and attitudes to the development of environmental programmes among Western Australian SMEs.

Previous research on SMEs and sustainable development

As the ISO 14000 series sets the achievement of sustainable development within a voluntary self-interest framework, its success will rely on adoption by a critical mass of companies. Experience of previous business behaviour may provide a guide for their likely success. Information measurement and reporting is an integral part of any EMS, and is indicative of the importance placed on the environment by a company. There is no Australian literature for SMEs on this topic; however, Deegan and Gordon (1996) and Guthrie and Parker (1990) report a lack of information disclosure by the majority of larger corporations.

There is also no previous Australian research in the area of SMEs and sustainable development. O'Laoire and Welford (1996) cite a British Chamber of Commerce survey of small firms, which found that it was legislation rather than any other factor that was driving environmental programmes. Welford (1994a) concludes that SMEs see themselves as having a negligible impact on the environment and, further, that pressure to improve environmental performance is a passing phase. Whether this is the same for Australian SMEs will be addressed through the results of a survey conducted in 1997.

Survey and results

Given the voluntary nature of the raft of environmental standards and the potential impact SMEs can have on the environment, in November 1997 a survey was conducted of 210 Western Australian businesses to determine the attitudes and perceptions of management towards environmental management.

The survey sample was taken from *The Yellow Pages Small Business Index* database, which defines a small business as a business with up to 19 full-time employees. A medium-sized business is defined as employing between 20 and 200 full-time employees. The general manager for each company was contacted by telephone to obtain an appointment for a return phone call to complete the questionnaire. This was designed for use over the telephone with the aim of obtaining responses to all the questions within ten minutes of conversation. A response rate of 80% was achieved which provided 169 usable questionnaires. In 15 cases, the manager asked for and completed the questionnaire by facsimile. The companies surveyed were from a cross-section of business (see Table 1). Construction producers and suppliers made up 33% of respondents, 18% were retailers and 14% general manufacturers, with the remainder predominantly from service-based businesses. The mean number of employees per company in the final sample was 25.

Cleaning, pest control, veterinarian products, tanning	11
Retailers	9
Primary food production	8
Secondary food processing	14
Construction (production and supply)	56
Textiles	9
Manufacturing	24
Services	12
Petroleum/rubber products	22
Wool	4

Table 1: RESPONDENTS' INDUSTRIAL SECTOR BREAKDOWN

Environmental impact

Of the companies surveyed, only 48% considered that they have a significant impact on the environment. Another 15% thought they did have an impact but it was minimal, while 37% thought they had no impact. This indicates that, in the sample, management had little appreciation of the impact of all commercial activities on the environment, for example in producing waste, using power and resources, etc.[2] Secondary food processors had the highest awareness (79%) of their impact, which was probably a reflection of their need to comply with state regulations on the disposal of wastes in this industry type. The group with the second highest awareness was textiles at 67%. This group includes paper and packaging manufacturers and reflects the environmentally sensitive nature of these industries' activities. Primary food producers were the next highest (63%) which is, again, probably a reflection on government regulations on waste and pesticide/fertiliser/water use. There was a surprisingly low awareness among manufacturers at 29% and those in the petroleum/rubber industries at 34%.

Awareness of general environmental standards

There appears to be a very low awareness of international environmental standards, with only 5% of respondents being aware of the ISO 14000 series. Some 46% were aware of other (mostly industry-specific) standards, while 47% were unaware of any standards applicable to their business. The awareness of standards enforced for specific industries was strongest among secondary and primary food processors and pest controllers.

Awareness of industry-specific environmental legislation or regulations

Only 36% respondents (60) admitted any knowledge of industry-specific environmental legislation, licence requirements or other relevant legislation, while a mere 8% (13) were aware of the Environmental Protection Act of Western Australia. Six of the respondents used occupational health and safety standards as examples of standards they were aware of. Most others quoted specific items such as licensing requirements, local council requirements for toxic material movement, effluent disposal requirements, fishery licences, and grease trap and noise regulations.

Attitudes towards environmental issues and business practice

Environmental issues were seen as a potential cost by 46% of respondents. Nearly the same proportion—44%—did not view the environment as a cost, however. The reason for this could be twofold. Some industries may view their organisation as being exempt from the need for such a programme, while others may view a programme as an integral part of their activities.

Contrary to the marketing of environmental standards, 69% of respondents did not see the environment as a marketing opportunity. In fact only 26% reported that they did accept the marketing benefits, although this attitude varied according to industry. Strangely, 75% of primary food producers did not see the environment as a marketing issue. More understandably, 92% of the services group did not see it as

2 Similar attitudes to environmental impact were found in UK SMEs (see Chapter 1).

a marketing opportunity, although industries such as transport and private hospitals, which are in this group, can have a high environmental impact. The cleaning, pest control, veterinarian products and tanning group was the only one to have a majority of respondents seeing this issue in marketing terms.

Only 26% of respondents saw the environment as an issue to be used for competitive advantage. It is interesting that groups with the lowest scores on this question were industries with high potential impacts such as primary food producers, manufacturers and construction. In only one group—the cleaning, pest control and vets—was there a majority that saw the environment as a competitive issue.

How critical are environmental issues to organisations?

The responses showed a majority who did not see environmental issues in general as critical (56% compared to 43%). Interestingly, while only 26% saw the environment as having an impact on competitiveness, 43% saw it as being critical to their organisation. The most significant discrepancy was with primary food producers, where 75% did not see the environment as a competitive element; however, 63% did see it as critical to their organisation. This finding could be a result of the direct importance the natural environment has on supply-side production in this industry, but not correspondingly on the demand side.

Is the environment important to clients?

The respondents generally did not feel their clients thought of the environment as an issue (59% as against 33%), although the cleaning, pest control, vets and tanning group had 73% of respondents saying it was important. This might have been caused more by occupational/domestic health considerations rather than strictly environmental aspects. Consistent with this view, 73% of this group thought the environment was critical to their organisation, but only 27% thought it represented a marketing opportunity.

It is significant that manufacturers and retailers do not view the environment as being important to their clients (the respondents had a 71% and 78% negative response rate respectively).

Significance of suppliers' environmental record

The majority (65%) did not consider the environmental record of suppliers as an issue. (This is in contrast with the practices of large companies in Europe and the US, where small suppliers are often vetted by their environmental record.)[3] In the cleaning, pest control, vets and tanning group 55% considered it important. This was probably caused by the stringent regulations covering this industry. Generally, it shows the low priority that these SMEs place on the environment. Overwhelmingly, respondents stated that their clients had *never* mentioned environmental factors when choosing their company to trade with (83%).[4] There seems to be little motivation for organisations to make these issues a priority.

---

3 See Welford 1992 for examples of this compliance requirement.

4 The disinterest of customers in SMEs' performance is commented on by Hillary in Chapter 10.

Conclusions

The goal of sustainable development will require the involvement of SMEs, given their call on resources in the production of goods and services and the size of the workforce they employ. The behaviour of SMEs and the attitudes of their management towards the environment will play an important role in determining whether sustainable development can be achieved.

The SMEs reported here display a lack of awareness to the existence of local or industry standards, let alone the emerging international standards. Further, the respondents, in the main, do not seem to echo the views of the originators and promoters of the international standards—that is, in terms of how they view the environment in general or environmental programmes as a management issue. Specifically, the majority of respondents do not view the environment or environmental programmes as a potential source of competitive advantage, a marketing issue, of importance to their customers or as a factor when they are purchasing from their suppliers. Whether it is demand- or supply-driven, it is not an issue for the majority of respondents.

These responses cannot, however, necessarily be interpreted as a reflection of the importance of the natural environment to the respondents. To a degree this view is backed by the proportion of respondents who see the environment as critical to their organisation, but who do not see it as a consideration for their customers or as a means of differentiating their suppliers. This is also backed up by the proportion of respondents who answered that they would be willing to reduce profits to improve the state of their local environment. It is much easier to express good intentions than actually have to put them into practice; it should be recognised that concern for the environment will not necessarily mean adoption of environmental standards.[5] Whether this reflects a lack of awareness of the standards or a reasoned assessment of their potential impact is difficult to say.

Adoption of standards imposes a certain private cost with the promise of uncertain long-term private and public benefits in the form of the goal of sustainable development. As always for the SME, it is a question of balance. If they assess the voluntary standards as being of little practical benefit either to the environment or to their bank balance, what should they do? Do they ignore them and save the private compliance costs but run the risk of losing potential business if their clients and competitors do decide to adopt the standards?

Standards are but one of a range of mechanisms that can be used to engage SMEs in sustainable development. It would seem clear that voluntary mechanisms that incur private costs have less chance of being successful in the absence of tangible benefits. A standards-based approach, coupled with direct subsidies to cover private costs of adoption, would face better prospects of success, as would a reward-based system. The use of standards may be a necessary step towards achieving sustainable development. They are not, however, sufficient by themselves.

5 A fuller discussion of this issue is presented in Chapter 3.

6
Small business, sustainability and trade

Lynn Johannson

> The truth is he's not dead but only ignored.
>
> Poète, mon enfant, tu me chantes en vain (Saul 1997: 409).

Small and medium-sized enterprises (SMEs) are big business in the global economy. Their health and wealth is critical to trade and sustainability. To date, however, their interests have been largely ignored, especially in the area of trade. That is a serious mistake, because, without bringing the needs and concerns of SMEs into the development of trade *and* sustainability, neither will be successful.

SMEs are the economic backbone of most national economies. The exact percentages may differ slightly, but for most of the globe they lie in the same ballpark—the majority. If we look at 1995 world statistics, the combined gross domestic product (GDP) of the world was about US$29 trillion, 55% of total world income was in Asia Pacific Economic Co-operation (APEC) member countries and 45% in non-APEC member countries. The 40 million SMEs throughout APEC economies account for well over 90% of all enterprises, employing from 32%–84% of the workforce, contributing 30%–60% of the GDP and accounting for 35% of exports in the region (APEC 1997). These figures are paralleled in non-APEC economies.

According to *The Economist*, innovation is the primary engine of economic growth. SMEs are creative and resourceful—it is in their hands that most innovation takes shape. Clearly, the health and wealth of SMEs are very important to their respective national economies and are also instrumental to economic growth. Their role in sustainable development is no less critical. That it why it is so troubling to note that SMEs have not been adequately involved in trade discussions around sustainability. Nor have they been sufficiently informed about internationally accepted business systems established for greening companies and, in turn, trade. ISO 14001 is the core specification standard for environmental management systems (EMSs) which is on its way to becoming a key screen in purchasing and trade transactions.

The ISO 14000 series of standards were developed to bridge the gap between trade and sustainability and reduce the potential for non-tariff trade barriers developing across political boundaries around environmental issues. However, unless things change, ISO 14001 could well be a barrier to SMEs trading in the global village. The standard was designed to be flexible and to act as a model for a common system. No matter where an organisation operates in the world, it would be able to apply a common approach to managing its local environmental relationships. ISO 14001 does not prescribe specific outcomes such as levels of toxic materials or waste generation; the organisation decides what environmental aspects it must be concerned with, because of associated significant impacts, through a systemised management framework.

ISO 14001: one standard for the global village

In theory, the standard presents any organisation, regardless of size, sector or geographic location, with a globally common approach to manage its own site-specific environmental concerns, and influence those outside its boundary of control. This can occur through purchasing relationships in a supply chain, loan arrangements with financial institutions, or activities involving the community. ISO 14001 does not dictate *how* the elements of the standard are applied by the organisation, only that they are considered within the business context. Its framework enables a consistent approach to managing an organisation's significant impacts, whatever they may be. ISO 14001 is a 'what' document, specifically designed to meet the needs of any organisation, including those of an SME.

There are five elements to the core standard, as shown in Figure 1. Similar to ISO 9001, the standard for quality management, ISO 14001 is a process standard. Its requirements relate to management systems, not specific product lines. An organisation that uses ISO 14001 can select any of the elements as ideas for self-improvement or implement all the requirements of the standard. If the latter is undertaken, the organisation can self-declare its conformity to ISO 14001, or certify/register[1] to the standard by employing a third party to audit its EMS and verify that the requirements are in place.

Conceptually, with its 'one-size-fits-all' style, the standard has a lot of merit. There is no question that ISO 14001 provides an organisation with the opportunity to better itself (see Box 1). An EMS aligned to ISO 14001 can lead to process efficiencies and may result in performance improvements. Overall, the standard is relatively straightforward.

1 Certification and registration are interchangeable terms; their use is tied to regional market cultural preferences.

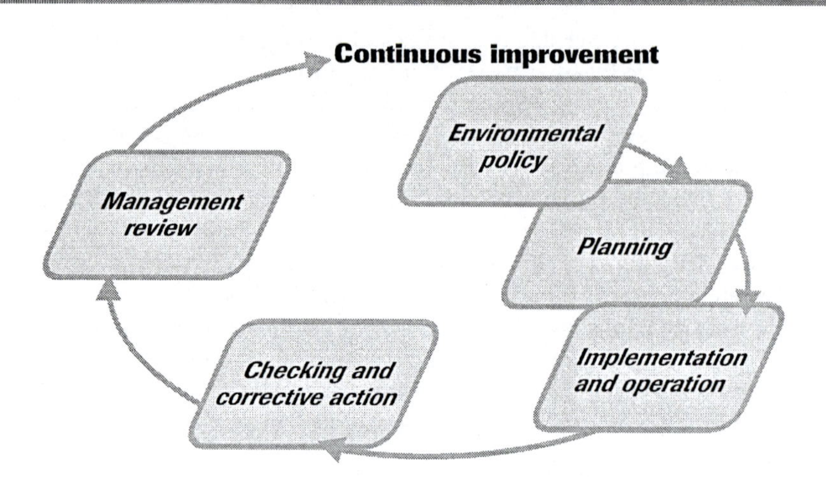

Figure 1: THE FIVE ELEMENTS OF ISO 14001, SUPPORTED BY A COMMITMENT TO CONTINUOUS IMPROVEMENT

Some of the benefits that can accrue to an organisation using ISO 14001 include:

- Better-trained employees lead to improved attitudes towards the business.
- Reduced costs due to the elimination of outdated or unnecessary equipment and/or revised streamlined processes.
- Through process improvements, decreases in resource use occur and there is the opportunity to minimise the creation of waste.
- Cost savings are realised due to lower insurance premiums—a well-managed company poses a lower risk to the insurer and an improved perception of creditworthiness to banks and other lenders can be seen.
- More confident shareholders interfaced with better stock performance; evidence suggests that companies that manage their environmental issues well show a slightly higher stock value. As these companies continue to improve and gain experience, this trend may continue to grow, making them more competitive.
- Defensible legal positions for due diligence in times where conflict arises.
- Improved credibility with customers valuing environmental ethics as part of their supplier selection criteria can move organisations onto a preferred list; the potential to reduce reporting requirements is being considered in some jurisdictions, bringing the potential for greater efficiencies for all parties.

Box 1: BENEFITS OF ISO 14001

However, the standard has not been marketed effectively to SMEs, if at all. When information has appeared, it has seldom been shown or heard as an opportunity for improving a business for competitive advantage in domestic markets or for global trade opportunities. This needs to change. This chapter addresses the challenges in attracting the attention of SMEs, introduces some mechanisms for advancing ISO 14001 with SMEs and larger entities as a better business tool and opportunity, and gives consideration to ISO 14001 as one of the tools for engaging SMEs in achieving sustainability.

ISO 14001 as a barrier to trade

Any country not involved in the ISO Technical Committee (ISO TC 207)[2] process may face disadvantages as it is not privy to the negotiations that developed and influence the continuous improvement process around the standard. These nations will find it difficult to keep up with improvements in the standard. Additionally, companies from those nations may find their product or service at a disadvantage in the export market if their customer's country has adopted ISO 14001 as a national standard.

Less developed countries that *are* members of the ISO process may, nevertheless, be unable to build the necessary infrastructure, or may face potential frustrations if they cannot afford to retain a national expert. For transnational corporations (TNCs), while a lack of local support may be inconvenient, corporate resources can simply be moved to where they are needed. In addition, TNCs can afford to hire the services of consultants and registrars/certifiers to ensure that their remote plants and foreign divisions conform to the standard.

Who else may find the standard a barrier rather than a boost? SMEs are potentially at a disadvantage in both developed and less developed countries for a number of reasons, the most basic being that the information is simply not available or not easily available to them through national channels. SMEs that have access to the Internet may be able to overcome this problem. In the opening half of 1999, the Canadian Federation of Independent Business (CFIB) indicated that 61% of business owners said they were connected to the Internet—almost double the number measured just two years before (Mallett 1999). This figure is likely to be far smaller in less developed countries.

Research conducted in Canada and in several other member countries in the ISO forum (TC 207) shows that SMEs were typically *not* involved in the standards development process. Governments, non-governmental organisations (NGOs) or TNCs spoke on their behalf, often without prior consultation. Organisations such as the CFIB are helpful as they act as a conduit between SMEs, governments and TNCs. Because of the importance of SMEs in the global economy, it is critical that their needs and wants are an integral part of the standards development process, especially as ISO 14001 is tied up to trade in practice.

2 ISO TC 207 is the Technical Committee responsible for the ISO 14000 series of standards. Its members are from 71 countries.

SMEs are key to trade and sustainability

To understand the significance of SMEs' adoption of ISO 14001 it would be useful to have an insight into their role in trade and sustainability. In Canada, there are estimated to be between 951,000 and 2.2 million SMEs; 99% of registered Canadian businesses are enterprises with 100 people or less, 97% are companies with 50 or less, 83% have less than 20 and 74% have less than 5 (see Fig. 2). SMEs account for 58% of Canada's GDP and represent the business majority. In terms of international export trade, even the smallest SME can play a part, although the dollars involved per firm and the numbers of SMEs in Canada involved are not extensive (see Fig. 3).

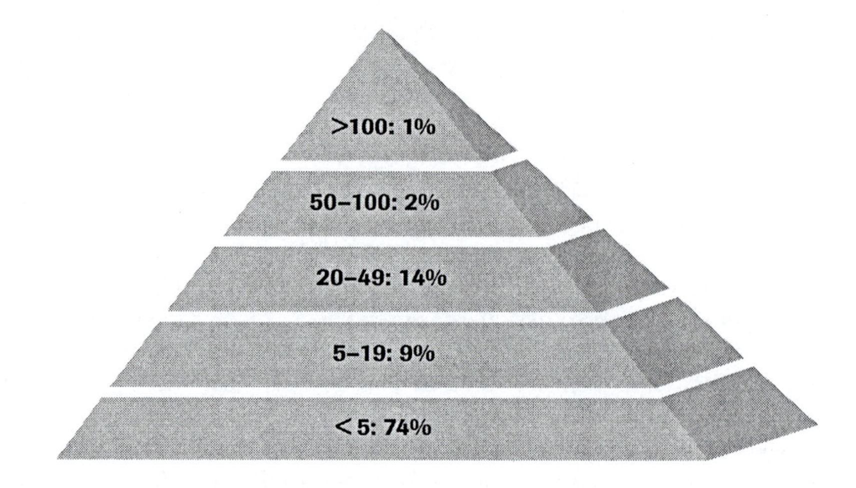

Figure 2: NUMBER OF EMPLOYEES, PERCENTAGE OF BUSINESS BY SIZE

Source: Statistics Canada 1998

For SMEs, trade must bring a positive, consistent cash flow whether it is transacted across a fence, the imaginary line of a political border, or over an ocean. In Canada, new markets for an SME may mean shipping to the next province or to the other side of the country, a distance of 5,500 km. The smaller the firm the more likely it is that business is conducted locally or regionally. Smaller firms typically serve their communities first.

SMEs do not have to 'grow up' to become exporters; they may launch their company with an eye on, and an interest in, the export market from day one. Their interest is related to the products or services they offer and whether there is a market niche. Few very small firms capture export revenue; as the firm grows in size, so too does its export revenue (see Fig. 4). In 1995, SMEs directly exported CDN$21 billion worth of goods and services—accounting for 8.5% of total Canadian exports. This trade involved 86,000 SMEs, representing 97% of all Canadian companies involved in exporting, although SMEs were fewer than 10% of the total registered Canadian

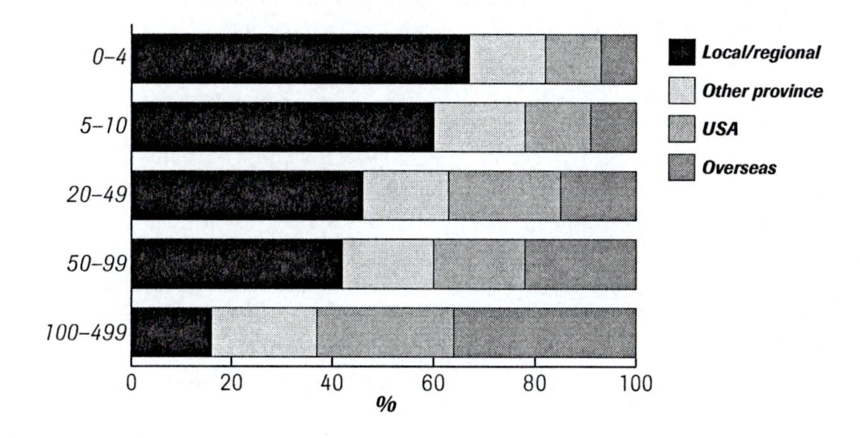

Figure 3: CANADIAN SMES' FURTHEST MARKETS, BY FIRM SIZE

Source: CFIB 1997

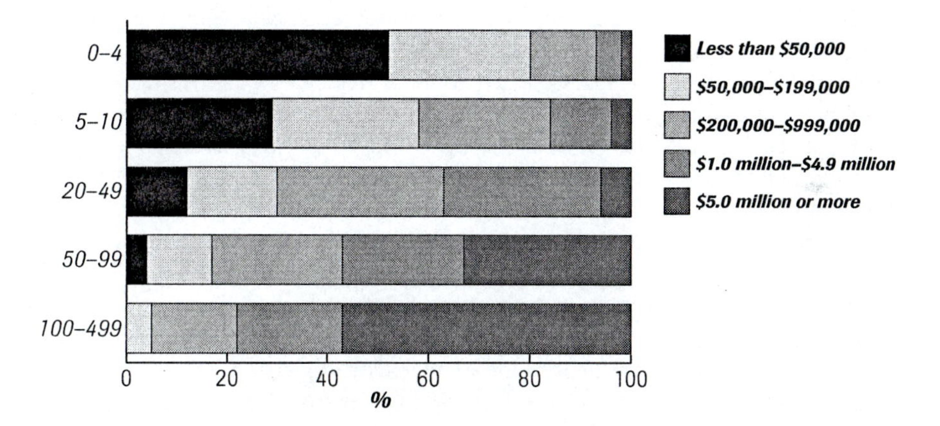

Figure 4: VALUE OF SALES EXPORT, BY FIRM SIZE

Source: CFIB 1997

businesses in 1995. However, SMEs are also involved indirectly in trade as components of, or support services in, the exports of larger companies. In fact, as of 1997, one in three jobs in Canada was tied to trade. This considerably reinforces the importance of SMEs in trade from a Canadian perspective.

SMEs are likewise critical to the success of a sustainable future, even though they may not perceive their role to be central. SMEs typically think of themselves as being insignificant in relation to environmental issues. It is true that an SME is unlikely to cause an environmental catastrophe or disaster of the magnitude of the *Exxon Valdez*, for example. Typically, much of their activity only has a local impact and may fall under the volumes or tonnages set as a threshold in regulations. However, everything an SME does, just as everything a householder does, has an environmental impact. Hence, it is the cumulation of seemingly trivial, everyday actions that adds up. As a group, SMEs can incrementally have a critical role in a polluted future or a sustainable one.

Barriers to participation

So what is preventing SMEs from participating in a global solution? There are several facets to the problem. While SMEs are big business, and not just in Canada, they do not exercise their clout as a collective bargaining unit either in national forums for standards development or international trade policy. Their views are seldom sought on policy issues, including trade and sustainability. Hence, a 'power élite' of big business and government makes decisions that impact SMEs without really understanding the resource constraints of SMEs, nor their dogged focus on cash flow. In the smallest firms, every hour away from the business is equivalent to 3% of production time.

As stated earlier, ISO 14001 was developed to bridge the gap between countries on the issues of trade and sustainability. However, adoption of the standard is in the hands of governments and TNCs or large national firms. In two surveys on Canadian SMEs, conducted in 1997 and 1998, results indicate that less than 1% of SMEs had even heard of ISO 14001—fewer than 10,000 companies. Discussions in the ISO forum (TC 207) indicate that the situation in other countries is similar.

Few countries are actually marketing or advertising the existence of the standard to SMEs. Much of the investment in the standard has, understandably, gone into its development. So, after four to five years of investment, those governments and organisations that have been involved typically believe they should see some return. However, without a targeted marketing campaign, it will be several more years before information about the ISO 14000 series filters down to SMEs. Standards by themselves are seldom seen as hot market opportunities. Instead they may be viewed as an added layer of cost that also brings increased exposure to government interference. More often than not, SMEs in Canada perceive environmental and ISO standards to be derived from government and, in fact, almost any contact with government is perceived by SMEs as an intrusion. Combined, these factors do not present the standard as a value-adding opportunity. If a standard is really valuable to SMEs and valued by its developers, it needs to be supported by action, not passive statements, and marketed as an opportunity.

Those countries that are engaged in activities to promote the ISO 14000 series tend to focus on training opportunities. Canada, Argentina, Colombia and Austria[3] are four nations where training support is being provided to SMEs. While training is valuable as part of the tactical approach needed to support the development of infrastructure, it is not an adoption strategy. Nor is training as a stand-alone activity going to be successful. SMEs need to see a real demand in the market for environmental management systems in order to take advantage of training.

For SMEs, environmental issues that are deemed significant are community issues under the 'brown agenda', i.e. local end-of-pipe problems requiring local solutions: for example, the 'blue box' curbside recycling programme or composting opportunities to manage waste. Local issues require local solutions; how does an international standard fit into this picture? There are global issues that require the attention of each and every SME, such as climate change and ozone depletion. While these issues continue to be perceived as big-company or national issues, it is the seemingly insignificant contribution of each SME that can combine to perpetuate the problem or add to the solution. Who or what is relating the input of each SME to its role in these global issues?

As there are no highly visible market drivers, the bodies that represent SMEs, such as business associations, do not show a strong interest in ISO 14001. This is true even for trade-related associations. The reason is that their members are not telling them they are interested. In today's market, associations focus on member-driven interests. If their members have no interest in an issue, they tend to ignore it because the membership won't pay for it. So the bubble of ignorance is reinforced.

There is yet another series of barriers, associated with communication, that relates to the adoption of the standard from a bottom-up perspective. Some of these are more perceptual but are potentially no less challenging to adoption. People have a tendency to perceive ISO 14001 as being more complicated a process than it actually is, which can turn them away. It can also lead to far more documentation than is truly necessary, thus adding volume to an EMS without necessarily adding value. This does not make for a good selling proposition. There are also vested interests that will play a part in keeping a veil of mystery around the standard, a mist to cover its

3 Based on the Austrian experience, costs for training, documentation and verification to the Eco-Management and Audit Scheme (EMAS) run to about US$100,000 (Johannson 1997). When these costs were converted to Canadian dollars, they represented 26% of the annual revenue of the smaller Canadian SME. If this level of investment were the true cost of certification to ISO 14001, it would preclude at least 74% of Canadian SMEs. In some cases, grants for training have helped. There is a support programme in Austria whereby the government provides funds to hire consultants to help small business with implementation, including certification, to EMAS. SMEs received up to a maximum of US$50,000 to undertake EMAS implementation and registration. The breakdown of cost sharing for EMAS related activities is: 50% for companies with <20 employees; 40% for companies with <50 employees; 25% for companies with <250 employees; and 15% for companies with <500 employees. In Austria, the smallest firm on record being certified to EMAS was a three-person cheese factory. Compare the size of funds that would have to be set aside for Canadian companies with 20 employees or less to provide this level of support. It would require at least CDN$38.27 billion. And that amount pertains only to registered businesses (corporations) in Canada. If other companies were to be included, such as limited partnerships or sole ownerships, the funds would have to be more than double that amount.

flexibility and adaptability to any size or any sector. Sometimes service providers offering EMS consulting support are the source of this concern. Additionally, not all the entities involved in the development of ISO 14001 are interested in its success or its continual improvement.

It is true that the requirements of ISO 14001 will be widely interpreted, and misinterpreted. Part of this misinterpretation is due to the fact that the standard suffers from being developed by the international consensus process. As such, it does not read easily or well. Tools such as sector guides or small business guides can be developed that are clearly written and sensitive to the culture of the target market. Furthermore, ISO 14001 was written in English, the international language of business. This obviously causes problems for non-English speakers and is a concern for the developing world where environmental stewardship is even less likely to figure as a priority issue.

Another consideration for SMEs is the issue of regulatory requirements, although regulators may view ISO 14001's primary benefit as a means to enhance compliance. While the standard may be used for selective improvements, the ISO support process is geared towards full adoption and favours third-party certification/ registration. This brings another challenge for SMEs to the surface. While organisations can self-declare to the standard, only certifications or registrations are monitored.[4] By and large, SMEs are unlikely to certify due to cost and resource limitations—time, knowledge, skills and staff. (see Table 1).

Cost of registration (based on ISO 9000)					
Organisational size/sales	**Initial document review**	**Follow-up reviews**	**Number of employees**	**Total cost per employee**	**Minimum total cost as a percentage of sales**
Small: $500,000	$6,000	$2,500	1–49	$174–$8,500	1.70%
Medium: $5,000,000	$10,000	$5,000	50–100	$150–$300	0.30%
Large: $50,000,000	$18,000	$7,000	101+	$0.21–$248	0.05%

Table 1: THE COST OF CERTIFICATION IS NOT ALWAYS SUPPORTED BY MARKET DEMAND OR SEEN AS ADDING VALUE TO AN SME.

Source: E2M 1998

Self-declaration, while a stated viable option in ISO 14001 (and an unstated option in ISO 9001), is not well supported by ISO and is not seen as being as attractive as certification by some sceptics of the ISO process. Aside from the self-declaration versus certification dilemma, there remains the issue of attracting a broader and deeper market. As long as no one is marketing EMSs to SMEs, the small business link to sustainability will remain poorly connected.

4 There is a new process emerging out of Canada that will change this.

A final problem is the absence of immediate market drivers for ISO 14001. There needs to be an obvious business benefit for SMEs to move down the standards route. Without pressure from customers or higher up the supply chain from larger businesses, there is no commercial imperative for SMEs to improve their environmental management. Of all the challenges that may exist, this is perhaps the single most critical one.

The implications of the FTAA

SMEs are critical to sustainability but are largely ignored by big companies and governments. This, coupled with communication and awareness barriers, keeps SMEs outside much of the debate and development on sustainability. The 'power élite' recognises that environmental issues must be addressed. Evidence of its importance is in the number of multilateral agreements (there are over 100). A specific example of an agreement that ties trade and environmental/sustainable development issues together is the Free Trade Area for the Americas (FTAA), the extended version of the North American Free Trade Agreement (NAFTA).

At the Summit of the Americas held in Miami in December 1994, 34 countries acknowledged that 'Sound environmental management is an essential element of sustainable development.' The Miami Declaration, signed by these countries, *guarantees* sustainable development. Action plans include it as part of the free trade negotiations, with the FTAA expected to be completed in 2005. ISO 14001 is seen as a stepping stone to reach that objective.

Now consider the Miami Declaration as part of the FTAA objectives and time-frame, against the example of Canada. Fundamentally, it requires acceptance that the presence of an EMS/ISO 14001 will lead to sustainable practices or conditions. If this outcome is desired, the next hurdle is to have a majority of companies adopt ISO 14001. Those more likely to adopt are TNCs or large national companies.

Using 1997 figures from Statistics Canada to weight the potential success of the FTAA objective, all Canadian TNCs and large national companies would need to adopt the standard—that is, all organisations with over 100 employees. It would mean, regardless of local market conditions or sector interests, that 9,520 firms would have to implement ISO 14001. However, these companies only represent 1% of the incorporated number of business in Canada. As of autumn 1999, Canada had less than 200 companies registered to ISO 14001. Ontario's Minster of the Environment announced at this time his intention to have 1,000 companies registered to ISO 14001 by 2003. To be successful, it would require a substantial buy-in from the business community. Time will tell whether this is a success story or not: 97% of Canadian firms have fewer than 50 employees—approximately 923,360 registered businesses. Most of these comprise fewer than five people, in many cases working from home. Add to this the 1.2 million sole ownerships, one-person businesses that also add to the health and wealth of the national economy. Trends in job development show that it is in the smaller companies where new jobs are being created, so SMEs as a group are growing and, with it, their importance. Here lies the real challenge—the CFIB has shown that organisations with 50 employees or fewer tend to operate with informal management systems. Are they expected to add a formal EMS to an informal core?

Benefits of ISO 14001 for SMEs

Does this mean that ISO 14001 is not relevant to SMEs? No; the advantage of the standard is its flexibility and non-prescriptive nature. What a TNC may need to establish a consistent systems approach is going to be considerably different from an EMS in a five-person operation. Yet both can benefit from an EMS. For an SME, the whole EMS may be recorded on one poster, in a few pages, or by using software tailored to the pockets of SMEs. The documentation requirements of an SME are not likely to be onerous.

For an SME, there is one very simple motivation to spur change—the bottom line. From a strategic perspective, to encourage SMEs to adopt ISO 14001 without creating trade restrictions, means that a push–pull mechanism must exist. Making an EMS an opportunity to play in the global game of trade has to be supported by the SMEs' domestic market buying 'green'. SMEs need to see evidence that improved environmental management is valued by their customers and is, even, a condition for doing business.

How should this message be sent? Initially, EMS/ISO 14001 cannot be an ultimatum demanded by TNCs or by governments. It has to be marketed as a business opportunity. An on-line marketing opportunity has emerged that allows companies to post their conformity to ISO 14001, or EMAS and other standards such as those in the ISO 9000 series, Responsible Care, etc. It was initiated in Canada with the support of the CFIB and the Canadian Institute of Chartered Accountants. It is developing connections between customers (those businesses posting their conformance) and client groups (those entities that wish to associate with better-managed businesses). It was designed to reflect the intent, impartiality and flexibility of the standard and the resources of SMEs from all sectors, anywhere in the world. The registry[5] also offers other business opportunities. Attached to the site is a service that will be fulfilled by accountants. It will provide clients of self-declared customers with additional confidence through a 'Review Engagement Report', which parallels a financial review statement using GAAP (generally accepted accounting principles). Accountants who are able to offer this service will be listed on the site.

The site also overcomes some of the communication barriers as the information listed will enable a business-to-business connection 24 hours a day 365 days a year. Companies can post on-line, by fax or mail-in. While there are sites that track the statistics of certification/registration and some registrars list their own customers, the registry goes further. It is global, and it is as advantageous to SMEs as it is for TNCs. It is a one-stop shop. By making the market connection to adopters of ISO 14001, it compliments the development work of ISO process and brings the potential for added

5 *www.14000registry.com*

business to the SME in any regional economy. The registry will not solve all challenges around ISO 14001 and greening trade, but it is a positive step that connects EMS to green procurement opportunities.

We know in which countries ISO 14001 is most popular—Japan, Korea, Taiwan and the EU member states—but this picture has been based on certifications. As this is mostly a 'big-company' count, we do not know how many SMEs are using the standard unless they choose to certify to it. There is evidence that a few big companies are requiring their suppliers to adopt ISO 14001 but this has not yet reached the mainstream. These requests are usually international in scope, which brings SMEs back to the issue of trade and sustainability.

As organisations begin to 'buy green', SMEs will recognise that environmental management is favoured by local and international markets. The adoption of ISO 14001 would then offer an immediate business opportunity. In addition, it may be the case that a banker will favour an EMS as improving the creditworthiness of the SME for a loan, or an insurer may provide a modest reduction in insurance costs resulting from better risk management gained through the implementation of ISO 14001. The registry can serve to connect SMEs to these interested parties for mutual benefit.

Is it important for SMEs to buy into ISO 14001 today? Yes. The GlobeScan Syndicated Survey of Experts 1998-1 (GlobeScan 1998) indicated that, after the removal and/or redesign of subsidies, environmental management systems (e.g. ISO 14001) and the greening of the tax system are seen by GlobeScan experts as being the next most effective tools for achieving sustainability. By virtue of their importance in national economies as we merge into one global economy, SMEs are critical to achieving sustainability. We ignore them at our peril—environmentally and economically. We will succeed with their innovation, health and wealth.

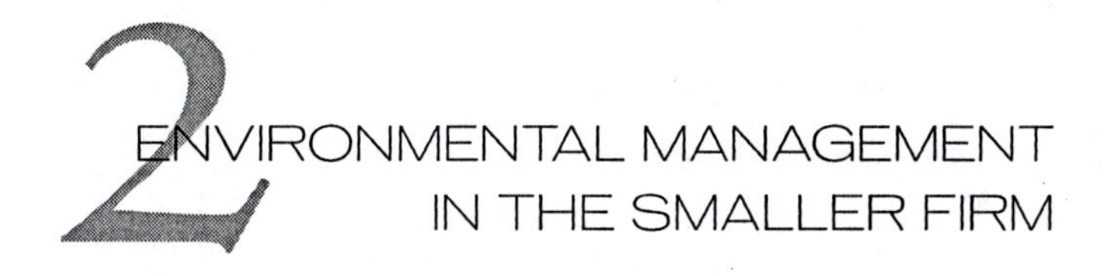

2 ENVIRONMENTAL MANAGEMENT IN THE SMALLER FIRM

7
Environmental management tools

Some options for small and medium-sized enterprises

Richard Starkey

This chapter briefly surveys the various tools available for the management of environmental performance and assesses their relevance to SMEs. Before looking at specific environmental tools, however, it is worth mentioning briefly the assumption made by those who advocate voluntary environmental management initiatives: namely, that it is possible for firms to improve their business performance by improving and/or publicising their environmental management and performance. In other words, firms are not viewed as maximisers of competitive advantage.

It should be pointed out that this is not a universally held assumption. Take, for example, the issue of cost minimisation. There are those who believe that the behaviour of firms by and large corresponds with that described by the neoclassical economic model, arguing that firms tend to be cost minimisers, perched on or near their efficiency frontier and picking up pretty much instantly (to use that well-worn analogy) any $10 bills that are lying around (e.g. Sutherland 1991; Cairncross 1994; Palmer *et al.* 1995). There are others, however, who argue that firms are not the omniscient cost minimisers portrayed by the neoclassical model and that they are characterised by 'limited knowledge of the ever-changing technical opportunities for cost saving and suffer from organisational inertia and control problems reflecting the difficulty of aligning individual, group and corporate incentives' (Porter and van der Linde 1994). They argue that, in the messy 'real' world of imperfect knowledge and mixed motives, dollar bills with a value sometimes substantially greater than $10 are just waiting to be snapped up—as numerous case studies of companies both large and small show.

In *Factor Four*, von Weizsäcker *et al.* (1997) describe the case of an energy- and waste-saving contest that took place at the Louisiana Division of Dow Chemicals. Staff (never higher than supervisor level) were asked to suggest projects that would save energy or reduce waste and would pay for themselves within one year; those that were judged most promising were implemented. Over the 12 years that the contest was held the average return on investment was 204%. As the authors state:

> Dow is one of the world's largest and most sophisticated chemical companies—a leader in a cut-throat industry noted for penny pinching. Dow's competitors would hardly say that Dow is stupid or lazy. Yet Dow has made the astounding discovery that there are $10,000 and $100,000 bills lying all over its factory floors—and that the more of them it picks up, the more it finds.

While sceptics may argue that such case studies do not constitute proof that 'win–win' opportunities pervade the economy (Palmer *et al.* 1995), it does seem highly likely that, if such opportunities exist in large leading-edge companies such as Dow, they will be common elsewhere in the economy.[1] That said, the tools available for achieving competitive advantage through environmental management are discussed below.

A brief survey of environmental management tools

In the early 1990s, the International Organization for Standardization (ISO) recognised the need for standardisation in the field of environmental management tools and, in 1993, it set up a Technical Committee (TC 207) to produce standards relating to the following environmental management tools:

- Environmental management systems (EMSs)
- Environmental auditing
- Environmental indicators
- Life-cycle assessment (LCA)
- Environmental labelling

This section examines these tools and three others used by businesses, namely:

- Environmental policies
- Ecobalances
- Environmental reporting

Each of these tools is described briefly below and its relevance for SMEs examined. The section also contains some brief case studies of SMEs that have used these tools.

1 Indeed, Jackson and Clift (1998) have argued that 'the evidence for profitable pollution prevention is extensive and convincing'.

Environmental policy

An environmental policy is a document that sets out a company's overall aims and intentions regarding the environment. Increasingly, SMEs will find it useful to have such a policy. Large customers are increasingly concerned to see evidence of environmental management on the part of their suppliers and an environmental policy can go some way to providing this. According to Groundwork's 1998 survey of SMEs (Smith and Kemp 1998; see Chapter 1):

> Just under two-thirds of respondents said they addressed environmental issues either through a formal policy (23%) or a business plan (38%). A third have neither, and this rises to 58% for micro-sized companies.

In order to write a relevant policy, a firm needs to make an appraisal, and have some understanding, of its environmental aspects.[2] And, in order to meet the aims set out in the policy, it will need to take some action, possibly utilising one or more of the tools described below.

Environmental management systems

A system can be thought of as 'a number of interrelated elements functioning together to achieve a clearly defined objective' and hence an environmental management system (EMS) can be described as a number of interrelated elements that function together to achieve the objective of effective environmental management. Recently, a common model for an EMS has been formulated by ISO which standardises the elements that an EMS should contain. The model has been designed to be applicable to organisations of all types and is set out in the standards ISO 14001 (ISO 1996a) and ISO 14004 (ISO 1996b).

ISO 14001 is a specification standard which means it consists of a set of requirements, in this case for establishing and maintaining an EMS. By demonstrating compliance with these requirements to the outside world, a firm can signal that it has a credible management system in place. One way in which a firm can demonstrate that it has met the requirements of the standard is by 'self-declaration' which means that the firm checks its own compliance. However, a firm may feel it carries more weight with the outside world if its compliance with the requirements of the standard is checked by an independent third party. If this third party is satisfied that the requirements of the standard have been met, then 'certification' can occur: certification being defined as the 'procedure by which a third party gives written assurance that a product process of service conforms to specified requirement' (ISO/IEC 1996).

As its title suggests, ISO 14004 consists of general guidance on a broad range of management systems issues. It contains guidance on all the elements of the ISO 14001 management system, as well as on additional relevant EMS topics. The standard

2 An environmental aspect is an 'element of an organisation's activities, products or services that can interact with the environment' (ISO 1996a).

provides useful guidance for a firm wishing to establish an EMS that conforms with the requirements of ISO 14001, but is equally useful for those organisations who wish to establish a system but do not (yet) wish to seek certification.[3]

The second EMS specification in use is the EU Eco-Management and Audit Scheme (EMAS). EMAS is an EU Regulation (as opposed to a standard) and only companies in EU member states may participate. The EMS requirements contained in EMAS are very similar to those contained in ISO 14001 but, unlike ISO 14001, EMAS also requires firms to produce an environmental statement and to have their EMS and statement verified by a third party as meeting the requirements of the Regulation (i.e. self-declaration is not possible).

So are these various EMS documents relevant to SMEs? The authors of ISO 14001 and 14004 certainly think their standards are. ISO 14001 states that 'it has been written to be applicable to all types and sizes of organisations and to accommodate diverse geographical, cultural and social conditions' (ISO 1996a) and ISO 14004 states that 'it can be used by organisations of any size. Nonetheless, the importance of SMEs is being increasingly recognised by governments and business. This guideline acknowledges and accommodates the needs of SMEs' (ISO 1996b).

In March 1998, 5,000 certificates had been issued worldwide, around 720 of those in the UK (ENDS 1998). A number of these UK organisations are SMEs but, as no organisation in the UK currently keeps a record of how many of these certifications are of SMEs, the exact number is not known.[4] EMAS is also regarded by its authors as being relevant to SMEs—Article 13 of the Regulation promotes their participation in the scheme—and, in the UK, 19 of the 67 companies with sites registered under EMAS are SMEs.

Loudwater, a printing firm based in Watford, UK, has gained great benefits from its EMS. Its core business is printing greeting cards and in 1996 it became registered under EMAS—the first small company (i.e. less than 50 employees) in the UK to do so. One of Loudwater's managing partners describes the benefits as follows (UK Competent Body for EMAS 1996):

> We have saved in excess of £20,000 by reducing our waste, cutting down on our energy consumption and by recycling or selling our unavoidable waste. We are winning new business all the time, particularly from blue-chip companies which previously would not have considered us. The fact that we can prove our green credentials through EMAS and demonstrate our ability to work within the framework of environmental legislation in a wider Europe gives us that competitive edge.

While the writers of ISO 14001 and EMAS consider their documents suitable for use by SMEs, others are less sure, regarding these schemes as too complex and bureaucratic for companies with fewer employees and less formal management structures. The first rather obvious point to make is that, given that a number of SMEs have become certified to ISO 14001 and/or EMAS, the above view does not accurately reflect the reality

3 The use of ISO 14004 is considered by Hillary in Chapter 10.
4 An estimate is presented in Chapter 10.

for *all* SMEs. However, even if it did for *most* SMEs, why is this a problem? Surely those SMEs that perceive a formal system as advantageous will develop one and those that don't, won't?

There is a problem, if, as has been argued, there could be something better for SMEs than a choice between ISO 14001/EMAS or nothing. The argument goes that the idea of a certifiable/verifiable EMS is worthwhile but what is needed is a simpler, more straightforward version for SMEs as this alternative will be much more relevant to most SMEs. However, in my view, there is nothing in the requirements of ISO 14001 that necessities the development of a complex and bureaucratic system. It is possible to meet its requirements by developing an effective system that is both simple and user-friendly. As Annex A of ISO 14001 states (ISO 1996a):

> . . . the level of detail and complexity of the environmental management system . . . and the resources devoted to it will be dependent on the size of an organisation and the nature of its activities. This may be the case in particular for SMEs.

There is a great deal of truth in the remark made to me by a consultant who had been employed by a local authority in the UK specifically to help SMEs set up management systems. His view was that bulging manuals and bureaucratic, unwieldy systems were caused not by the requirements of the standard but by the inappropriate implementation of the standard by consultants.

Environmental auditing

An environmental audit is a tool for checking whether a firm is doing what it should be doing—or, to put it slightly more technically, for evaluating whether particular environmental activities, conditions, management systems and so on conform with audit criteria. For instance, a legislative compliance audit checks that those activities of the firm covered by environmental legislation (i.e. what it is doing) actually comply with that legislation (i.e. what it should be doing/audit criteria). A waste management audit will tell a firm whether its waste management practices (i.e. what it is doing) conform with the industry sector best-practice guidelines it has committed itself to following (i.e. what it should be doing/audit criteria). Hence, any SME that needs to check its environmental management activities will use this tool.

Environmental indicators

Environmental indicators allow a firm to make measurements related to its environmental performance. Indicators can be used within an EMS to check that a firm has met the targets it is required to set for itself, but can equally well be used in firms that have not developed an EMS. ISO 14031, the international standard on environmental performance evaluation (EPE) (ISO 1999a), sets out guidance on how organisations can develop and use indicators to evaluate their environmental performance. The standard defines three categories of indicator: management performance indicators (MPIs) which provide information about management's efforts to influence environmental performance; operational performance indicators (OPIs) which provide information about the actual environmental performance of an organisation's operations; and environmental condition indicators (ECIs) which allow an organisation to assess its impact on the environment by measuring the condition of the environment over time. The standard makes it clear that the standard can be used by all organisations 'regardless of type, size, location and complexity' (ISO 1999a).

One SME that has developed a set of indicators in response to ISO 14031[5] is the Malaysian firm Perussahaan Pelindung Getah (PPGM). PPGM employs approximately 100 staff and manufactures medical rubber gloves, producing about 8 million per month (ISO 1999b). The company has set itself the following environmental targets and has developed indicators to track its progress towards meeting them (see Table 1):

- Total compliance with legislation
- Zero public complaints regarding its operations
- Minimal adverse environmental effect
- Rejected gloves to be less than 5% of the total produced
- The level of extractable protein in the gloves to be less than or equal to 0.1 milligram extractable protein per gram of rubber (the occurrence of residual soluble proteins in latex gloves has the potential to cause allergic reactions in sensitised individuals)

Ecobalances (also known as mass balances or input–output analysis)

An ecobalance records the various raw materials, energy, resources, products and wastes entering, held within and leaving a company over a specified period of time. In other words, it provides a record of a company's physical inputs, stock and outputs. Once a company knows exactly what is coming in and going out, it can begin to assess the particular environmental impacts of those inputs and outputs. An ecobalance therefore enables a firm to undertake the comprehensive environmental review of its activities required by ISO 14001 and EMAS. It also provides relevant data for operational performance indicators so that progress toward meeting targets can be tracked.

Ecobalances are used widely by German SMEs as a tool for undertaking environmental reviews and establishing environmental information systems. Some of the SMEs using the ecobalance have a formal EMS while others do not, and the tool is used by smaller SMEs as well as larger ones, including one company, Möbelwerkstätte Schmidt (see case study in 'Environmental reporting' section below) with just 11 employees.[6]

The Augsburg Worsted Yarn Spinning Mill, a German SME founded in 1836, uses the ecobalance tool to analyse its material and energy flows in order to identify areas

5 ISO 14031 is a Final Draft International Standard (FDIS), due to be published as a full standard in December 1999. PPGM piloted the FDIS.

6 Personal communication with R. Rauberger, Institut für Management and Umwelt, Germany, 16 April 1999.

Indicators for EPE	Basis for selection of the indicator
Management performance indicators (MPIs)	
Annual total cost of implementing environmental programmes	For the evaluation of management commitment (i.e. useful public relations material)
Number of environmentally related complaints received by PPGM per year	For evaluation against the environmental performance target on zero public complaints
Number of effluent samples analysed monthly not complying with regulatory standards	For evaluation against the environmental performance target on total compliance with regulations
Operational performance indicators (OPIs)	
Number of pieces of gloves rejected in relation to the total number of pieces of gloves produced per month	For evaluation against the environmental performance target on controlling rejects in order to reduce wastes
Extractable protein level of glove measured in milligrams of extractable protein per gram of rubber	For evaluation against the environmental performance target on eliminating the potential cause of protein allergy. (This information is useful to the US FDA.)
Quantity of zinc in kilograms discharged to the receiving watercourse per month	For evaluation against the environmental performance target on minimising wastes
COD load in kilograms discharged to the receiving watercourse per month	For evaluation against the environmental performance target on minimising wastes
Quantity of dried sludge in kilograms produced per month	For evaluation against the environmental performance target on minimising wastes
Environmental condition indicators (ECIs)	
Incidence of protein allergy associated with the use of rubber gloves by sensitised individuals (i.e. number of official reports per year)	For evaluation against the environmental performance target on eliminating the potential cause of protein allergy
Changes in the quality of surface water upstream and downstream of the factory's effluent discharge point. This indicator is based on the test for inhibition of oxygen consumption by activated sludge as carried out in accordance with the ISO 8192 procedure. The specific indicator is the percentage of change in EC 50 value where EC 50 is defined as the concentration which inhibits the oxygen consumption by 50%	For evaluation against the environmental performance target on ensuring the environment is not adversely affected by PPGM's operations. (This information will be useful to the water treatment plant operator.)

Table 1: INDICATORS FOR ENVIRONMENTAL PERFORMANCE EVALUATION SELECTED BY PPGM

Source: ISO 1999b

for improvement in its operations. The most important data resulting from the ecobalance is converted into relevant environmental indicators (see Table 2).

Environmental indicators	Unit	1993	1994	1995
Heavy-metal-free dyes	%	35.2	35.3	40.0
Re-utilisation of spinning bobbins	%	–	3.5	8.5
Re-utilisation of transport packaging	%	–	8.3	11.9
Loss of primary raw material	%	6.7	5.4	4.8
Energy consumption	MWh	89,285	82,422	73,865
Water consumption for dyeing	m^3	249,670	241,450	219,010
Ecotex 100 standard approved products	%	50	90	98
Residual waste	kg	158,014	102,598	81,658
Temporary limit value excesses	number	–	3	5
Environmental cost savings	£	–	–	100,000

Table 2: ENVIRONMENTAL INDICATORS USED BY THE AUGSBURG WORSTED YARN SPINNING MILL

The company's indicators reveal that since 1994 it has realised a number of cost savings by improving its resource efficiency. For example, the spinning bobbins and transport packaging returned to the company by customers are cleaned, checked and repaired by members of a local community project for later re-use by the company. This re-use saves the company around £10,000 per year. Technical improvements to machinery and the increased motivation and awareness of the staff regarding environmental matters have led to significant reductions in the loss of raw materials. This loss was reduced from 6.7% in 1993 to 4.8% in 1995 which resulted in savings of over £40,000. In total, over £100,000 per year has been saved by measures implemented on the basis of the company's ecobalance.

◻ *Life-cycle assessment*

Life-cycle assessment (LCA) is a tool for identifying and assessing the various environmental impacts associated with a particular product. LCA takes a 'cradle-to-grave' approach, looking at the impacts of the product *throughout its life-cycle*—i.e. from the raw materials acquisition (the 'cradle') through its production and use to its final disposal (the 'grave'). LCA allows manufacturers to find ways of cost-effectively reducing the environmental impact of a product over its life-cycle and to support their claims about the environmental impact of their products. LCA is a complex process for which SMEs will almost certainly not have the in-house expertise. LCA is also expensive to carry out and so, while used by some SMEs (see example below), is likely to be of interest to only a limited number.[7]

7 It is possible to reduce the cost and complexity of an LCA by streamlining the process, but this must be done with care to ensure accurate results (see Weitz and Sharma 1998).

Climcon A/S, a Danish SME, employed consultants to carry out an LCA on a product it was developing using grant money provided by the Danish Environmental Protection Agency. The company had developed a prototype for a vehicle air conditioning unit that used ceramic semiconductors. (Air conditioning in passenger vehicles has significant environmental impacts. Not only are ozone-depleting or greenhouse gases commonly used as cooling agents, but an air conditioning unit increases the overall energy consumption of a vehicle by as much as 47 litres of petrol per 10,000 km driven—equivalent to approximately 10% of vehicle energy consumption.) Initial testing suggested that the unit could be produced at a cost equivalent to, or lower than, conventional units and would be considerably more energy-efficient—meaning lower running costs and improved environmental performance. The LCA showed that the energy consumption was only one-third that of a conventional unit, equivalent to as much as 1,000 litres of petrol over the lifetime of a standard car. In addition, the materials from which the unit was built were also shown to have a lower life-cycle impact than those used to build a conventional unit.

Environmental labelling

Environmental labelling schemes award an environmental label to those products that are judged to be less harmful to the environment than others within the same product group. Firms that wish their products to be considered for a label must apply to the scheme organiser. To be awarded a label, a product has to meet a set of environmental criteria drawn up for its product group by the scheme organiser. The criteria relate to the complete product life-cycle and are drawn up using LCA. They are set so that only a certain percentage of products within a group, say 20%–30%, can meet them. Hence, environmental labels can be used as marketing tools as they signify that a product is one of the least environmentally harmful products in its group.

There are currently nine national/regional environmental labelling schemes within the EU as well as an EU-wide eco-labelling scheme. Although most labelling schemes do not keep an official record of SME participants, staff from various schemes report that SMEs do take part. The Austrian labelling scheme, which does keep records, reports that of the 80 companies whose products have labels, 20 have between 50 and 249 employees and 45 have fewer than 50 employees.[8]

Environmental reporting

Having undertaken various environmental management initiatives to improve its environmental performance, a firm may wish to communicate the results of these to the outside world. One way of doing this is by publishing an environmental report. Issuing an environmental report can improve a firm's public image and lead to improved relationships with stakeholders. To date, it is mainly large companies that have issued such reports, although some SMEs have also done so.

8 Personal communication with J. Raneburger, Federal Ministry for Environment, Youth and Family, Austria, 26 April 1999.

Möbelwerkstätte Schmidt is a small-scale German carpentry workshop with 11 employees and an annual turnover of about £400,000. The impact of its products on both the environment and on human health has long been a prime concern of the company. The company has sought to improve its environmental performance (and at the same time generate new market opportunities) by producing furniture from solid wood (rather than chipboard) and by using natural oils and waxes to treat its products. In 1995, the company conducted a full input–output-type mass and energy balance in an effort to further improve its environmental performance. As a result, it was able to improve control of various materials and substances used in its processes. The company published the results of this work in a 12-page environmental report which was distributed to customers and is publicly available. In addition, it made the report available on the Internet.[9] As a result of its environmental management activities, the company has significantly reduced its environmental impact, reduced its costs through waste avoidance and gained new customers (Starkey 1999).

Comments

While voluntary environmental management initiatives are not particularly prevalent among the SME population, the previous section has shown that the various tools listed can all be used constructively by SMEs. However, as SMEs are by no means homogeneous—varying in size, complexity and activity undertaken, it is doing nothing more than stating the obvious to say that not all of these tools will be relevant to *all* SMEs *all* of the time. What is important is that each SME is in a position to choose the most appropriate tool or tools for its particular environmental management needs.

For this to happen, it is necessary for service providers to be fully conversant will the full spectrum of possible environmental management options in order to help SMEs come to the correct conclusion about which form of action is the most (competitively) advantageous. It may be that the most advantageous form of action is no action, or is to deal with environmental issues on an ad hoc basis, for instance with discreet energy-efficiency or waste-reduction projects. Or it may be that a highly systematised approach developing a formal EMS that includes life-cycle assessment of particular products[10] is the most advantageous. On the other hand, some firms may find a simple set of indicators is all that is required, while others may find that obtaining an environmental label for their products might be particularly beneficial. The challenge for service providers is in enabling their SME clients to make an informed and appropriate choice.

9 *http://members.aol.schmids40/*

10 Obviously, it is unlikely that many SMEs will, in the near future, be looking to get involved in a formal LCA of their product(s), but the appropriate advice and guidance should be there for those companies to whom it might be beneficial.

8

Size matters

Barriers and prospects for environmental management in small and medium-sized enterprises

Agneta Gerstenfeld and Hewitt Roberts

This chapter focuses on small and medium-sized enterprises (SMEs) in the UK. It is based on the belief that business and industry are significant causes of global environmental damage and thus the implementation of environmental management systems (EMS) will, at least in part, improve corporate environmental performance and thus improve the global environmental state of affairs.

However, as EMSs are not, at present, being adopted by SMEs, for one reason or another, there lies an inherent problem in this seemingly simple solution to improving corporate environmental performance.

By addressing the reasons why EMS uptake in the SME sector has been weak, and by analysing the needs of the SME community, it is possible to develop a support mechanism that may increase the implementation of EMSs by SMEs.

To increase the rate of EMS implementation in the SME community, a successful support programme for SMEs must be inexpensive, co-operative, locally based, flexible, unique and accessible. Furthermore, an effective programme must provide training, legislative compliance support, and clear, concise, dependable sector-specific information and support.

Environmental impact of the SME sector

The majority of the world's output is actually produced by SMEs (Hutchinson and Hutchinson 1995). SMEs are a vital part of the economy, contributing the largest proportion of growth (CBI 1996) and their economic significance is undeniable. It has been stated that 'SMEs are an integral and dynamic part of a healthy economy' (CBI 1996), 'they provide innovation, competition, flexibility in the labour market, and are a crucial source of job creation'.[1] In short, although the role of the large and often multinational corporation is considerable, without SMEs the present fabric of our social and economic systems would simply unravel (Hillary 1995).

However, the fact remains that there is a general misconception that environmental impact is restricted to large manufacturing companies. Undoubtedly, larger manufacturing firms do have an often considerable impact on their environment, but it is a simple fact that all firms, large and small, have an effect on the environment. According to a recent survey, individual SMEs are likely to 'have little effect on the environment but collectively their sheer numbers mean their environmental impact could be quite significant'.[2]

Perhaps one of the most serious problems we face is that, like individuals, small businesses are unable to measure the effect of their activities on the environment and, therefore, have no motivation to change. Compounding this is the fact that the environmental problem has yet to be defined with sufficient clarity to enable SMEs to deal with it (Hutchinson and Hutchinson 1995).

Definition of the SME sector

According to the EU definition, the SME sector comprises three sub-groups consisting of a 'micro' or 'very small' category, a 'small' category and a 'medium' category.

EU definition	*Size*
Micro	1 to 9
Small	10 to 49
Medium	50 to 249

For the purposes of this chapter, we shall not make distinctions within the sector as a whole, but will regard the SME sector as bound by an upper limit of 250 employees.

Environmental management and the SME sector

Issues of importance to SMEs

If improved environmental performance is, in fact, the goal of corporate environmental management and if, as we have seen, the current approach is escaping the SME sector, then it is rational to conclude that the sector needs additional assistance to increase its uptake of environmental management. What is important, therefore, is the

1 Between 1985 and 1989, enterprises employing less than 20 people created over one million jobs (Hillary 1995).

2 Interview taken from ENDS 1995.

development of a mechanism that will activate increased environmental management in the SME sector. In designing assistance or support programmes for SMEs, it is crucial to accept that they are what they are, understand their limitations and respect them. It can hardly be said more succinctly than by recognising that small firms know what type of support services they prefer. They want company-specific advice provided by a local adviser and access to improved technology. They need: assistance in getting started in the process of addressing company environmental issues; support services that are delivered locally and fit within the business context; mechanisms that bridge the gap between SMEs and the available support systems and pump-priming finance to initiate environmental management actions in SMEs (Hillary 1995).

From Table 1, it is easy to gauge not only the issues of concern, but what weight they each hold respectively. Thus, in developing an assistance programme, consideration of these concerns would, of course, be necessary.

Issue of importance to UK SMEs	% of respondents considering this an issue
Legislative compliance	55%
Industrial standards compliance	48%
Environmental protection	39%
Insurance requirements	30%
Customer pressure	29%
To improve business efficiency	27%
Employee pressure	22%
Investor pressure	3%

Table 1: IMPORTANCE OF ISSUES TO UK SMEs

Source: Hutchinson and Hutchinson 1995

While the implementation of environmental management in SMEs certainly is not impossible, the standardised approach offered by the EU Eco-Management and Audit Scheme (EMAS), and the EMS standard of the International Organization for Standardization (ISO) ISO 14001 appear to be the domain of larger companies.

Whether SMEs are not being pushed by the same pressures, pulled by the same incentives or are simply unaware of environmental management altogether, 'the smaller business sector has failed, so far, to change its practices and adopt the management solutions offered' (Hutchinson and Chaston 1995).

This is confirmed by the results of a UK survey in which it was concluded that only 6% of SMEs have an environmental policy, only 4.6% of have an environmental management system in place and 27% feel that an environmental audit would be irrelevant to their business. Some 70% of SMEs had no clear intentions of even starting the environmental management process.[3]

The reason for the SME sector's categorical disinterest in environmental management seems to be expressed in one of two ways. First, SMEs are aware of environmental management but hold a strong belief that it offers nothing for them or, second, they are simply unaware of the issue and the entire concept of environmental management genuinely escapes them.

In support of the first scenario, it is the opinion of some that SMEs seem to concentrate their attention on everyday operational issues rather than producing an holistic strategic plan for the environment.[4] Conversely, there is the opinion that 'small firms tend not to have awarded a high priority to environmental issues and are typically unaware of general environmental issues or the increasing demand for environmental quality in the market place' (Griffin *et al.* 1995). An extension of this feeling is the theory that awareness in the SME sector is directly related to the size of the enterprise, with 'micro businesses being the least aware and medium-scale businesses being the most' (Hillary 1995). As a result, and for whatever reason, interest from the small and medium-scale business sector is 'relatively low' (Line and Vogt 1996).

As many SMEs presently consider the environment to be peripheral to their business practices[5] and as the management strategy of most SMEs tends to be biased towards immediate, critical incidence situation management, long-term intangible environmental benefits will typically receive a low priority (Hutchinson and Hutchinson 1995). There is the possibility that the present approach to environmental management is simply out of reach for SMEs and will remain so for the time being. Research shows that the concerns of the smaller businesses are different in kind from those of the larger firms and, therefore, blanket solutions will not work to resolve this problem now or later. Furthermore, 'with neither the perceived financial incentive nor the legal force to require change, SMEs lack the incentive to seek effective means to minimise their impact on the environment' (Hutchinson and Hutchinson 1995).

Why environmental management?

In short, there are three reasons for SMEs to implement environmental management. The first, and of ultimate importance, is that increased implementation of environmental management will improve corporate environmental performance and consequently improve the prospect for global sustainability.[6] However, while this reason is praiseworthy and essential for long-term survival, for the average small shopkeeper or medium-scale widget-maker it may not be of immediate concern. Thus, the second and third reasons for implementing an EMS, i.e. the 'stick'—avoiding the business costs

---

3 Survey conducted by Hutchinson and Chaston in 1993–94 with Plymouth Business School to assess current attitudes towards BS 7750 in the SME sector; 600 SMEs were surveyed in the Devon and Cornwall area.

4 See Chapter 3.

5 Additionally, '24% list no benefit from pursuing positive environmental actions and only 5% believe that customer satisfaction is a benefit of positive environmental actions' (Hillary 1995).

6 See the discussions in Chapters 5 and 6.

of not implementing environmental management, and the 'carrot'—reaping the immediate and long-term benefits of doing so, are more likely to be of prime concern.

However, even though SMEs play such a vital role in our socioeconomic systems, are experiencing mounting pressures to adopt EMSs, and are collectively responsible for a considerable percentage of our past and present global environmental impacts, we are still in a situation where there are serious barriers and deterrents preventing the SME sector as a whole from embracing environmental management. Clearly, the solutions currently offered to overcome this do not satisfy the concerns of SMEs or eradicate the barriers experienced by them. Consequently, for improved environmental performance, a different approach and better support and information dissemination mechanisms are needed.

This conclusion is proven by the results of a recent survey that found that a majority (66%) of SMEs felt that environmental issues were either 'quite' or 'extremely' important to their business but 'only one in five companies are convinced that positive environmental action would generate cost savings' (ENDS 1995). Clearly, 'the SME sector is ready and willing to take more care of the environment but at the moment it seems they are unable to change their practices';[7] SMEs see no bottom-line reason for doing so and do not know what to do if they did.

Ultimately, SMEs need a solution offering support that concentrates more on the day-to-day running of small companies, increases the availability of advice and reduces the cost of consultancy through more detailed, less generic, developments for each industrial sector based on best practices and not on strategic management systems. The fact remains that companies are increasingly recognising the opportunities offered by improved environmental performance. For SMEs, 'the first steps towards this change can prove daunting, yet it is these firms which hold the key to our success in ensuring a sustainable future' (Hillary 1995).

Pressures driving the implementation of environmental management in SMEs

Legislative pressures

One of the clearest and least contestable pressures experienced by SMEs is that created by present legislation—be it the fear of fines, liability or closure: 'what little action SMEs are taking appears to be driven by legislation' (ENDS 1995).

Business-to-business pressures

While it is irrefutable that, for many SMEs, legislative compliance is a matter of business survival, and is likely to be their single largest motivator, business-to-business pressure is another dominant factor behind the implementation of environmental management.

7 From a mail survey conducted in conjunction with Hutchinson and Chaston 1995. Furthermore, as this survey was compared to similar surveys for different geographical regions of the UK, it was concluded that the results of the paper could be applicable to the rest of the UK.

While some maintain that 'legislation is by far the most significant driving force in persuading firms to effect an environmental improvement' (Griffin *et al.* 1995), others hold that business-to-business pressures are by far the most prevalent.[8]

Stakeholder pressures

For the SME, the enforcing authorities, local residents, insurers, bankers, employees, customers, interest groups and the general public are increasingly the source of pressure to improve environmental performance. Furthermore, while some of the aforementioned pressures are felt equally by large and small firms alike, SMEs, unlike their larger counterparts, are much more transparent and thus susceptible to and influenced by the concerns of these stakeholders.[9]

Barriers to implementing environmental management in SMEs

One characteristic of the SME sector that is of undeniable relevance is its diversity. Thus, when addressing the shortfalls of the present state of affairs with respect to SMEs there will be barriers, issues and difficulties that will be common to some SMEs but not all. These will depend on the firm's size, location, sector, organisational structure and so forth. This discussion is meant to identify problems that are resident within the SME sector generally but not necessarily universally common across the sector as a whole. By identifying problems, whether experienced unanimously throughout the sector or not, a solution is more likely to be of merit to a majority of the sector.

When one considers the practical barriers to environmental management implementation for SMEs, the obstacles seem to be as numerous as the sector is diverse. However, like the definition of 'SME' itself, while there are many propositions and differences, there are, too, a number of widely repeated and common barriers. In simplest terms, lack of money, time, experience, access to information, support and a general misconception and lack of interest in the both the standards and the environment are echoed throughout most surveys, questionnaires and academic papers on the subject.

Lack of training and awareness

There seem to be two 'educational' issues that arise when dealing with environmental management implementation in the SME. First, there is a general lack of awareness about the environment and the benefits of improved environmental performance and, second, there seems to be a serious lack of expertise and understanding about actually

8 See Chapter 3 for a discussion on attitudes to legislative drivers.
9 See Chapter 10 for a discussion on stakeholder drivers.

implementing environmental management. A programme of support for SMEs must therefore provide general information and assistance to encourage the next step of more complete assistance and education. Education must obviously then be delivered in a fashion that encourages its own perpetuation and thus addresses the benefits of the steps taken and the advantages of taking the next step.

Initially, therefore, it is very important to provide 'awareness' about costs and benefits of continuing environmental education. 'The belief that pollution prevention pays is not widespread among SMEs' (van Wijngaarden 1995) and thus to address it early in the process of education would encourage continued participation. Similarly, a 'report on the Aire and Calder Project in Yorkshire shows that smaller companies have yet to be convinced about the benefits of cleaner technology and waste minimisation' (*Environmental Manager* 1995). Thus, if there is going to be increased uptake of EMS by SMEs through a process of initial awareness and education, it is imperative that there is not only an increased understanding and appreciation of the issues at hand, but also the provision of self-perpetuating incentives to continue.

Subsequent to awareness training, and in light of a number of clear indicators, SMEs need specific training on environmental management and how to implement it in their businesses and realise the benefits from doing so.

> Training will need to be concise and of direct relevance. It will need to incorporate practical, competence-based management skills and delivered largely through distance learning in the workplace. It should lead to a recognised qualification and, most importantly, it must be affordable (Kemp *et al.* 1996).

Lack of legislative support

One fact about the SME sector that is resoundingly clear is that there is an alarming lack of knowledge and application of the environmental legislation by which they are bound.[10] Clearly, and in conjunction with the training needs addressed above, the SME sector needs assistance to become aware of relevant legislation and abide by it. In the UK, surveys have found that the majority of SMEs have only achieved 50% legislative compliance.[11]

Obviously, the 'general awareness of environmental legislation and regulation is low in SMEs' (Hutchinson and Chaston 1995) and they are 'unreceptive or unable to interpret the relevance of environmental legislation to their business' (Hillary 1995). Additionally, with a forecast for considerable increase in the volume and severity of environmental legislation, a further issue confronts SMEs. Not only are they unaware or unreceptive to present laws and regulations but they need to be made more aware of the likely impact of anticipated legislation and international agreements on their activities. This is echoed in the alarming fact that 'regardless of forthcoming laws, 40% feel that an increase in environmental legislation would have no impact on their businesses' (Hillary 1995).

10 See Chapters 2 and 3.

11 See footnote 3, page 109; see also Chapter 1 for the 1998 survey results.

In short, smaller companies in particular need support to understand what they must do to comply with regulations and thus, in the first instance, a programme should be designed to get the environmental management process started and, in the second, to increase the understanding and application of the law.

☐ *Lack of sector-/industry-specific support and solutions*

As has been identified in the preceding material, SMEs perceive the environmental management process as generic and biased to larger companies. As a result, there is also a wish for an environmental management standard that is uniquely catered to SMEs and respects the individuality of each and every one. However, as the objective is not to redesign a standardised EMS for SMEs, it is instead necessary to educate the SME with reference to the present approach to EMS implementation, but in doing so allow them to create something that is inherently unique to themselves.

Clearly, for an EMS to be effective and ultimately improve the environmental performance of an individual company, large or small, it must, by its very nature, be wholly unique to that company as 'the environmental management strategy which a business adopts will depend on its own market situation and environmental pressures as well as the internal resources available' (Coopers & Lybrand and BiE 1995) Ultimately, then, a support mechanism for SMEs must allow them to learn and implement in their own language and using their own frames of reference. A support system cannot merely provide generic EMS implementation—support that is then subsequently altered or watered down by the SME—it must cater to the development of an EMS for the SME from the onset.

☐ *Expenses involved*

SMEs' management of daily affairs is often based on critical incident management with a day's work generally spent dealing with issues and incidents in an ad hoc and reactive manner. In this situation, management's opportunity costs are very high and, thus, 'unless environmental issues are a bottom line pressure, SMEs' focus tends to be on survival and maintaining competitiveness' (Line and Vogt 1996). Simultaneously, significant management resources will be required for EMS implementation, detracting from other parts of the business and thus environmental management receives little or no attention in most SMEs. Simply,

> where there are no bottom line or business-to-business pressures, typical perceptions are that the time and costs associated with implementation of a more structured approach to environmental management are prohibitive and thus environmental management receives low priority in relation to other business pressures (Line and Vogt 1996).

☐ *Lack of relevant information*

Evidently, even if an SME has overcome the initial resource barriers and summoned the necessary courage to engage in environmental management, they are by no means in the clear. An obvious problem is that small businesses do not have access to environmental information, or the expertise to introduce complex management systems without assistance (Welford and Gouldson 1993), and seeking answers to the ongoing questions is time-consuming and very expensive (van Wijngaarden 1995).

The issue of environmental management information dissemination for SMEs is central for the success of every environmental programme on national, European or international level. It is also of uttermost importance to recognise the fact that information in itself is not enough to trigger change. Information will be a powerful tool for change only when it is communicated to a target group, turned into knowledge and put into practical use.[12] This may seem evident, but communication remains a problem. Therefore, it is important not only to provide information, but also to know the target group and to determine what delivery vehicles should be used to achieve the best results.

As confirmed by a survey of certified UK companies (Robinson 1996), 100% of the UK SME respondents needed external assistance, be it in the form of consultants, software, checklists or 'official' advice to develop their system. However, as a similar survey concludes, while 70% of respondents are concerned about their environmental impact, only 34% have sought advice from support services and 46% of SMEs did not even know of any support services available (Hillary 1995).

Even though lack of information is frequently cited as one of the major impediments to improving environmental performance, one must note that there is yet another aspect to be considered: the need for *relevant* information. Smaller firms often find themselves in one of two possible situations:

- Smaller firms do not, for a number of different reasons, have access to environmental management information.
- Smaller firms have access to an overwhelming flood of data but are unable to identify the relevant information.

In both cases it is unlikely that the SME will be able to improve its environmental performance. Communication and information screening are therefore central concepts when developing and disseminating environmental information for SMEs.

Understandably, 'poor knowledge or understanding of a particular issue often introduces a confidence gap', but 'where these issues are addressed and it can be demonstrated that the nature and extent of the environmental risks are clearly understood, the confidence gap decreases' (Line and Vogt 1996). Therefore, in developing a programme to assist and streamline information requirements, accessibility to that support, and knowledge of its existence, is certainly necessary. Important also is the fact that 'few SMEs seek the help available because the support services do not match their needs' and it is often claimed that 'regulators present contradictory messages' (Hillary 1995).

Consequently, SMEs need a mechanism that provides accessible, clear and ongoing advice and information that is relevant and understandable to them.

---

12 See Ludevid Anglada's discussion on this point in Spanish SMEs in Chapter 4.

Environmental management standards are ill-suited for SMEs

When examining the present environmental management standards (ISO 14001 and EMAS) through the eyes of an SME, three clear incongruities appear. First, a common concern is that the standards are strategic approaches to management and SMEs generally do not approach management strategically. Second, standards are generic and SMEs are specific and, third, standards are market-based instruments that rely on market-based pressures which presently are not felt extensively by the SME sector.

Thus, if we can accept that standardised EMSs are more suited to those organisations with an existing strategic management structure, and if we can accept that SMEs generally do not employ strategic management structures, there is a fundamental problem for SMEs when it comes to the implementation of the present standards. In light of the fact that the EMS approach is generic and the SME sector is so diverse in nature, it cannot possibly provide an environmental management panacea for all companies without seriously compromising its effectiveness. The needs, requirements, abilities and limitations of the business community as a whole are irregular and disparate, while the standards, by definition, are not. This of course leads to a situation where the standards are an attempt to be all things for all people at the risk of providing little for a few.

A further problem for SMEs is that the present environmental standards are 'market-based instruments; they are voluntary codes and rely on the assumption that the market will reward companies for participating' (Hillary 1995). Thus, if there are no immediate market rewards—which appears to be the case for many SMEs—the instruments are flawed as mechanisms to improve environmental performance for 99.8% of UK businesses (DTI 1999).

Ultimately, it has been shown that 'a voluntary standard of this type is not tailored to the needs of the SMEs in any realistic way' (Hutchinson and Hutchinson 1995) and thus the EMS approach, in its present form, is seriously deficient when one considers the SME.[13]

Other barriers

While the aforementioned barriers are considerable and often prohibitive, they are by no means the only hindrances to the implementation of environmental management in an SME. Some claim that 'the lack of evidence of a substantial uptake by SMEs to the development of environmental management is often related to perception of the requirements involved', while others hold that the problems lie in a 'general fear of the unknown and a common perception is that small companies in particular would rather not know about potential problems than be aware of them' (Line and Vogt 1996).

Other issues of definite consequence are a lack of commitment from top management (ENDS 1996) and, in the case of EMAS, the level of detail of information

13 See Chapter 7 in which Starkey suggests an alternative view that ISO 14001 is appropriate, unaltered, for SMEs.

to be released within the environmental statement would be a task well beyond the ability of most SMEs who are simply not capable of preparing statements of that nature about the internal workings of their companies (Coopers & Lybrand and BiE 1995).

In short,

> many of the barriers preventing SMEs from adopting a more formalised approach to environmental management relate to their perceptions of the costs, effort and commitments involved, their perceptions of the regulators and their role and the potential for environmental management to impact on the commercial aspects of the business (Line and Vogt 1996).[14]

Solutions for SMEs

A solution should be:

Inexpensive

It has been mentioned time and again that the SME sector is comprised of a diverse incomparable lot. However, while this is essentially true, there is one thing that is generally universal among the SME community—lack of cash. Whatever the viewpoint, the facts remain the same. A support programme must respect that 'SMEs need help to facilitate their growth and survival as they lack the financial resources to tackle new pressures such as environmental regulations' (Hillary 1995).

What transpires is fairly clear. Not only is there a disproportional expense/turnover bias in favour larger firms—as the costs of implementation do not rise in a linear correlation to a firm's size—but SMEs simply struggling with cash flow, financing difficulties, bad debts and late payments are not at financial liberty to spend unless it is absolutely essential. Consequently, 'supporting organisations need to develop strategies which deal with the internal weaknesses in SMEs and deliver affordable company specific advice' (Hillary 1995). Financial support through funding is also essential in this context.

Sensitive to the limitations of SMEs

Tight cash flow translates directly to a lack of human resources. As a result, an SME generally will just not have the human resources required and thus the SME sector is consistently ill-equipped to the mounting environmental pressures it faces. Evidently, not only do SMEs lack human resources, they also 'lack the time and money to investigate their environmental performance, read all the literature or access the high cost consultancy/conference support network'. Furthermore, while simply not having the 'time' is a considerable limitation, a second and equally limiting factor to consider when designing support programmes is that smaller businesses do not have the internal skills to respond to these new demands, and at the pace required, and

14 See Hillary's discussion, in Chapter 10, on the internal and external barriers to EMS uptake by SMEs.

thus support must not only be provided to ease resource deficiencies but must also provide expertise (Hillary 1995).

One thing that makes SMEs similar is that they are all different. They have different products and services, they have varying *raisons d'être*, and they typically have special clients in niche markets. Simply, they 'have special needs of their own and are all faced with the challenge of one or more of a group of special characteristics'. They have unique characteristics that include: short track records; heavy reliance on niche markets; lack of specialist skills; low cash flow; small asset base; the need to make changes in structure and ownership at various stages of growth' (CBI 1996).

Consequently, it is of no use if environmental management efforts require extensive internal operational experience, considerable market rewards, expertise, or heavy finance, which is, in a nutshell, where we are today.

Co-operative

In developing an effective support mechanism for SMEs, co-operation plays a fundamental role. In broad terms, 'environmentally sustainable development is dependent upon co-operation within an environmentally aware and dynamically active SME sector' (Hutchinson and Hutchinson 1995). True, also, is the fact that the co-operation must be specific.

Ultimately, and respecting that 'individual change in SME practices is unlikely to occur in isolation' (Hutchinson and Hutchinson 1995), a programme for success must not only be developed in the spirit of co-operation, but it must be inherently co-operative itself and promote co-operation among its participants.

Locally based

As is surely the case with almost any programme for sustainability, it must be locally based. Whether it be buying vegetables or developing government policy, the key to sustainability is generally always local. This fact does not differ when considering SMEs. SMEs are usually closely linked with community culture, their employees are generally local, their ownership is typically local and often traditionally associated with the heritage of the area.

Similarly, and as has been exposed by the incomparable success of both the Aire and Calder Project and Project Catalyst (both locally based waste minimisation initiatives), a programme that is locally based will provide a number of auxiliary benefits. It will encourage the combination of resources to make use of the maxim 'a problem shared is a problem halved', and it will be enable the exploitation of economies of scale that such co-operation brings (Shayler 1996). Furthermore, the successes of one SME will have visible and encouraging spin-off effects on another and the programme will achieve multiplied or synergistic results collectively far greater than if they were achieved individually.

Ultimately, 'an effective approach will require support mechanisms at local level, providing knowledge, skills and resources to sustain the strategies' (Hutchinson and Hutchinson 1995).

User-friendly

Clearly, it is important to provide training, support and implementation programmes that not only avoid the facetious use of esoteric terms themselves, but also provide easy guides or practical translations of the many easily confused and erroneously interchangeable terms, such as 'review' and 'audit', or 'impacts and effects'.

For example, SMEs generally see the standards as bureaucratic, confusing and esoteric and frequently feel like the victims of environmental red tape (ENDS 1995). The SME community simply needs a hand through the bureaucracy, jargon and legal language on which these standards are built. Support must be user-friendly both in terms of jargon and accessibility—easy to understand and easy to use.

Flexible

When assisting firms to improve their environmental management, flexibility is a crucial issue. For instance, if a company provides time and flexibility in improving environmental performance, its approach is more likely to be innovative and preventative. If, however, the improvement is required rapidly, then the 'end-of-pipe' reactive approach is chosen (Griffin *et al.* 1995). The same applies when implementing environmental management or providing support services that assist implementation.

From the aforementioned material, it is clear that SMEs know what they want and what they do not want. Considering that legislative compliance and cost savings are their greatest concerns, many 'felt that it might only be appropriate to complete a basic review of environmental performance, with no subsequent development of any formal management system in the short term' (Coopers & Lybrand and BiE 1995). Consequently, developing a support programme that necessitated commitment for the entire implementation and certification process would almost certainly be unwelcome in the SME community.

Conclusions

Experiences from ongoing and previous European environmental management projects and initiatives show clearly that the needs of SMEs are different from the needs of other sectors of industry and that a variety of obstacles have to be overcome to achieve a substantial improvement in their environmental performance. To be successful, a support programme for SMEs must be inexpensive, co-operative, locally based, flexible, unique and accessible. Furthermore, an effective programme must provide training, legislative compliance support, and provide clear, concise, dependable sector-specific information and support.

9

What are 'appropriate' systems for assessing environmental risks and performance in small businesses?

Simon Whalley

If we pause for thought, practical common sense tells us that small and medium-sized enterprises (SMEs as defined by the European Commission),[1] despite the convenience of the label, are not a homogeneous section of the business community. In attempting to get a perspective on the relevance of their impact on overall environmental performance (whether at local, national or even global level), it is important to note that, although they form by far the largest business segment, the focus of environmental concern has only recently shifted in their direction. Consequently, all the mechanisms for establishing environmental probity have been designed around methodologies more appropriate for larger companies and not easily adaptable to the multifaceted circumstances of smaller businesses.

As a result of all this, when it comes to responding to stakeholders' concerns, many hard-working and successful businesses find it galling to be treated as if their individual actions and opinions do not count, while at the same time having a finger of blame pointed at them for their collective contribution to environmental degradation without

1 Definition contained in the Annex to Commission Recommendation of 3 April 1996 concerning the definition of small and medium-sized enterprises: SMEs have fewer than 250 employees and either an annual turnover not exceeding ECU 40 million, or an annual balance sheet not exceeding ECU 27 million. Within this 'small enterprises' have less than 50 employees and turnover of less than ECU 7 million or an annual balance sheet less than ECU 5 million. 'Micro enterprises' have less than ten employees. In all cases, no more than 25% of the enterprise must be owned by a 'large' business.

any real evidence or hard facts to support it. Their sense of grievance is heightened when they are further criticised for failing to adopt what many of them feel is a 'big business' solution to environmental issues. Although their exclusion from the debate has been mostly a function of their circumstances (lack of resources for non-essential activities), there exists a strong sense of a continuing failure by 'officialdom' to recognise and respond to their particular needs—without imposing a 'second-class' system on them.

Concentrating on size is, however, only another misreading of the situation and indeed for 'small' it would be better to substitute 'independent-minded' or 'owner-managed' as these are more accurately descriptive of the principal wealth-creating sector in the UK. Owner-managed business (OMB) culture (or rather 'cultures' since they are not homogeneous) is not the same as the culture of a small unit in a large company. This is often difficult for those immersed in the culture of a large business, institution or government office to recognise or accept. It is not 'tidy': individual managers have complex, non-linear responsibilities; records and systems are informal (sometimes chaotic!); professional advisers and consultants are frequently as much a problem (because of their cost) as the problems they are meant to solve. Worse, for those who are seeking neat answers to environmental questions, every OMB has its own culture—even within sectors—because they are much more dependent on proprietoral influences and local conditions than large companies which can impose corporate 'standards' across even very different activities and international boundaries. Despite this individuality there are some common themes at work which are not even remotely appreciated outside the world of the 'small' business. This chapter will attempt to put this into perspective using a particular theme to address the issue of 'appropriateness' of environmental management systems (EMSs) for small business.

The double bottom line

The theme that, perhaps, most clearly distinguishes an OMB is the 'double bottom line' effect (see Fig. 1). All OMBs contain elements of 'visionary' and 'reactionary' motivations in their management. For those on the visionary wing (certainly until they outgrow the small business sector—by which time they have often either moved away from owner-management or transferred to the reactionary style), the bottom line is usually survival and they will only consider change when they are fully convinced that it will not increase the pressures on cash flow and management time. For extreme reactionaries, on the other hand, the bottom line is control—often they would rather see the business collapse than give away any control—and acceptance of any outside influence is to give away control. As a result, they are often unwilling to consider any change. Although the extremes are surprisingly common, most are somewhere in the middle and, balancing survival with control, are willing to accept change only reluctantly and have very little interest in growth. Attempting to impose external values on such situations is often counter-productive. Only long-term

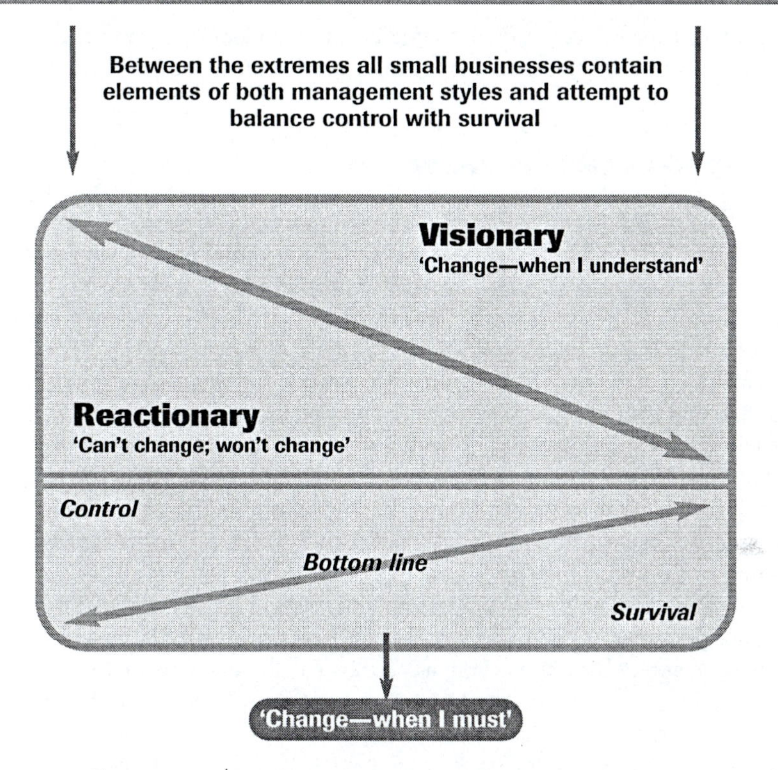

Figure 1: OMB/SME BUSINESS MODEL: THE DOUBLE BOTTOM LINE

education, through appropriate stakeholders, to create cultural change is likely to be effective.

Wholesale cultural change in the environmental field is beyond the resources of any agency or government department if attempted across the board with several million small businesses.[2] For this reason it is essential to begin to identify what is 'appropriate' for small business in terms of management response to environmental issues. All those involved should remember that the move to establish environmental management as a discipline was not, as it was with quality standards, in response to customers' needs for codification of existing 'best practice', but rather to attempt to create a better climate (literally perhaps!) for the future. It is essential to avoid the trap of 'waiting for evidence' on how business is coping before attempting to grapple with the needs of OMBs. Delay will result in the gap between the environmental

2 Every estimate seems to vary, but the Federation of Small Businesses (FSB), using Department of Trade and Industry (DTI) statistics, reckons on 3.65 million accounting for 99.8% of the total number of business in the UK (DTI 1999).

management capacity of small and large business widening beyond the ability of governments, agencies or standards organisations to provide a bridge.

Proportionality of response

The very scale of the apparent 'problem' has a mesmerising effect. Recent surveys such as that by Groundwork in 1998 in the UK (and somehow the very number of surveys itself has created its own distortion of the perceptions) have shown that increasing numbers of small businesses are aware that there is concern among their stakeholders that a problem exists.[3]

The first step is to try to identify which businesses or business categories need to respond in any formal sense (i.e. perhaps by implementing an *appropriate* EMS). In this there is little point in considering including sole traders who are, by definition, not able to put a formal 'management system' into place. This removes two-thirds of the numbers and leaves around 26,000 'medium' businesses, 72,000 'small' businesses and 980,000 'micro' businesses in the UK. These categories are here defined according to the EU definition.

Identification of the categories of business that might be 'at risk' from stakeholder or regulatory action on the environment has been often discussed. It is usually rejected as being too complex because, without a common baseline and with no generally agreed definition of what constitutes 'environmental risk', no risk management mechanism can be effectively applied. There are some steps being taken to address this: the International Organization for Standardization (ISO) is debating the definition of 'risk' (and its subdivisions, including environmental risk) and a number of commercial products are being produced that attempt to give business a management tool to help evaluate its environmental risk exposure. None of these can yet be described as 'definitive'. However, there is enough evidence available in current UK and EU legislation to enable the allocation of 'high risk' to many business process categories. For example, in the February 1999 edition of *Environment Action* Barry Keyworth, a team leader with the Environment Agency (EA) dealing with process industry regulation, is quoted as saying: 'There are about 2,000 IPC (Integrated Pollution Control) processes in the country (UK) which must be re-authorised by the Agency. With IPPC (Integrated Pollution Prevention and Control), that number could go up to about 7,000!'

Perhaps rather more subjectively 'low risk' may be allocated to all those categories with very little impact on the environment for which there is little prospect of any deleterious consequences from the normal pursuit of their business practices. The remainder can be considered 'medium risk'. When such a simple mechanism is applied to a list of process categories, 76% of business processes are found to be in the high- and medium-risk area. Although this is not directly indicative of numbers of businesses in these categories, if these proportions are applied with a balancing factor (see Table 1) to the numbers of small businesses, a guess at the possible numbers that might be affected can be made.

---

3 See the survey results in Chapter 1.

Business size	Total no. of UK SMEs*	'High' and 'medium' risk (by process)	Illustrative balancing 'factor'†	No. of businesses 'at risk'	% of total
Medium: 50–249 employees	26,000	76%	50%	9,880	38%
Small: 10–49 employees	72,000	76%	20%	10,944	15%
Micro: 1–9 employees	980,000	76%	10%	74,480	8%
Total	1,078,000			95,304	9%

* Figures exclude 'sole traders'
† Percentage of 'normal' risk (illustrative only)

Table 1: PUTTING SME ENVIRONMENTAL RISKS IN PERSPECTIVE: AN ILLUSTRATION OF POSSIBLE OUTCOMES

Different factors of risk for each size-range of business will need to be used to do this. *For illustrative purposes only* it has been assumed here that medium-sized businesses are only half as likely to involve inherently 'risky' processes as large business, that small businesses have a 20% chance and micro businesses only a 10% chance of being affected. These assumptions have not been tested and are not put forward with any purpose except that of being illustrative of a possible outcome. Research into any correlation between size (or culture) and environmental risk would be required before an adequate indicator could be drawn up.

From these figures it is perhaps reasonable to estimate that the total numbers of environmentally 'at risk' SMEs in the UK might be around 100,000 and certainly not more than 750,000—with the balance of probability (and practicality) inclined to favour the lower figure.

If a mechanism such as that illustrated here—though with rather more robust methodology—was used to identify the focus of environmental concern, it would be possible to allocate resources, both in the commercial and regulatory sectors, more effectively. This could also reduce the burden on businesses when inappropriate demands for environmental response damage productivity and even threaten survival. It is imperative for some research to be undertaken to test and quantify the appropriate scale of need for environmental management (whether systemised or not) in small businesses. Without it, everyone will continue to be bemused by the sheer numbers of small businesses and fail to come up with any sense of proportion.

How appropriate are existing EMSs?

Moving on from scale, it is clearly necessary to determine whether the EU's Eco-Management and Audit Scheme (EMAS) and ISO 14001—the international EMS standard—are appropriate in their current form for OMBs or whether amendments or better interpretation is needed. The research so far undertaken in the UK and Europe has been inconsistent so that a cumulative conclusion is difficult. Arguably, also, the results are inherently biased because of the way in which the work has, of necessity, had to be focused on the small numbers of businesses who have so far embarked on this approach. Before going any further into this, however, it is important to define what is meant by 'appropriate'. The dictionary defines this as meaning 'suitable'. There is some debate over the alternative use of the word 'relevant' instead of 'appropriate' but this is defined as meaning 'concerning'—a rather weak word. There is also a greater sense of immediate importance about 'appropriate' and for current purposes it seems to be the most useful word to use.

The importance of cultural aspects in the response of small business to issues such as the environment has already been discussed. There has been considerable attention given to the relative importance of the 'carrot' and the 'stick' in environmental decision-making for SMEs. It can, however, be seen from the studies in this area that, in fact, almost all response is reactive and can be categorised very simply as:

When asked = opportunity
When demanded = threat

The more complex responses of large business are simply irrelevant to the small OMBs. Further, perception in such matters is everything and the source of the enquiry (stakeholder, regulator) will determine the balance between perceived opportunity and threat. There is very little incentive—and totally insufficient information—for any unrequired and unasked initiative in this area. The level of activity among stakeholders will, therefore, determine the response of SMEs. In the 1998 Groundwork survey the correlation between those that enquired about environmental performance and those considered to have the greatest influence was very close. A total of 40% of SMEs had received an enquiry and these sources of enquiry were ranked as follows:

Local authority (regulator)
Customers
Insurers
EA (regulator)
Suppliers
Trade association
General public
Environmental group
Investors/shareholders
Lawyers
Competitors
Bankers

Since Groundwork also determined that six out of ten of the businesses surveyed had been asked for environmental information through their supply chain, it is clear that there will be increasing pressure to respond. However, at the small business level, social and community pressures will act on an individual manager's decision-making in a way that is not normally considered in the structure of business surveys. This aspect did, however, come out of a study published by the Centre for Hazard and Risk Management at Loughborough University on behalf of the Economic and Social Research Council (Petts *et al.* 1998b).[4] This determined that pressure for change frequently came from employees, but that management often separated moral concern for the environment from actual performance in the business—even where there was a perception of a need for compliance. This left the business vulnerable and dependent on being reactive to individual threats. In most OMBs there is never enough time for 'planning ahead'.

Although there is clearly a need for research and comparative analysis of the whole range of studies already undertaken, a picture seems to emerge of only the very beginning of any awareness level of EMS appearing in most small businesses. The weakness in the data is, however, compounded by the bias already alluded to. This arises partly from the 'double bottom line' effect (see Fig. 1), which results in the participants in most initiatives coming mainly from the 'visionary' wing of the sector, and partly from the fact that most of the small numbers with actual experience of implementing EMSs will have had some form of financial or practical assistance or sponsorship. These two factors have distorted even the limited experience so far gained and reinforce the view that reliance on 'results' is unsatisfactory and that it is important to incorporate simple common sense and an understanding of OMB culture in considering requirements for any form of EMS.

In order to address this, another factor in the culture of small business needs also to be considered—the lack of dedicated resources for 'non-essential' activities. This means that the business principle has to take on responsibility for all 'new' areas of management. Any such activity dilutes attention from essential core activities and is regarded with suspicion. In order to impress the need for any environmental response, let alone any formal management 'system' onto a small business, even material that is perfectly understandable has to be put into the context of 'minimum available time' for the business to be able to assimilate the essential elements. Any material that does not immediately strike a chord will, at best, be filed for attention when there is time to think about it—a time that probably never comes!

The Loughborough study identified that even legal compliance (which was theoretically accepted as necessary) was often ignored if the benefits were not accepted as being deliverable. The case for demands for compliance with any standard that cannot demonstrate tangible, short-term, benefits, and whose proponents cannot immediately and succinctly put it into the context of an individual small business, is proportionally weaker still. The same, of course, applies to OMB response to any related activity (e.g. a 'waste minimisation' programme).

The challenge now is to ensure that any mechanism for improving environmental performance in small business:

- Is used appropriately
- Applies only to businesses 'at risk' or that will otherwise benefit from it

4 See Chapter 3 for a fuller discussion of the study.

- Is written, and explained, in an appropriate way
- Has its benefits 'sold' clearly and simply
- Is cheap and easy to implement
- Has appropriate support mechanisms in place to give advice and assistance

Taking these six points and comparing current 'performance' in the implementation of both EMAS and ISO 14001, the following comments might be made on behalf of small business:

- Jury still out, but the mistakes made in demands for compliance with the quality standard (ISO 9000) are still fresh in the memory.
- No consistent definition of risk; therefore inconsistent application of criteria.
- Most small businesses have relied on outside help—so reports of 'no problems with the wording' may be premature. The 'explanation' is, however, still far from simple and inclusive of more informal systems.
- Attempts made so far in the UK (e.g. the ill-fated Small Company Environmental and Energy Management Assistance Scheme [SCEEMAS]) have been under-resourced and failed to address OMB cultural issues.
- 'Could do better'.
- There is still a long way to go on this. Some sort of Highway Code for better business performance on the environment might be possible when the issue of 'what is appropriate' is better defined.

How to meet this challenge?

What is clearly important is to continue to collect and analyse data on implementation of existing standards and other mechanisms. The data collected so far on EMS, for example, has not revealed anything SMEs did not already know—i.e. 'formal management systems are not easy to apply to small business'. It is also clear that, for small business, understanding the need is as important as understanding the legislation or the standards.

When, in their respective ways, the EU and the ISO took on the challenge of environmental management, their aims were 'creation of best practice' rather than 'codification of existing best practice'. In this they entered new territory where it was no longer enough to rely only on existing mechanisms for development and delivery.

To address the challenge of small business requires attention at all levels. It is not sufficient to address only one facet, particularly as the core of the problem is not standards, waste management or even environmental legislation. All of these issues only impact on a small business when the owner-manager is convinced that the threat to the company's bottom line (control/survival) is greater if the issue is ignored than it would be if it was addressed. There may be a need (though this is still debatable) to address the 'words' used in all these areas in order to make what is said more accessible to those who have no time for jargon and puzzling out complex concepts. It is also important to engage smaller business in the process of review and alignment between ISO 9000 quality management standards and the ISO 14000 environmental management standards—for small businesses cannot afford to go through the process two or—if health and safety issues are also concerned—three times. The development of new guidance documents and/or awareness programmes needs very careful thought to avoid producing 'deliverables' that actually deliver nothing because they are never considered by those at whom they are targeted. At this level, 'opportunities' and 'savings' are debased currency. Operating at the margins allows nothing for such luxuries.

Any new material produced to assist SMEs should address not just the interpretation of the standards in the context of small business, but clearly, and simply, explain when and why a small business needs to consider any formal systems—based on appropriate reality checks. The issue of definition of appropriate levels of documentation and the formality of systems needs to be considered, with ongoing dialogue with those setting certification standards and guidance (e.g. the International Committee on Conformity Assessment—ISO/CASCO) on the direction given to certification bodies. The circumstances in which self-declaration of compliance with standards is appropriate needs clarification and promotion to both stakeholders and small business. As awareness is, perhaps, the biggest hurdle, dialogue with national standards organisations and governmental agencies on improved ways of promoting appropriate use of the standards and alternative mechanisms, focusing only on the areas of greatest risk, also need to be developed and maintained.

What is required most, however, is positive and sensitive leadership, commitment to finding a practical and deliverable solution and an open mind to the complex nature of the smaller, owner-managed business sector.

10
The Eco-Management and Audit Scheme, ISO 14001 and the smaller firm

Ruth Hillary

Small firms make up the vast majority of businesses in Europe and the UK. In 1996, around 90% of European businesses were classified as small and medium-sized enterprises (SMEs)[1] (CCEM 1997) and in 1998 there were 3.7 million businesses in the UK, of which 99% were small businesses employing less than 50 people and only 25,000 were medium-sized, employing between 50 and 249 people (DTI 1999). The environmental impact of small firms is not known, either at national or regional levels. It is often, and widely, quoted that, as a sector, SMEs could contribute up to 70% of all industrial pollution (Hillary 1995). The author found no evidence to support this assertion, however. Probably one of the reasons so little is known in environmental terms about the sector—and so much is speculated about it—is because of its size and diversity.

The heterogeneous nature of the SME sector makes it difficult to generalise about its environmental impacts and strategies. The environmental issues facing a sole trader or partnership will have little similarity to those of a firm employing 249 people; and yet they are lumped together in the SME sector. The lack of knowledge about the impacts of the sector, and the recognition of its importance in helping to ensure a

1 The EU defines SMEs based on employee numbers, turnover or balance sheet total and ownership. An SME has less than 250 employees and either:
- An annual turnover not exceeding ECU 40 million; or
- An annual balance sheet total not exceeding ECU 27 million; and
- Is an independent enterprise, i.e. 25% or more of the capital or voting rights cannot be owned by a larger enterprise.

healthy economy, has stimulated a growing interest in SMEs. In the EU, engaging SMEs in environmental improvements is viewed as a vital part of the drive towards sustainable development (CEC 1992).

As part of a broader strategy to provide businesses with tools to manage their environmental impacts more effectively and contribute to sustainable development, the EU developed the Eco-Management and Audit Scheme (EMAS) Regulation (CEC 1993) on the back of the British environmental management system (EMS) standard BS 7750 (BSI 1994). BS 7750 generated interest in EMSs in the international standards world, eventually being superseded by the international EMS standard ISO 14001 (BSI 1996a). The voluntary regulation and the EMS standards sought to provide organisations with the means to develop systematic approaches to environmental performance and to complement legislation. All purport to be relevant and applicable to small and medium-sized firms.

This chapter presents the findings of two studies that sought to investigate EMAS and the EMS standards in the SME sector. The first was a pan-EU assessment of the implementation of EMAS undertaken for the Commission of the European Communities (CEC) (Hillary 1998).[2] Its purpose was to investigate the current implementation practices of EMAS in the different member states. It provided data on EMAS-registered sites, their use of EMS standards, EMAS implementation periods, the support needed to participate in the scheme and the benefits of participation. The second study—a review study—evaluated 33 different studies that investigated the practical implementation experience of SMEs with EMSs and the attitudes of smaller firms to the environment (Hillary 1999). The aim of the evaluation was to identify the barriers, opportunities and drivers for smaller firms in the adoption of EMSs.

Methods employed

The pan-EU EMAS survey employed a telephone survey to gather in-depth objective information from four groups associated with EMAS implementation, i.e. competent bodies or administrative individuals, accreditation bodies, accredited environmental verifiers and EMAS-registered sites. The EMAS help-desk provided the contact details for each group.[3] The EMAS site list was dated 31 December 1997 and included a total of 1,211 sites in 12 member states.

The large number of EMAS-registered sites meant that interviewees were selected at random. Random selection criteria were developed to select a representative 10% sample from those member states where there are large numbers of EMAS-registered sites, i.e. Austria, Denmark, Finland, France, Germany, the Netherlands, Sweden and the UK.

A five-part confidential questionnaire—'Questionnaire on the Implementation Status of EMAS'—was developed as the investigative tool to achieve the study's

2 *europa.eu.int/comm/emas*

3 *www.emas.lu*, the former EMAS help-desk.

objectives. The questionnaire was pilot-tested in October 1997 and modified accordingly. The majority of questions were unprompted and received spontaneous responses from interviewees. The questionnaire was in English and was orally translated into German, French, Italian and Spanish where necessary. The interview schedule spanned four months. EMAS-registered sites were interviewed during February 1998.

A 'standardised analysis database' in Excel 5.0 software was developed for the rapid collation and analysis of survey data. Questionnaire responses were codified and entered into the database.

The second study analysed 33 different studies at UK and EU level, dating from 1994 to 1999 (see Appendix, page 146). Study reports were identified from: academic and government sources; support organisations, e.g. green business clubs; expert individuals, e.g. members of the British Standards Institution (BSI) Small Firms Panel; non-governmental organisations (NGOs); consultants; and companies. Four standardised tables were developed to formalise the analytical process. The table headings fell into three categories: study details including author, publication date, title, etc.; content of study (including scope and issues covered, limitations and quality of data[4]); and study findings (including internal and external barriers, benefits and disbenefits, stakeholder pressures and drivers). Each report was read and analysed and relevant details entered into the standardised tables.

There is a scarcity of quality studies into SMEs and their adoption of formal EMSs. This analysis is of 33 studies. Others have been identified and no doubt more exist, but in the time-frame available all reasonable efforts were made to obtain the key reports. Not all the selected studies examined the formal EMSs of EMAS and ISO 14001 (BS 7750 is included as some studies prior to and around 1996 refer to it [BSI 1994]). Some studies looked at SME environmental awareness and attitudes towards the environmental issues they face, and provided an insight into the reasons why EMSs are not adopted.

EMAS and ISO 14001

Improving the environmental performance of SMEs is important, irrespective of their total—as yet unknown—impact, because they are a vital part of the enterprise society that collectively can contribute to sustainable development. One means of bringing about improved environmental performance is through the adoption of EMSs. The two formal EMSs in the marketplace are EMAS and ISO 14001. Common to both is the need for an organisation to implement a number of management system stages to

4 A: High-quality data, reliable findings due to sample size or in-depth nature of study and method of research and/or material directly related to EMSs and SMEs. B: Good-quality data, reasonably reliable findings due to moderate sample size or moderate depth of study, and/or material on the attitudes of SMEs to environmental issues and EMSs. C: Moderate-quality data, low sample size or shallow study and method or research and/or material on attitudes of SMEs to environmental issues.

formalise the policies, procedures and practices that control an organisation's environmental aspects. EMAS has the added requirement of an environmental statement which publicly reports the environmental performance of a site. Both standards purport to be applicable to large *and* small firms. Indeed, ISO 14001 was reportedly written with the 'chip shop owner' in mind;[5] as a safeguard of its relevance to SMEs a working group was convened at the International Organization for Standardization (ISO) to investigate this issue (Dodds 1997). The EMAS Regulation is less certain of the unaided participation of SMEs and suggests that supporting measures should be introduced by member states to assist smaller companies to register.

The value of an EMS to the smaller-firms sector and the proposition that SMEs will be able to straightforwardly participate in EMAS and ISO 14001 meant that both of the studies discussed in this chapter sought to estimate the number of SMEs registered to the standards.

Estimation of the number of SMEs registered to EMAS

The official UK and EU lists of EMAS-registered sites are the most reliable source of data on registration numbers (see Table 1). The UK competent body asks companies registering sites to EMAS to supply enterprise size data.[6] The EU list holds no such data.

EMAS figures and information	UK	EU
Percentage of companies registered to EMAS that are SMEs	24%	18%
Number of SMEs registered to EMAS	17	541
Number of large enterprises registered to EMAS	52	2,464
Number of enterprises registered to EMAS size not known	2	N/A
Total number of EMAS-registered sites	71	3,005
Source of information	UK EMAS competent body*	EC EMAS help-desk†
Type of information	Official	Official
Date of information	15 September 1999	1 September 1999

* *http://www.emas.org.uk/siteregister.htm*
† *http://europa.eu.int/comm/emas*

Table 1: ESTIMATE OF SME REGISTRANTS TO EMAS IN THE UK AND EU

5 Personal communication with O.A. Dodds, 1996.
6 *www.emas.org.uk/siteregister.htm*

Estimation and degree of accuracy in the UK and EU estimates

There is a high degree of accuracy in the estimate of number of SMEs with sites registered to EMAS in the UK (17). Since the UK EMAS competent body has been housed at the Institute of Environmental Management and Assessment it has collected enterprise size data on EMAS-registered sites. Of the 71 sites registered on 15 September 1999 only two companies could not be identified by size.

There is a moderately high degree of accuracy in the estimation of the number of sites of SMEs registered to EMAS in the EU. The European Commission does not hold size data on enterprises that have sites registered to EMAS. The calculation of the number of SMEs is based on the estimation in the pan-EU EMAS survey undertaken in 1998 (Hillary 1998). The study completed telephone interviews of EMAS-registered sites and collected size data. Sites were classified by size using the EU definition of SMEs.[7] There were no registered sites in Greece, Luxembourg and Portugal. In eight countries—Austria, Denmark, France, Finland, Germany, the Netherlands, Sweden and the UK—population sizes were too large to enable all sites to be interviewed so a minimum representative sample of 10% was taken. All EMAS-registered sites were interviewed in Ireland, Italy and Spain.

The survey's use of randomised sample data of EMAS-registered sites allowed for extrapolation to the EMAS population as a whole. This gives a sound estimate of the number of SMEs registered to EMAS and is the most accurate data currently available in the EU. National variation can be wide, however—for example, the UK has 24% of its registered sites from SMEs[8] while in Denmark over 50% of registered sites employ less than 100 people (Andersen 1998).

Estimation of the number of SMEs adopting ISO 14001

There are no official UK or EU lists of ISO 14001-certified organisations. Reliable commercial, but unofficial, sources of UK ISO 14001 numbers have been found (see Table 2).

Neither the EGA Environ Environmental Consultancy list[9] nor the BSI list[10] hold size data on organisations. The EGA Environ Environmental Consultancy list,

---

7 See footnote 1, page 129.

8 This figure is considered by some in the UK government and the EU to be encouraging. It is not. Large firms, which comprise less than 1% of all UK firms, account for 76% of all EMAS-registered sites while the vast majority of firms, the SMEs, account for only 24% of EMAS-registered sites. A MORI survey undertaken in 1998 (Smith and Kemp 1998) determined, from a sample of 300 SMEs, that 3% were registered to EMAS and 9% intended to register. This survey still presents an over-optimistic picture of EMAS uptake by the general population of SMEs, however, as the sample population was biased towards medium-sized companies (See Table 1 in Chapter 1, page 29; see also below, 'EMAS and ISO 14001 registrations as a percentage of UK businesses', page 134).

9 *www.eagenviron.co.uk*. EGA Environ Environmental Consultancy collects data on ISO 14001 certifications from UKAS-accredited certification bodies. (UKAS does not hold a publicly available list of ISO 14001 certified organisations.) The company names of the organisations certified by BSI are not held on EGA's list.

10 *www.bsi.org.uk* enables individual company details to be accessed if the name of the company is known.

ISO 14001 figures and information	UK
Percentage of companies certified to ISO 14001 that are SMEs	25%
Number of SMEs certified to ISO 14001	254
Total number of organisations with ISO 14001 certifications from UKAS-accredited certification bodies	1,014
Sources of information	EGA Environ Environmental Consultancy*, BSI†
Type of information	Commercial/unofficial
Date of information	13 September 1999

* *http://www.eagenviron.co.uk/*
† *www.bsi.org.uk*

Table 2: ESTIMATE OF THE NUMBER OF SMEs WITH ACCREDITED ISO 14001 CERTIFICATES IN THE UK

which holds the total number of certificates to ISO 14001, does not include the names of organisations certified by BSI. These need to be obtained directly from BSI as its web list can only be searched if the certified organisation's name is known.

Estimation and degree of accuracy in the UK

The estimation of the proportion of companies certified to ISO 14001 in the UK that are SMEs (25%) has a moderate-to-low degree of accuracy. No size data is collected on organisations with ISO 14001 certifications from UK Accreditation Service (UKAS)-accredited certification bodies. No single UK list holds all ISO 14001-certified organisations. ISO does not collect size data in its surveys (see ISO 1998).[11]

The data limitation means that the method employed to estimate the proportion of SMEs certified to ISO 14001 must rely on survey results and trends in EMAS uptake. A MORI survey, discussed at length by Smith and Kemp in Chapter 1 (see also footnote 8, page 132), determined from a sample of 300 SMEs that 3% of those interviewed were registered to EMAS and that 7% were certified to ISO 14001. These figures are very high for the general population of SMEs. This may be partially explained by the survey's bias towards medium-sized companies (see footnote 8, page 132); however, they do suggest that ISO 14001 is significantly more popular among SMEs in the UK than EMAS, possibly with over twice as many registrants. The current number of UK SMEs registered to EMAS is 17 (see Table 1); this might suggest that around 35 SMEs are certified to ISO 14001. However, a review of the company names held on the EGA and BSI lists of certified ISO 14001 organisations tells us that this figure is much too low. Furthermore, considering the ratio of SMEs to large firms with sites registered to EMAS (24:76), and, given the greater popularity of ISO 14001 in the marketplace and the preference SMEs have for the standard, if they show any interest in EMSs at all (Hillary 1999),

11 *www.iso.ch*

it is suggested that the percentage of SMEs certified to ISO 14001 will be higher than those registered to EMAS. Therefore, it is estimated that the proportion of companies certified to ISO 14001 that are SMEs is, at a minimum, 25%.

EMAS and ISO 14001 registrations as a percentage of all UK businesses

UK government statistics indicate that there were approximately 3.66 million businesses in the UK in 1998 (DTI 1999). Over 2.33 million of these were 'size class zero' businesses: those made up of sole traders or partners without employees. Small businesses, including those without employees, accounted for over 99% of all businesses. Only 25,000 businesses were medium-sized (50–249 employees) and there were only 6,660 large businesses (250 or more employees).

Using this data, it is calculated that 0.03% of all UK businesses are registered to either EMAS or ISO 14001 and that 0.007% of SMEs are registered to one of the initiatives. If businesses of 'size zero class' are excluded, the percentages increase so that 0.08% of all businesses and 0.02% of SMEs are registered to EMAS or ISO 14001. The numbers of registrations represent a minuscule proportion of total businesses, and it is likely that the actual numbers are lower than the figures presented, as joint registrations have not been identified.

EMAS and ISO 14001 together

The estimates for the number of SMEs registered to EMAS or ISO 14001 may conceal some duplication in the registrations whereby an enterprise is registered to both. The pan-EU survey identified that organisations adopted both EMAS and ISO 14001 (Hillary 1998). Just under half (47%) of all registered sites interviewed were also certified to ISO 14001. However, the majority (85%) of registered sites of small-sized enterprises did not adopt ISO 14001 (see Fig. 1).

ISO 14001 can be used as the EMS component of EMAS. However, the data in Figure 1 indicates that small firms, by and large, do not go for joint registration. Medium-sized enterprises are more likely to do so. Of those firms not registered to ISO 14001, 100% of small firms and 71% of medium-sized firms have no intention of obtaining certification to ISO 14001 in the future. In the revised version of EMAS that is due to be adopted in 2000, ISO 14001 is included verbatim in the EMAS text (CEC 1998). This is to avoid confusion and potential competition between the two initiatives and to enable companies to use the international standard as the EMS for EMAS.

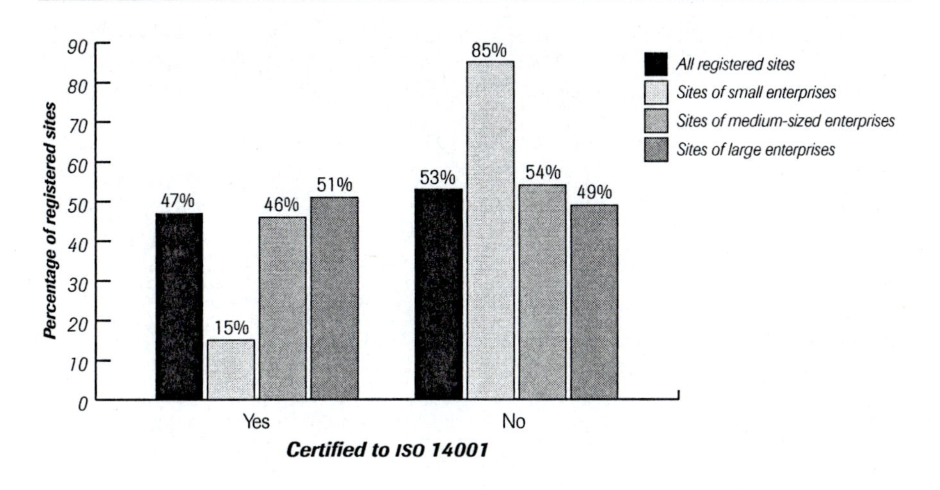

Figure 1: PERCENTAGE OF EMAS-REGISTERED SITES CERTIFIED TO ISO 14001

Source: Hillary 1998

Internal and external benefits of SMEs adopting EMSs

Numerous internal and external benefits are expected from the implementation of formal EMSs such as EMAS and ISO 14001. In the review study, 22 studies identified benefits from EMS implementation experiences of SMEs (Hillary 1999). All but one of these drew their findings from practical experience of SMEs implementing EMSs.

Internal benefits

Internal benefits are positive outcomes from the implementation of an EMS which relate to the internal operation of an SME. The internal benefits identified from the review study are shown in Table 3 and are grouped into the following three categories:

- Organisational benefits
- Financial benefits
- People benefits

Numerous organisational improvements and efficiencies are achieved in the SME from the adoption of an EMS and are not solely related to the EMS: i.e. spin-off management benefits arise from the implementation of an EMS. For example, quality systems are improved (INEM 1999),[12] the overall quality of management rises (NALAD 1997), training

12 *www.inem.org/emas-toolkit*

1. Organisational benefits	2. Financial benefits	3. People benefits
• EMS enhanced quality and Investor in People systems • ISO 14001 possible to combine with quality systems (ISO 9000 series of standards) • Quality of management improved • Improved quality of training • Improved working conditions and safety • Improved quality of environmental information • Legal compliance is documented and can be demonstrated • Encourage innovation • Review and improve procedures • Stimulate process, transport, raw materials and packaging changes • Demonstrate environmental responsibility • Provide a strategic overview of environmental performance	• Cost savings from material, energy and waste reductions and efficiencies • Improved economic condition of SME	• Increased employee motivation, awareness and qualifications • Improved employee morale • Enhanced skills and improved knowledge in SME • Creates a better company image among employees • Provides a forum for dialogue between staff and management

Table 3: INTERNAL BENEFITS CATEGORIES AND EXAMPLES

is introduced where previously there was none (INEM 1999) and innovation is encouraged (WWF and NatWest Group 1997).

The range of financial savings and payback periods for investments generated in SMEs adopting EMSs is as diverse as the sector itself and is frequently mentioned (e.g. Elliot *et al.* 1996; Court 1996; NALAD 1997; ETBPP 1998). The pan-EU survey of EMAS sites identified cost savings as the top benefit cited by both large, medium-sized and small companies from the implementation of EMAS (Hillary 1998). However, SMEs place cost savings as the second-ranked benefit behind better image.

Communication channels, skills, knowledge and attitude are all improved in SMEs adopting EMSs (NALAD 1997; Hillary 1998; INEM 1999). EMS implementation opens up new interactions between staff and management and provides intangible benefits such as enhanced morale, which is seen as very important for both small and medium-sized firms (Hillary 1998; Smith and Kemp 1998).

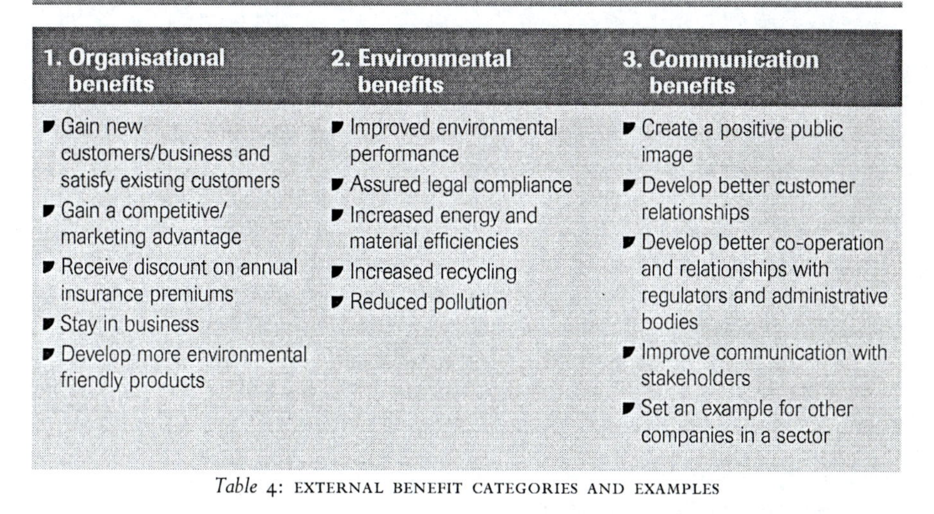

1. Organisational benefits	2. Environmental benefits	3. Communication benefits
Gain new customers/business and satisfy existing customers	Improved environmental performance	Create a positive public image
Gain a competitive/ marketing advantage	Assured legal compliance	Develop better customer relationships
Receive discount on annual insurance premiums	Increased energy and material efficiencies	Develop better co-operation and relationships with regulators and administrative bodies
Stay in business	Increased recycling	Improve communication with stakeholders
Develop more environmental friendly products	Reduced pollution	Set an example for other companies in a sector

Table 4: EXTERNAL BENEFIT CATEGORIES AND EXAMPLES

External benefits

External benefits are positive outcomes from the implementation of an EMS that relate to the external interactions of an SME. The external benefits identified from the review study are shown in Table 4 and are grouped into three categories:

- Commercial benefits
- Environmental benefits
- Communication benefits

SMEs found numerous financial, competitive and business rewards from adopting formal EMSs. For many, ISO 14001 and EMAS are living up to the claims of their promoters. Key benefits for SMEs are the attraction of new business and customers and the ability to satisfy customer requirements (Hillary 1995; Elliot *et al.* 1996; Business in the Environment/Coopers & Lybrand 1995; MORI 1994). These benefits are closely linked to customers in their role as the paramount driver for the adoption of EMSs in the SME sector. However, the pan-EU EMAS survey indicated that, for SMEs, gaining more customers and greater customer satisfaction was not a paramount benefit of EMAS implementation (Hillary 1998). Only one small firm cited this as a benefit and 23% of medium-sized firms considered it a benefit.

Alongside the commercial benefits, SMEs found positive outcomes in terms of improved environmental performance (NALAD 1997; Hillary 1997a), assured legal compliance (University Bocconi 1997; NALAD 1997) and energy and material efficiencies (Smith and Kemp 1998; WWF and NatWest Group 1997; Elliot *et al.* 1996). The most frequently cited improvements were those relating to reduced energy consumption and waste minimisation. Coupled with these benefits, SMEs found that their image was enhanced and dialogue and relationships with stakeholders improved

(Elliot *et al.* 1996; Hillary 1995; MORI 1994). Indeed, improved image was cited in the pan-EU EMAS survey as the most important benefit for SMEs implementing EMAS and this became more important as the size of the firm decreased, i.e. 38% of medium-sized companies and 54% of small companies cited it as the first benefit from the adoption of EMAS (Hillary 1998).

SMEs saw distinct benefits of ISO 14001 over EMAS (Baylis 1998; Hillary 1997a, b). The two key ones are that ISO 14001 does not have an environmental statement requirement and that ISO standards are well-known and accepted concepts.

A caveat needs to be added to the overly positive image given by SMEs of formal EMSs. Some of the studies in the review study were uncritical in the presentation of their findings (e.g. INEM 1999; Smith and Kemp 1998; WWF and NatWest Group 1997). These case studies were often adjuncts to survey results and presented positive messages about the adoption of EMSs by SMEs.

Disbenefits of SMEs adopting EMSs

Few studies (seven) in the review study identify disbenefits (Hillary 1999). The probable cause of this is that many reports present best-practice case studies and seek to 'sell' EMSs to the SME sector.

Disbenefits are negative outcomes or non-materialisation of benefits from the adoption of EMSs. The disbenefits identified in the review study are shown in Table 5 and are grouped into the following three categories:

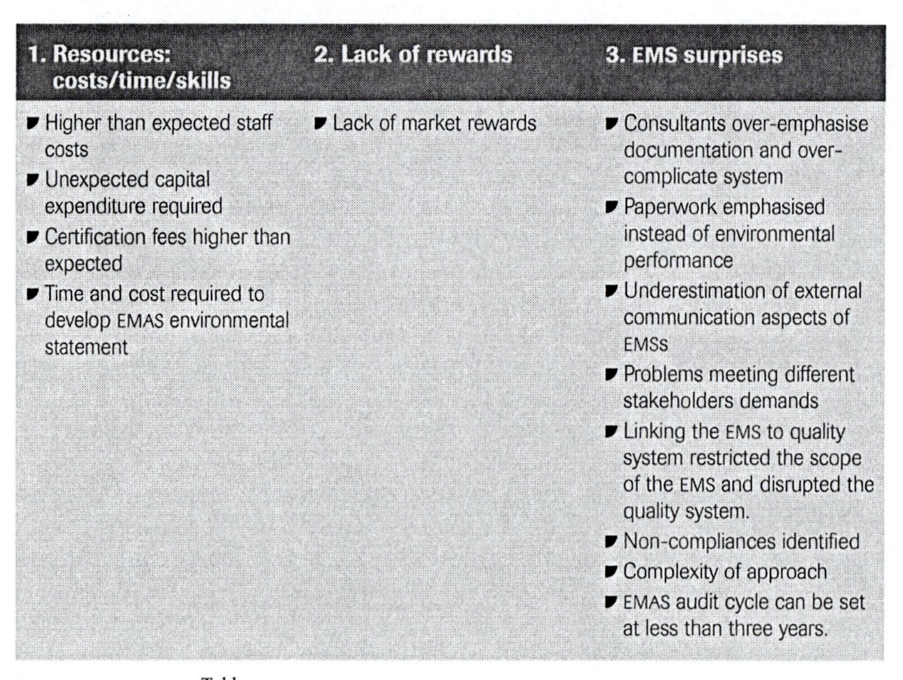

1. Resources: costs/time/skills	2. Lack of rewards	3. EMS surprises
• Higher than expected staff costs • Unexpected capital expenditure required • Certification fees higher than expected • Time and cost required to develop EMAS environmental statement	• Lack of market rewards	• Consultants over-emphasise documentation and over-complicate system • Paperwork emphasised instead of environmental performance • Underestimation of external communication aspects of EMSs • Problems meeting different stakeholders demands • Linking the EMS to quality system restricted the scope of the EMS and disrupted the quality system. • Non-compliances identified • Complexity of approach • EMAS audit cycle can be set at less than three years.

Table 5: DISBENEFITS OF IMPLEMENTING EMSs FOR SMEs

- Resources
- Lack of rewards
- EMS surprises

SMEs found that more resources than expected, in terms of costs, time and/or skills were required for EMS implementation (Goodchild 1998; University Bocconi 1997). SMEs also faced implementation surprises that had impacts on resources. Components of the EMS failed to meet their expectations. For example, they found problems meeting stakeholder expectations (University Bocconi 1997), or the EMS did not integrate smoothly into their quality systems (Hillary 1997a; University Bocconi 1997), or they underestimated EMS requirements—for example, that the audit cycle in EMAS could be less than three years (Hillary 1997a).

A major source of irritation for SMEs, surfacing in a number of studies, is the cost of certification/validation (Goodchild 1998; Hillary 1997a, 1998; KPMG 1997b). They were also aggrieved by the cost and quality of consultants advising them (KPMG 1997b; NALAD 1997; Hillary 1997a). It is apparent that some have been misadvised and developed bureaucratic and ineffective systems (KPMG 1997b). Such cases feed back into the general impression some SMEs have of the inappropriate nature of formal EMSs for smaller firms, a perception that was identified as an internal barrier in some studies (Hillary 1997a; Court 1996).

Identification of non-compliance was viewed as a double-edged sword, being a benefit if the SME could readily rectify the cause of the non-compliance and a disbenefit if action could not be taken because of lack of, or unwillingness to allocate, resources (Hillary 1997a; Elliot *et al.* 1996). In the latter case, an attitude of 'ignorance is bliss' seemed to exist.[13]

Some studies cited SMEs' dissatisfaction with the fact that benefits had not materialised as expected (MORI 1994; Hillary 1998). It is apparent that SMEs have been 'sold' the benefits of EMSs on grounds such as cost savings, and when they fail to materialise the SMEs feel cheated.

Internal and external barriers to the adoption of EMSs by SMEs

Internal and external barriers to EMS implementation are extensively covered in 28 study reports in the review study (Hillary 1999). Ten of the reviewed study reports do not discuss EMS implementation but instead detail SME attitudes that act as barriers to EMS adoption.

---

13 See Petts's extensive analysis to compliance among management and non-management in SMEs in Chapter 3.

Internal barriers of SMEs adopting EMSs

Internal barriers are obstacles that arise within the firms and prevent or impede EMS implementation. The internal barriers identified in the review study are shown in Table 6 and are grouped into the following categories:

- Resources
- Understanding and perception
- Implementation
- Attitudes and company culture

1. Resources	2. Understanding and perception	3. Implementation	4. Attitudes and company culture
▸ Lack of management and/or staff time for implementation and maintenance ▸ Inadequate technical knowledge and skills ▸ Lack of training ▸ Multifunctional staff easily distracted by other work ▸ Loss of environmental champion ▸ Lack of specialist staff ▸ Transient workforce ▸ Requirement for capital expenditure	▸ Lack of awareness of benefits ▸ Lack of understanding of EMAS environmental statement or value of reporting ▸ Lack of knowledge of formalised systems ▸ Uncertainty and concern over possible de-registration (from EMAS) for minor breaches of legislation ▸ Perception of bureaucracy ▸ Perception of high cost for implementation and maintenance ▸ Confusion between ISO 14001 and EMAS and how they relate	▸ Implementation is an interrupted and interruptible process ▸ Inability to see relevance of all stages ▸ Internal auditor independence difficult to achieve in a small firm ▸ Doubts about ongoing effectiveness of EMSs to deliver objectives ▸ Difficulties with environmental aspects/effects evaluation and the determination of significance ▸ Uncertainty about how to maintain continual improvement	▸ Inconsistent top management support for EMS implementation ▸ Management instability ▸ Low management status of person spearheading EMS implementation ▸ Resistance to change ▸ Lack of internal marketing of EMS ▸ Negative view or experience with ISO 9000 standards rubs off on ISO 14001's acceptance

Table 6: INTERNAL BARRIERS TO EMS IMPLEMENTATION

Human rather than financial resources are the major barrier impeding EMS implementation and are frequently cited in the studies (e.g. Poole *et al.* 1999; Goodchild 1998; Charlesworth 1998). Lack of human resources and the multifunctional nature of staff become increasingly important as the size of the company decreases, not only to the implementation of the EMS, but also to its maintenance (Hillary 1997; University Bocconi 1997).

SMEs are largely ill-informed about EMSs, how they work and what benefits can be derived from their implementation (Charlesworth 1998; Smith and Kemp 1998; Baylis *et al.* 1997; KPMG 1997b). As such, ISO 14001 and EMAS hold relatively little interest for the sector. Furthermore, EMAS has a public reporting component that frightens SMEs (KPMG 1997b; Hillary 1997a) and ISO 14001 has an added disadvantage among small companies who may have had negative experiences with the ISO 9000 quality standards (Hillary 1997a).

Negative corporate attitudes toward EMSs and an unfavourable company culture, often cited by SMEs, conspire to create a climate that deprives the EMS implementation process of support (Rowe and Hollingsworth 1996; Hillary 1997a; NALAD 1997). Inconsistent top management support is frequently cited as a factor in the stop–start approach that SMEs, and small firms in particular, may take to the implementation of an EMS (Charlesworth 1998; Hillary 1997a).

Implementation in SMEs is an interrupted and interruptible process that may lose momentum and resources. Practical difficulties, such as how to achieve internal auditor independence and how to determine environmental aspects and assign significance, also scupper implementation (Hillary 1997a, b; CEC 1997a; University of Bocconi 1997). In the pan-EU EMAS survey, the environmental review and the EMS elements took SMEs the most time to implement, were cited as the most difficult to understand and were identified as the elements that required additional guidelines (Hillary 1998). Fear of de-registration for minor breaches of legislation also make EMAS an unattractive proposition for many firms (Elliot *et al.* 1996).

◻ *External barriers of SMEs adopting EMSs*

External barriers are obstacles that arise outside the firms to prevent or impede EMS implementation. The external barriers identified in the review study are shown in Table 7 and are grouped into the following categories:

- Certifiers/verifiers
- Economics
- Institutional weaknesses
- Support and guidance

SMEs face inconsistencies in, and barriers from, the certification system for ISO 14001 and verification system for EMAS (KPMG 1997b; NALAD 1997; University Bocconi 1997; Goodchild 1998). EMAS verifiers and the verification process come in for greater criticism, whereas little comment is made about the certification process for ISO 14001. This does not mean the ISO certification system is perfect—more that it has not been investigated and presented in the analysed studies. SMEs found cost of certification to be a problem. In the pan-EU EMAS survey, it was found that small firms were charged the most per day for their verification to EMAS (ECU 1,085/day) and large firms the least (ECU 878/day) (Hillary 1998).

Many SMEs experience insufficient drivers for EMS adoption and are uncertain about the market benefits of EMSs (MORI 1994; Baylis *et al.* 1997; Court 1996; Business in the

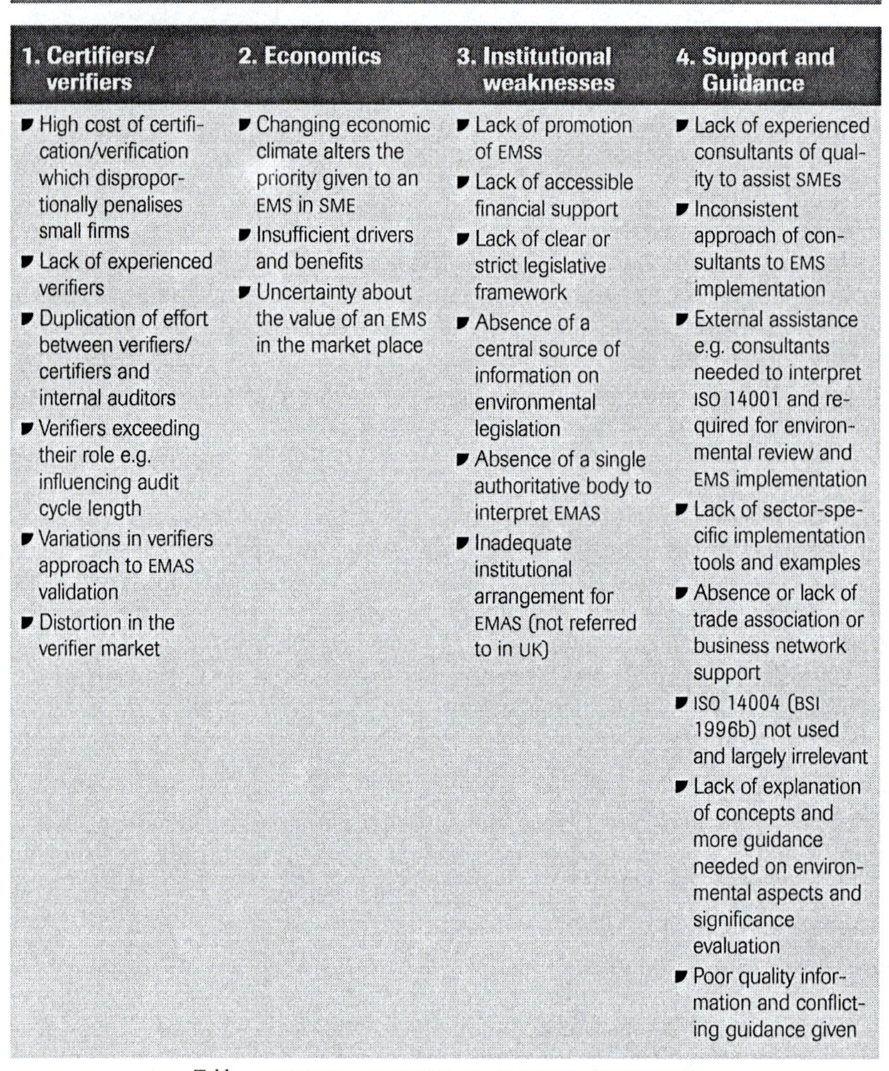

1. Certifiers/ verifiers	2. Economics	3. Institutional weaknesses	4. Support and Guidance
▪ High cost of certification/verification which disproportionally penalises small firms	▪ Changing economic climate alters the priority given to an EMS in SME	▪ Lack of promotion of EMSs	▪ Lack of experienced consultants of quality to assist SMEs
▪ Lack of experienced verifiers	▪ Insufficient drivers and benefits	▪ Lack of accessible financial support	▪ Inconsistent approach of consultants to EMS implementation
▪ Duplication of effort between verifiers/ certifiers and internal auditors	▪ Uncertainty about the value of an EMS in the market place	▪ Lack of clear or strict legislative framework	▪ External assistance e.g. consultants needed to interpret ISO 14001 and required for environmental review and EMS implementation
▪ Verifiers exceeding their role e.g. influencing audit cycle length		▪ Absence of a central source of information on environmental legislation	▪ Lack of sector-specific implementation tools and examples
▪ Variations in verifiers approach to EMAS validation		▪ Absence of a single authoritative body to interpret EMAS	▪ Absence or lack of trade association or business network support
▪ Distortion in the verifier market		▪ Inadequate institutional arrangement for EMAS (not referred to in UK)	▪ ISO 14004 (BSI 1996b) not used and largely irrelevant
			▪ Lack of explanation of concepts and more guidance needed on environmental aspects and significance evaluation
			▪ Poor quality information and conflicting guidance given

Table 7: EXTERNAL BARRIERS TO EMS IMPLEMENTATION

Environment/Coopers & Lybrand 1995). The pan-EU EMAS survey indicated that just under half (49%) of all sites did not believe the market had rewarded them for achieving EMAS registration (Hillary 1998). Medium-sized enterprises were less critical of the scheme with 46% stating they had received market rewards, whereas 54% of small enterprises stated they had not received any market benefits. In addition, the changing economic fortunes faced by SMEs alter their priorities and push the environment to the bottom of the list, further depressing interest in ISO 14001 and EMAS (Hillary 1997a; ETBPP 1998).

Shortcomings in the institutional framework that facilitates the operation of EMAS and ISO 14001 inhibit SME uptake of the two initiatives (CEC 1997b; University Bocconi 1997; KPMG 1997b). Some of these shortcomings apply to EMAS: for example, the lack of a competent body or the absence of accredited verifiers (KPMG 1997b; Hillary 1998), and are not found in the UK, but others, such as the absence of a single body to interpret EMSs and the absence of a central source of information on legislation, do apply (Hillary 1997a; CEC 1997b).

SMEs appear to need support and guidance, in particular for the environmental review and environmental aspects and significance evaluation, but experience problems gaining consistent quality information and experienced consultants of good quality (Abrams 1998; KPMG 1997b; NALAD 1997; Court 1996). ISO 14004 is rarely used by SMEs (Baylis 1998). The lack of sector-specific guidance and material tailored to different sizes of firms, especially very small firms, is a frequently referred to external barrier (Poole *et al.* 1999).

Stakeholder pressure and drivers to adopt EMSs

SMEs are subjected to a variety of stakeholder pressures related to their environmental performance and their adoption of EMSs. The review study sought to identify which stakeholders influence SME behaviour in the adoption of formal EMSs.

In the review study, 22 studies identified stakeholder pressures experienced by SMEs, of which 16 identified key stakeholders as drivers for EMS adoption and six identified the stakeholders that are influencing SMEs' attitudes towards environmental performance (Hillary 1999). One of the main drivers for the adoption of a formal EMS is the customer (e.g. Goodchild 1998; Hillary 1998; Baylis 1998; MORI 1994). Table 8 shows stakeholders identified in the review study and listed in order of priority.

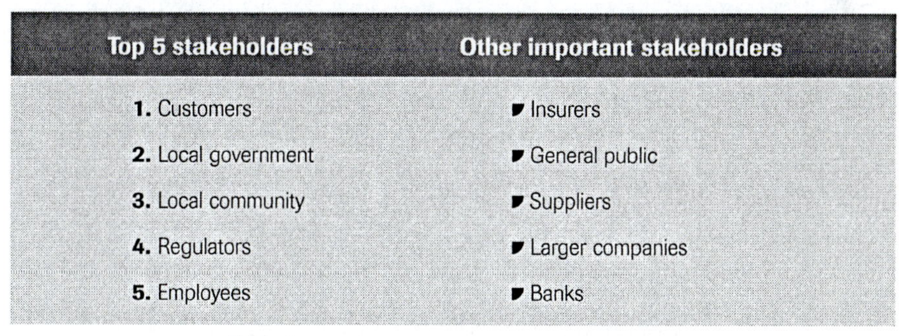

Top 5 stakeholders	Other important stakeholders
1. Customers	Insurers
2. Local government	General public
3. Local community	Suppliers
4. Regulators	Larger companies
5. Employees	Banks

Table 8: STAKEHOLDERS DRIVING EMS ADOPTION

Customer and supply-chain pressures are also prominent in driving SMEs' environmental improvements (Charlesworth 1998). The regulatory authorities, however, exert greater influence on the general environmental performance of SMEs, in particular medium-sized enterprises, than do customers (Poole *et al.* 1999; Hillary 1998; Smith and Kemp 1998; Hillary 1995).

The pan-EU EMAS survey identified the three main stakeholder audiences to which site environmental statements were targeted (Hillary 1998). Customers were the paramount audience for the EMAS statement for both small and medium-sized companies. Just under half (46%) of all SMEs cited local government and regulators as key audiences for their EMAS statements; however, small firms (62%) were more interested in this stakeholder group than medium-sized firms (38%).

Conclusions

Extensive benefits accrue to SMEs adopting formal EMSs and this is widely reported in the analysed studies. These benefits may also arise in small firms that implement non-formal EMSs. Disbenefits also exist, although there are less of them and fewer studies give them coverage. Some studies are uncritical of the practical experiences of SMEs with EMSs and do not address disbenefits; this may be why they are under-represented, although there may genuinely not be that many. The studies show, however, that internal and external benefits are real, valuable and demonstrable and can be duplicated in other SMEs. Nevertheless, the large majority of SMEs remain unconvinced of the need to tackle environmental issues (Smith and Kemp 1998; Hillary 1995).

Small and medium-sized firms face internal and external barriers when seeking to address their environmental issues and adopt and implement EMSs, but it is the internal barriers that, initially, have the more significant role in impeding progress. Negative company culture towards the environment and the disassociation between positive environmental attitudes of personnel and taking action cause the uptake of environmental performance improvements and EMS adoption to stumble at the first hurdle.

On top of this general culture of inaction on the environment, SMEs are also very sceptical of the benefits to be gained from making environmental improvements. In many cases, especially for the smaller organisations, low awareness and the absence of pressure from customers (the most important driver for environmental improvements and EMS adoption), along with insufficient other drivers, mean that few efforts are made to address environmental issues. SMEs also face the problem of locating, and having the time to locate, good-quality advice and information.

Once a smaller company has embarked on EMS implementation the process is often interrupted and resources are frequently diverted to core business activities. It is the lack of human resources, not financial ones, that SMEs find most difficult to secure and maintain for EMS implementation; this is particularly the case for micro firms. The more multifunctional the staff, as is common in micro and small companies, the more likely it is that the process of implementation will be interrupted. Some studies indicate that SMEs, once on the route to certified EMSs, face inconsistency and high charges in the certification system and poor-quality advice from consultants.

Customers are the key driver for the adoption of EMSs and have influence far beyond any of the other stakeholders cited in the analysed studies. Paradoxically, customers also show lack of interest in, or are satisfied with, SMEs' current environmental performance. Micro enterprises, in particular, found their customers to be uninterested in their environmental performance. This may be because the customers, like the micro firms themselves, consider micro firms' environmental impacts to be negligible. Legislation and the regulators are more important drivers for general environmental improvements in SMEs than customers.

The SME sector[14] is not homogenous; it is diverse and heterogeneous. Studies that seek to investigate the sector and draw conclusions about it are, to some extent, comparing not just apples and pears but all the fruit in the fruit bowl. This chapter's conclusions also suffer from this limitation. It is recommended that future research consider parts of the sector either as sub-groups by size, i.e. micro, small and medium-sized, or by industrial sector.

14 The EU definition for SMEs is the standard used in the analysis for this chapter. However, the sector is not a homogeneous group of companies and the definition for SME is a very blunt instrument when understanding the variety of enterprises in the sector.

Appendix: Studies analysed in the review study

No.	Author and year of publication	Title
1.	International Network for Environmental Management (INEM) 1999	*EMAS Tool Kit for SMEs*
2.	Association of Independent Entrepreneurs (ASU) 1999	Reported in *EMAS Tool Kit for SMEs* (INEM 1999)
3.	Österreichischer Kommunalkredit 1999	Reported in *EMAS Tool Kit for SMEs* (INEM 1999)
4.	Swiss Consultants 1999	Reported in *EMAS Tool Kit for SMEs* (INEM 1999)
5.	M. Poole *et al.* 1999	*The Environmental Needs of the Micro-company Sector and the Development of a Tool to Meet those Needs*
6.	E. Goodchild 1998	*The Business Benefits of EMS Approaches*
7.	R. Hillary 1998	*An Assessment of the Implementation Status of Council Regulation (No 1836/93) Eco-Management and Audit Scheme in the European Union Member States*
8.	J. Petts *et al.* 1998b	*Business Attitudes to Environmental Compliance*
9.	A. Smith and R. Kemp 1998	*Small Firms and the Environment 1998: A Groundwork Report*
10.	K. Charlesworth 1998	*A Green and Pleasant Land? A Survey of Managers' Attitudes to, and Experience of Environmental Management*
11.	R. Baylis 1998	*Summary of SME Responses to the SME Panel Survey*
12.	Environmental Technology Best Practice Programme (ETBPP) 1998	*Attitudes and Barriers to Improved Environmental Performance 1998*
13.	J.Q. Merritt 1998	*EM into SME won't go? Attitudes, Awareness and Practices in the London Borough of Croydon*
14.	S. Abrams 1998	*An Analysis of the Drivers and Barriers for Improving Environmental Performance of Small and Medium-Sized Enterprises in the Industrial Painting Contract Sector with Wider Applications*
15.	KPMG 1997b	*The Environmental Challenge and Small and Medium-Sized Enterprises in Europe*
16.	R. Hillary 1997a	*The Eco-Management and Audit Scheme: Analysis of the Regulation, Implementation and Support*

No.	Author and year of publication	Title
17.	National Association of Local Authorities in Denmark (NALAD) 1997	*Guide for the Promotion of Cleaner Technology and Responsible Entrepreneurship*
18.	University Bocconi IEFE 1997	*EMAS Pilot Projects: Third Progress Report*
19.	CEC DG XXIII 1997	*Euromanagement: Environment Pilot Action*
20.	R. Hillary 1997b	*UK National Co-ordinators' Report on Euromanagement: Environment Pilot Action*
21.	CEC DG XI 1997	*Lessons Learnt by SMEs in their Implementation of the EMAS Regulation*
22.	R. Baylis *et al.* 1997	*Environmental Regulation and Management: A Preliminary Analysis of a Survey of Manufacturing and Processing Companies in Industrial South Wales*
23.	WWF and NatWest Group 1997	*The Better Business Pack: Practical Management Guide*
24.	P. Court 1996	*Encouraging the Use of Environmental Management Systems in the Small and Medium-Sized Business Sector*
25.	The British Chambers of Commerce (BCC) 1996b	*Small Firms Survey: Businesses and their Communities*
26.	D. Elliot *et al.* 1996	*UK Business and Environmental Strategy: A Survey and Analysis of East Midlands Firms' Approaches to Environmental Audit*
27.	J. Rowe and D. Hollingsworth 1996	*Improving the Environmental Performance of Small and Medium-Sized Enterprises: A Study in Avon*
28.	Environmental Technology Best Practice Programme (ETBPP) 1996	*Attitudes and Barriers to Improved Environmental Performance*
29.	R. Hillary 1995	*Small Firms and the Environment: A Groundwork Status Report*
30.	Business in the Environment/Coopers & Lybrand 1995	*EC Eco-Management and Audit Scheme (EMAS): Position your Business*
31.	The Springhead Trust 1995	*Small Business Survey*
32.	MORI 1994	*Eco-Management and Audit Scheme: Research Study Conducted for the Department of the Environment*
33.	The British Chambers of Commerce (BCC) 1994	*Small Firms Survey: Environment: The View of the Small Firm*

11
Promoting cleaner production in small and medium-sized enterprises

Jonathan Hobbs

Small and medium-sized enterprises (SMEs) are not insignificant organisations. Their important role in economies throughout the world and their cumulative impacts on the environment make them essential contributors to strategies designed to promote sustainable development. This chapter discusses their participation in one such strategy—cleaner production.

On the one hand, SMEs provide employment and opportunity for millions of individuals. In the developing world 90% of all enterprises are considered to fall into this category. The global trend for larger enterprises to reorganise, downsize and outsource enhances SME numbers. They are, therefore, an increasingly important contributor to economies, development and society.

On the other hand, their cumulative contribution to pollution and global environmental concerns can be as significant as those of multinational corporations. It is even suggested that SMEs cause relatively more pollution than larger companies operating in the same sector because of their production techniques. SMEs commonly operate in some of the most problematic trades—metal finishing, leather tanning, dry-cleaning, printing and dyeing, brewing, food processing, fish farming, textile manufacture, etc.

There is considerable diversity behind the acronym 'SME'. The term encompasses everything from labour-intensive, low-skilled activities—mainly in developing countries—to highly skilled, individual and relatively sophisticated and well-managed companies, mainly in developed countries. The use of the term in this chapter refers primarily to the former.

Notwithstanding this focus on SMEs in developing countries, it must be recognised that the prosperity of the 600 million people in developed countries rests on unsustainably high energy, water and material consumption levels and excessive waste generation. So the principles inherent in sustainable development strategies apply to all enterprises.

SMEs and sustainable development

The potential contribution of SMEs to strategies promoting sustainable development—cleaner production, eco-efficiency, pollution prevention, green productivity, etc.—while critically important, is not being fully realised. The reasons for this are both external and internal.

External factors, such as policy and legal frameworks and institutional and administrative systems, fail to encourage and support attempts SMEs make in this direction. Resource pricing and inappropriate subsidies often fail to encourage frugality in resource management.

These, in turn, result in internal factors, such as general complacency, scepticism and ignorance. The status quo is accepted even if this includes recurring fines for pollution offences—stalling the search for corrective action.

SMEs are usually small in capital and staff resources, frequently dependent on simple, labour-intensive and outdated technologies, and focused on short-term quick profit—if not day-to-day survival. They often lack long-term strategic vision and are insensitive to market changes. This is not a profile conducive to sustainable development.

Paradoxically, however, on occasions, they can also be mobile and flexible and extraordinarily innovative.[1] The examples of case studies of SMEs' innovative successes are many and varied (Brugger and Timberlake 1994).

Lack of support

The neglect of SMEs by some critical influences on their behaviour, however, has left them ill-informed and ill-equipped to respond to mounting pressures to improve environmental performance.

Consequently, SMEs are generally characterised as:

- Unaware of the urgency and need for them to contribute to sustainable development strategies.
- Unconvinced of the cost savings and market opportunities that can accompany these 'win–win' strategies of environmental and economic improvement.

1 See Chapter 13 for a discussion of SMEs and environmental innovation.

- Out of touch with the changing desires of their principal stakeholders and end-consumers.
- Perceiving such strategies as technical, expensive and the concern of larger companies rather than themselves.
- Perceiving environmental degradation as a social issue largely the concern of government, not the private sector.

SME owners may be concerned about global climate change, loss of biodiversity, the build-up of toxic chemicals and stratospheric layer ozone depletion but are largely ignorant that their activities comprise significant contributions to these problems.

As a consequence, sustainable development strategies rate low on the list of SMEs' priorities. They remain unprepared to invest in such strategies, irrespective of the frequently favourable and prompt paybacks, as well as environmental improvements, that usually result.

This situation is compounded by the perceived poor rewards that potential sources of support and influence gain from focusing on SMEs. Pressure groups gain none of the prestige, headlines and publicity by targeting SMEs that campaigns exposing the environmental misdemeanours of high-profile multinationals brings. Consultants gravitate to the more lucrative returns that flow from advising large companies. Likewise, financial institutions are more inclined to support larger companies with investment capital. In spite of the implicit health, safety and social benefits that improved environmental performance brings, trades unions are not very active in working with SMEs on these issues either.

In addition, many SMEs are operating outside the formal sector of the economy and, therefore, outside the influence of legal and administrative systems. In some respects this is of little consequence because, even though policies and legislation promoting sustainable development strategies are increasing, the perennial problem in developing countries is one of ineffective enforcement. According to the Deputy Environment Minister of the Republic of China, Wang Xinfang,[2] 56,000 Chinese SMEs were closed down during 1997 for repeated pollution offences. The majority of these reopened shortly after the delivery of the closure notice.

Changing times

The chances are that this administrative failure is a temporary phenomenon. Enforcement of legislation will become more effective and penalties for persistent offenders more severe. Trends in regulatory mechanisms such as increasing waste disposal costs, increasing costs of pollution control technology, and the reduced availability of problematic chemicals will force a greater awareness of environmental strategies in SMEs—especially preventative ones. Attitudes to inefficiency and waste will change out of necessity and opportunities realised co-incidentally.

More direct influences are also emerging. Certainly supply chain linkages are starting to have an impact as larger companies embrace the spirit of sustainable development

2 Speaking at the *Fifth High-Level International Seminar on Cleaner Production*, Korea, October 1998.

through activities such as increased life-cycle responsibilities. The environmental performance of business partners of larger companies—suppliers, subcontractors, etc.—is under increasing scrutiny as larger companies accept accountabilities beyond their own factory gates.

Demands for improved environmental performance and more environmentally benign products are not prevalent in the marketplaces of developing countries, however. But the growing expression of 'green consumerism' in the markets of richer countries cascades back through international supply chains and trade routes. A back-street metal finisher in Pakistan using child labour to make door knobs, a sports equipment manufacturer in Thailand using endangered species components or a fruit grower in Kenya using a banned chemical to fumigate pests will destroy the credibility of a retailer in 'northern' markets if these practices are made known to ultimate customers. Exposure of poor and inappropriate performance of their suppliers is now a major liability to companies in Europe, North America, etc.

The reputations of large industry and the success of international voluntary and business initiatives are closely related to SMEs. One of the 16 principles of the long-standing Business Charter for Sustainable Development developed in the late 1980s by the International Chamber of Commerce (ICC) (ICC 1991) identifies the importance of these relations. Signatories to the ICC Business Charter must strive to get their contractors, subcontractors and suppliers to observe the Charter's 16 principles. The poor environmental performance of SMEs will negate the environmental successes of larger partners and encourage governments to enact ever more draconian measures for all. The Charter, in turn, influenced the business dimensions of Agenda 21, the environmental action plan adopted at the 'Earth Summit' in Rio de Janeiro in 1992.

The emergence in the late 1990s of international standards on environmental management systems has also had a significant effect on the environmental pressures on SMEs. These reinforce a company's responsibility to manage its activities throughout a product's entire life-cycle from procurement of inputs to reclaiming the redundant product components after use. Inherent in management systems standards is a commitment to 'continuous improvement'. This is not simply continuous improvement of the management system, as sometimes suggested, but also, and more importantly, performance. A management system, therefore, requires a strategy to fulfil this requirement just as, in the same way, environmental performance is likely to be enhanced by the existence of a management system.

SME response

The growing prevalence of ISO 14001 certification among SMEs in developing countries, particularly Asia, is going some way to indicate to existing and potential trading partners that environmental management is being addressed within the company. (APO 1997). It is also leading to the development of a cadre of consultants marketing their skills to this scale of business.

This is important because SMEs are unlikely to have staff and resources dedicated to exclusive management of the environment function, or to have research and development infrastructure that can be called on. Nor are they likely to have a real appreciation of their own situation and the strategies necessary to implement sustainable development, even if they are willing to do so.

SMEs are normally fully occupied with short-term survival and day-to-day operational and cash flow concerns. They have little time to devote to long-term considerations such as sustainability strategies. This is particularly acute in developing countries. As if it were not hard enough getting a successful business off the ground in complex and unsupportive business environments, the challenges of environmental management without supportive resources and infrastructure is daunting.

Yet, in spite of this, practical examples of SMEs taking steps to sustainability are growing worldwide. The United Nations Environment Programme (UNEP) regularly publishes examples of cleaner production initiatives in its 'Cleaner Production Worldwide' series and as a component of its International Cleaner Production Information Clearinghouse (ICPIC) programme. Furthermore, Van Weenen (1999) cites examples of products such as the use of solar dryers in food processing in Kenya and Uganda, biomass power plants used in recycling paper in India, self-powered lanterns and radios in South Africa and palm wood furniture in Fiji as indicative of this progress.

The strategies

A need has been recognised for a reorientation to more sustainable patterns of production and consumption. This needs to incorporate elements such as increased productivity and efficiency (doing more with less) as well as eco-effectiveness (doing not only things right but also the right things). These are neatly encapsulated in a term coined by the World Business Council for Sustainable Development (WBCSD) as 'eco-efficiency' (Schmidheiny 1992). Eco-efficiency is a win–win concept that targets the mutually interdependent ecological and economic goals required of sustainable development. Eco-efficiency is a strategy that should confer a long-term competitive advantage on a company while breaking the causal link between economic growth and environmental damage. The United Nations agencies prefer the term 'cleaner production'. The ingredients of the two strategies, however, remain essentially the same.

Strategies contributing to the reorientation of production and consumption patterns on to more sustainable paths are as relevant to SMEs as to any other enterprise. Cleaner production and eco-efficiency should be strategies with inherent appeal to all business people no matter the scale of operation. They are strategies promoting greater productivity, efficiency, cleaner technology, innovation, reduced liabilities and new market opportunities within a framework of social responsibility and sustainability. Getting this message across more effectively to SMEs—the engine of economic progress—will be a significant contribution to the reorientation to sustainable production and consumption.

Cleaner production can be defined as the continuous application of an integrated preventative environmental strategy applied to processes and products to reduce risks to humans and the environment. In production processes this includes conserving raw materials and energy, eliminating toxic materials and reducing the quantity and toxicity of all emissions and wastes before they leave the process. For products, the strategy focuses on reducing impacts throughout the entire life-cycle of the product from raw material extraction to ultimate use and disposal or recycling of the product.

In essence, cleaner production is a case of shifting environmental management away from end-of-pipe solutions of pollution control to anticipatory, preventative approaches that start by tackling the root causes of environmental degradation rather than the effects. Cleaner production is achieved by upgrading technology, good housekeeping, on-site recycling, substituting inputs, process change and modification, and/or product redesign. These options are identified after a cleaner production assessment. This relies on the calculation of material balances between inputs and outputs which will indicate waste reduction options.

The role of international agencies

The absence of supportive infrastructure and enabling environments for SMEs has encouraged many multilateral international agencies to attempt to fill the gap. The promotion of the cleaner production strategy is part of many agencies' SME programmes. Among others, The World Bank for Reconstruction and Development (WBRD), UNEP, United Nations Industrial Development Organisation (UNIDO), and the International Labour Organisation have recognised the importance of this strategy and include it in their programmes. Bilateral agencies do the same.

The activities of international agencies in promoting cleaner production can be categorised into three areas:

- Influencing national policy and administrative environments to encourage the accelerated implementation of cleaner production measures by industry.
- Raising awareness and providing access to support infrastructure—such as training, advice, demonstration projects, case studies, technology and information on cleaner production.
- Helping to overcome specific barriers to cleaner production—such as access to investment capital and putting up seed or incremental finance to initiate cleaner production actions.

UNEP has been active in all three areas with varying degrees of success. A Cleaner Production programme was initiated in 1990. The progress of this programme and the adoption of cleaner production throughout the world is monitored and subject to biennial review seminars.

In an attempt to influence policy environments UNEP launched an 'International Declaration on Cleaner Production' at the fifth biennial review seminar on cleaner production in Korea in 1998. This declaration is designed to secure the public commitment of public- and private-sector leaders to the cleaner production strategy. With political leadership the impetus for SMEs to adopt cleaner production will be greater. Over 150 signatories had committed to the declaration in the first six months and translations had been completed into 14 languages.[3] The declaration encourages the creation of national policies on cleaner production.

The Massachusetts Institute for Technology (MIT) in the US has developed a model approach to help national governments launch a comprehensive cleaner production policy package (UNEP 1994). This outlines a range of policy tools available for governments to catalyse industry commitment to this strategy through voluntary agreements, regulations, permitting systems, economic instruments and information and education support. Such measures need to be developed after an evaluation of the existing system to reveal any existing obstacles to cleaner production that exhibit a bias instead to end-of-pipe pollution control. These may be found in trade, industry, technology or innovation policies, resource pricing, tax systems, educational criteria or regulations.

An example of a successful effort to foster the adoption of cleaner production in a developing country has been the joint UNEP/World Bank project in China. In 1993, China's (then) National Environmental Protection Agency launched the promotion of a cleaner production programme. This encompassed cleaner production assessments in 29 companies—mostly SMEs. A total of 690 low- or no-cost options were identified for implementation. Of these options, 63% had a payback period of less than one year and all produced significant environmental benefits. In addition, all companies identified further higher-cost options. The World Bank provided loans to several companies to finance their implementation including an automated nickel-chrome plating line at Shaoxing General Bicycle plant and an integrated heat and water recovery system for MaAnshan Sulphuric Acid Plant.

The core element of any country's cleaner production programme will be the establishment of a cleaner production capacity-building centre—National Cleaner Production Centres (NCPCs). UNIDO, in partnership with UNEP, and with bilateral donor money have now established NCPCs in ten countries—Brazil, China, Czech Republic, Hungary, India, Mexico, Slovakia, Tanzania, Tunisia and Zimbabwe. More are planned—particularly in Latin America—with Swiss financial support. Each are provided with counterpart support in the formative stages by expatriate experts in cleaner production, funded by the donor agency.

The activities of these centres include setting up demonstration projects and conducting in-plant cleaner production assessments in SMEs in the hope that others will be inspired by the environmental and financial advantages. Education and training, providing training to managers, technicians and national consultants and disseminating information on cleaner technologies, good case studies and contacts are provided. NCPCs also provide policy advice.

Between 1995 and 1997, 110 in-plant assessments have demonstrated the efficacy of cleaner production practices and 190 national consultants now assist companies in implementing cleaner production techniques. Typical projects include toxic waste reduction in chrome-plating (Mexico), increased raw material use efficiency in pulp and paper manufacture (China), process optimisation in food processing (Zimbabwe) or textile dyeing (India), energy saving in soap production (Tanzania),

---

3 For the latest situation, see *http://www.unepie.org/Cp2/declaration/home.html*

re-use of chrome in leather tanning (Tunisia) and water savings in sugar refining (Philippines).

To provide further technical support various specialist sector-specific working groups have been established. Each are involved in further defining effective tools, techniques and policies for priority industry sectors such as tanneries (UK), textiles (India), metal finishing and food processing (Australia) and pulp and paper (Sweden/Thailand), as well as for specific parts of the cleaner production strategy such as sustainable product development (the Netherlands) or education and training (Denmark/Sweden).

A key determinant of success in SMEs is access to information, knowledge and skills. UNEP has provided an information clearinghouse function through the aforementioned ICPIC since 1990. Company case studies are collected illustrating cleaner production initiatives successfully carried out worldwide (UNEP 1995). Together with information on available contacts, expertise, publications, databases, training opportunities, etc. this is disseminated to a wide audience in electronic and printed format (UNEP 1999).

It is questionable, however, despite the wealth of information available, how much of this successfully reaches SMEs. Strategies to market this information more effectively to SMEs are needed. Key to these is the need to leverage the influence of information 'gatekeepers' such as trade associations, employer associations, business councils, capacity-building centres, productivity councils, chambers of commerce, rotary clubs, the media and others.[4]

With financial support from the European Commission and in conjunction with the WBCSD, UNEP has recently participated in a study investigating how to improve systems of communicating best practice in industry.

Financing cleaner production

Experiences in cleaner production programmes indicate that 20% reductions in pollution can be achieved with low- or no-cost measures in most SMEs. A further 10%–20% can be achieved with minor investments and pay-back periods of less than six months. Beyond these measures more substantial investment finance may be needed. But what 'substantial investment' means to an SME may be just an administrative burden hardly worth the effort to a large financier.

In 1999 UNEP, with financial support from the Norwegian government, embarked on a three-year study to investigate the financing strategies and mechanisms for promoting investment in cleaner production. This work is taking place in five pilot countries—Guatemala, Nicaragua, Tanzania, Vietnam and Zimbabwe. The objectives of the study are to evaluate access to cleaner production investment capital and formulate approval procedures and administration mechanisms.

As is the case with information provision, it is not the availability of the resource that is at issue. Financial resources exist, but it is the difficulty SMEs experience in mobilising these resources that needs to be addressed. Many SMEs lack collateral,

4 See Clark's review of the way information reaches SMEs in Chapter 14.

security and confidence and are unconvinced of the short payback periods that can be achieved by cleaner production investments.[5] For their part, lenders point out that small loans cost more per dollar loaned because the administrative costs are roughly the same no matter the size of the loan.

It is anticipated that this project will help the financial community better understand the technicalities, and thus more readily support cleaner production initiatives. It will also help cleaner production promoters produce proposals with appeal to financiers. In the process it is anticipated that new customised and streamlined ways of financing cleaner production investments will be found. In addition the project is designed to excite the interest of other financiers and SMEs in the benefits of cleaner production investments.

Conclusion

In spite of some progress in promoting the preventative strategy of cleaner production among SMEs, only the surface of the potential has been scratched. This reflects the scale of the challenge to be addressed.

Although the critically important role that SMEs play in addressing sustainable development is recognised, they are not yet fully effective participants in cleaner production strategies. The overwhelming majority of SMEs in developing countries continue to regard environmental issues as a threat and a cost rather than as a potential investment and opportunity.

The work of multilateral and bilateral agencies successfully compensates for the lack of interest in SMEs expressed by more traditional sources of support services. They, however, are too far removed from the day-to-day realities of SMEs to ever provide truly effective, direct support. Instead, they need to better appreciate and support the activities of the 'gatekeepers' to the SME sector. These speak the language of SMEs and are better at mobilising the larger-scale peers of SMEs in the business community from whom SMEs more readily take their cues for action.

Such intermediaries may be indirect, such as business or trade associations, or more direct partners of SMEs, such as customers involved in supply chain linkages and procurement of their products, or managers of the industrial estates where they are often located.

Among the strategies to be promoted in SMEs the preventative strategy of cleaner production has merit. The most effective way of coping with unsustainable practices is to prevent pollution and waste occurring in the first place rather than being forced to clean it up at the end—invariably at greater expense. Eliminating the source of the problem is always better than doctoring the effect.

But in a situation where any environmental measure is likely to be an improvement, and this is often the case in the environments in which SMEs are working, we should not limit our options to one strategy only. A strategy such as cleaner production should

5 A similar negative response is typical of SMEs' reaction to EMS success stories (see Chapter 10).

be seen as a contribution to a developing model of integrated environmental management—industrial ecology

Industrial ecology employs a systems approach. It draws on an analogy with natural ecological systems. In this model the relationships between SMEs and larger companies are more explicit and thus full of opportunity to better convey environmental messages between them. The need to get environmental messages more effectively to SMEs is critical. They are the engines of the required shift to sustainable production and consumption and ultimately sustainable development. While they remain on a dominant mission of short-term production, rather than fulfilling long-term needs in a sustainable way, we have a long way to go.

12
An approach to sustainable product development relevant to small and medium-sized enterprises *

John Holmberg, Ulrika Lundqvist and Karl-Henrik Robèrt

Manufacturing companies face a new challenge when it comes to product development. Over the past decade, in addition to existing technical requirements and economic constraints, concern over health and environmental issues has been an increasingly important factor in product development. Initially, such concerns have mainly been integrated into business activities by corporations that are large, with substantial resources for research and development, and that are close to their customers, e.g. retail companies. In turn, these companies have put demands on their suppliers. The process is now spreading to small and medium-sized enterprises (SMEs).

Most of the environmental burdens of a product are determined at the design stage. It is, therefore, meaningful to seek methods that can be helpful at this stage. Many tools and concepts have been developed to guide and support companies integrating environmental aspects in their product design, e.g. 'life-cycle assessment' (LCA) (Lindfors *et al.* 1995), screening and streamlined LCA (Todd 1996; SETAC 1997; Graedel 1998), and 'design for environment' (Ashley 1993; Graedel and Allenby 1995; Billatos and Basaly 1997). Other concepts that have been presented are 'environmentally adapted product development' (Ryding 1995; Hanssen 1997) and 'green design' (Sheng *et al.* 1995). Many of these tools are developed mainly for larger corporations. Their full implementation often turns out to be costly in resources and time and to

* Financial support by the Swedish National Board for Industrial and Technical Development (NUTEK) is gratefully acknowledged.

be essentially unworkable as routine tools for SMEs. Furthermore, these methods do not necessarily consider the *whole* concept of sustainability, but are limited to a more narrow perspective on environmental issues. Simon and Sweatman (1997) even suggest that 'design for environment' should, therefore, be distinguished from sustainable product design.

The next logical step in the evolution of product development is to broaden the concerns from health and environmental aspects to the wider field of sustainability, and to do so not only for large companies, but also for SMEs. The challenge of sustainability is to establish sustainable relationships between global society and the ecosphere as well as within global society itself. Thus, 'sustainable development' (WCED 1987) is a huge mission which involves a profound transformation of the societal metabolism. That global society is not sustainable means that different kinds of solid, liquid and gaseous wastes are steadily accumulating, resources and ecological functions are steadily diminishing, and the resource potential for both human health and the economy is systematically decreasing. At the same time, the Earth's population and the demand for services are increasing. This can be visualised as entering deeper and deeper into a 'funnel' in which the space to manoeuvre becomes narrower (Robèrt *et al.* 1997). For a firm wanting to invest skilfully, the crucial factor is to direct investment towards the opening of the 'funnel', rather than into its 'wall'. The 'wall' of the 'funnel' will superimpose itself more and more into daily economic reality in the following ways: environmentally concerned customers; higher insurance costs for risky activities; stricter legislation; higher costs and fees for resources as well as pollution; and tougher competition from competitors who invest themselves skilfully towards the opening of the 'funnel', i.e. towards sustainability.

The purpose of this chapter is to contribute to the discussion about how the perspective in product development can be broadened from environmental and health concerns to the field of sustainability. We will describe how principles for sustainability and the backcasting methodology (see below) can be integrated with a model for product development. We will also discuss what further research is needed to make this integration more effective for SMEs.

Strategic planning towards sustainability[1]

We have previously presented a method for strategic planning towards sustainability, developed together with the Natural Step Foundation, that has been successfully used by many companies, organisations and municipalities (Robèrt 1992, 1994, 1997; Holmberg *et al.* 1996; Holmberg and Robèrt 1997; Robèrt *et al.* 1997; Nattrass and Altomare 1999). The method is based on a framework of four non-overlapping principles of sustainability (Holmberg *et al.* 1996; Holmberg and Robèrt 1997) and on backcasting (Robinson 1990; Dreborg 1996). Rather than basing planning on today's trends (traditional forecasting), one tries to start the planning process by defining the

1 The text is this section is based on Holmberg 1998.

principles or conditions for a favourable outcome. Thereafter, the question is asked: 'What shall we do today to get there?' The framework helps to describe what parameters will change when the whole society has a metabolism that fits with the requirements set by sustainability. The reason for doing this must not solely be to inspire altruism. It is, rather, a method to identify early warning signals for when long-term investments based on present structure and trends lead to dead ends. Since the method uses the future sustainable situation as its starting point, it also helps to liberate beliefs about today's trends. It is, therefore, also a tool for supporting creativity, e.g. new product design and new business ideas (Holmberg 1998).

Figure 1 illustrates different steps in the method for strategic planning towards sustainability. The different steps will be further explained in the following text.

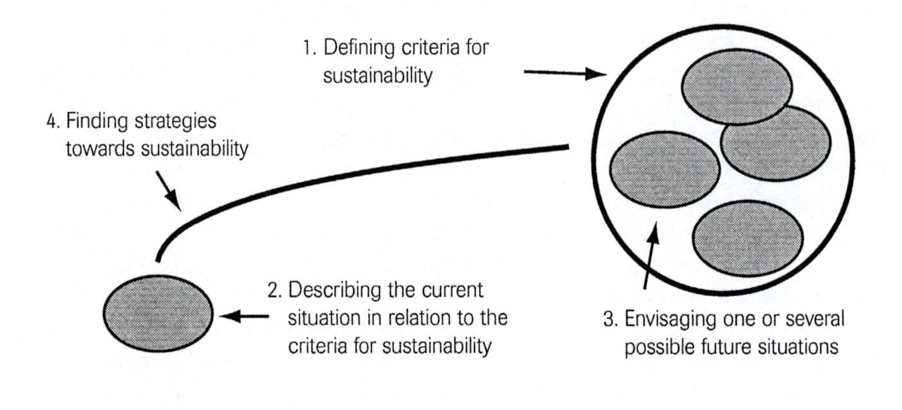

Figure 1: THE STEPS IN STRATEGIC PLANNING TOWARDS SUSTAINABILITY

Step 1: *Defining criteria for sustainability*

In this method, the starting point is criteria for a sustainable future. If the planning process does not take a set of properly defined criteria for sustainability as its starting point, but relies just on present trends, the course can be dangerously wrong. It can be compared with night navigation inside a belt of islands by just studying the lanterns of other boats. In order to avoid the rocks it is important to try to find relevant lighthouses.

The first version of the criteria for sustainability used in this method was developed by Karl-Henrik Robèrt and John Holmberg around 1990. The outcomes of several scientific workshops and review processes, as well as experience from their use in education, in scientific consensus processes and by corporations, municipalities and other organisations, have led to their continuous improvement, but they still have the same basic meaning (Robèrt 1992; Holmberg *et al.* 1996; Holmberg and Robèrt 1997).

In order for a society to be sustainable, nature's functions and diversity are not systematically:

1. Subject to increasing concentrations of substances extracted from the Earth's crust
2. Subject to increasing concentrations of substances produced by society
3. Impoverished by over-harvesting or other forms of ecosystem manipulation

and

4. Resources are used fairly and efficiently in order to meet basic human needs worldwide

Clarification of the principles

1. The societal influence on the ecosphere due to accumulation of lithospheric material is covered by the first principle. The balance of flows between the ecosphere and the lithosphere must be such that *concentrations* of substances from the lithosphere do not systematically increase in the whole ecosphere or in parts of it. The strategic question for planning is: 'Does our organisation systematically decrease its economic dependence on fossil fuels and mining to cover for losses, particularly of scarce elements in dissipative use that are already accumulating in the ecosphere?'
2. The societal influence on the ecosphere due to accumulation of substances produced in society is covered by the second principle. The balance of flows must be such that concentrations of substances produced in the society do not systematically increase in the whole ecosphere or in parts of it. The strategic question for planning is: 'Does our organisation systematically decrease its economic dependence on persistent substances foreign to nature, and dissipative use of natural substances that are currently accumulating systematically in parts—or the whole—of the ecosphere?'
3. The societal influence on the ecosphere due to manipulation and harvesting of stocks and flows within the ecosphere is covered by the third principle. It implies that the resource basis for (i) productivity in the ecosphere such as fertile areas, thickness and quality of soils, availability of fresh-water, and (ii) biodiversity are not systematically deteriorated by over-harvesting, mismanagement or displacement. The strategic question for planning is: 'Does our organisation decrease its economic dependence on activities which degrade productive parts of the ecosphere?'
4. The internal societal metabolism and the production of services to the human sphere is covered by the fourth principle. It implies that if the societal ambition is to meet human needs everywhere today and in the future, while conforming to the restrictions with regard to available resources given by the first three principles, then the use of resources must be efficient in meeting human needs. The strategic question for planning is: 'Does our organisation systematically decrease its economic dependence on using a large amount of resources in relation to added human value?'

In conclusion, in this first step, a company that wants to apply this method starts with the four principles. Often the process starts with a presentation of the principles, the rationale behind them, and this four-step backcasting procedure for the management team of a company. The principles are not presented in a prescriptive way, but as guiding instruments for discussion about sustainability. The result is a better understanding within the company about how the demands for sustainability in the global perspective will involve and influence the company.

Step 2: *Describing the present situation in relation to the criteria for sustainability*

In the second step, the present activities and competences are analysed. For each of the four principles a number of relevant questions can be asked in order to identify whether the current delivered products or services, production processes or other activities are in accordance with the principles or not. In this part of the analysis the flows that are critical in relation to the principles are determined and listed. Thereafter, possible indicators that can monitor those flows are considered: 'Which lithospheric flows, connected to our activities, are critical (principle 1)?', 'How could indicators to monitor those flows be designed?' (Similar questions can be asked for the other principles.)

In this inventory, activities that involve economic dependence on highly unsustainable patterns can then be studied in more detail. Often other more detailed inventory tools can then be used, for instance the LCA methodology, but they should then be related to non-overlapping principles which cover relevant aspects of sustainability. The reason is not to miss relevant aspects of sustainability and to make it possible to detect early warning signals. If they are not based on principles of sustainability, there is a great risk that we will solve today's problems by creating new ones. Unfortunately, a serious shortcoming of many inventory programmes is that they lack such principles of sustainability (Mitchell 1996).

Step 3: *Envisioning and discussing the future situation*

In this step, future possibilities are envisioned. Given the restrictions and possibilities set by the principles of sustainability (step 1), and the information regarding the present situation, e.g. regarding resource turnover, competence within the organisation and the services/utilities that are delivered (step 2), there are often a variety of future options. In this part of the analysis, these options are listed, whether they are realistic in the short term or not.

An essential aspect in this step is to focus on the aimed service rather than on the commodity, thereby avoiding a static view of the present activity or product. The product aims at delivering a service or a set of services which, in turn, might have the potential to fulfil human needs. The main idea with this step is to rid the mind of restrictions set by the present circumstances and open the mind for future options. Questions such as, 'do we sell trucking or logistics?' can open up avenues of business possibilities to list. It is important to remember that the perspective must then be large enough so that the starting point for the whole planning—the global 'funnel'—is not neglected. Will these measures, taken together, help society to change at a sufficient speed and scale to achieve sustainability without too much loss for humans and other species during the transition?

Step 4: *Finding strategies towards sustainability*

In the fourth step, strategies are identified that can link the present situation with the future sustainable situation. When identifying strategies to move towards sustainability, at least the following three points should be considered:

1. **Will each measure (e.g. investment or product design) bring us closer to sustainability?** Will it reduce our dependence on dissipative use of scarce elements and on dissipative use of persistent compounds foreign to nature etc?
2. **Is each measure a flexible platform for the next step towards sustainability?** Will it be possible to go from the actual investment to another that will bring us even closer to sustainability, or is the investment a dead end? This question is even more important than the previous. For instance: 'Can this new and efficient engine be developed to run on biofuels too, or is it stuck with petrol?'
3. **Will each measure pay off soon enough?** For instance: 'Is there already a market demand somewhere?'

LCA is a frequently used tool for the evaluation of the environmental impact during the whole life-cycle of a product, i.e. from the extraction of resources to the handling of waste.[2] Most LCAs are based on data from existing industrial systems as well as existing energy and transportation systems. In order to avoid a product design that can lead to a dead end, one must also analyse the product design in relation to the requirements of the future sustainable society. This means that tools such as LCA should be complemented with parallel assessments that assume that the product exists in a sustainable society. What parameters will change when the whole society has a metabolism that fits in the framework given by sustainability?

Sustainability aspects in product development

A systematic approach to product development is based on a structure of the process in different steps (Pahl and Beitz 1996). This approach may increase the efficiency and improve the result of the process. The efficiency can be further increased by the use of integrated product development which implies that the development of different types of activity such as marketing, design and production are carried on parallel to each other (Karlsson 1997; Magrab 1997).

2 See Chapter 7.

In this section, we describe the structure for the incorporation of sustainability aspects within the process of product development. We base the structure on the method for strategic planning towards sustainability described in the preceding section, and on the different steps in the product planning and design process, as denominated and defined by Pahl and Beitz (1996) (see Fig. 2).

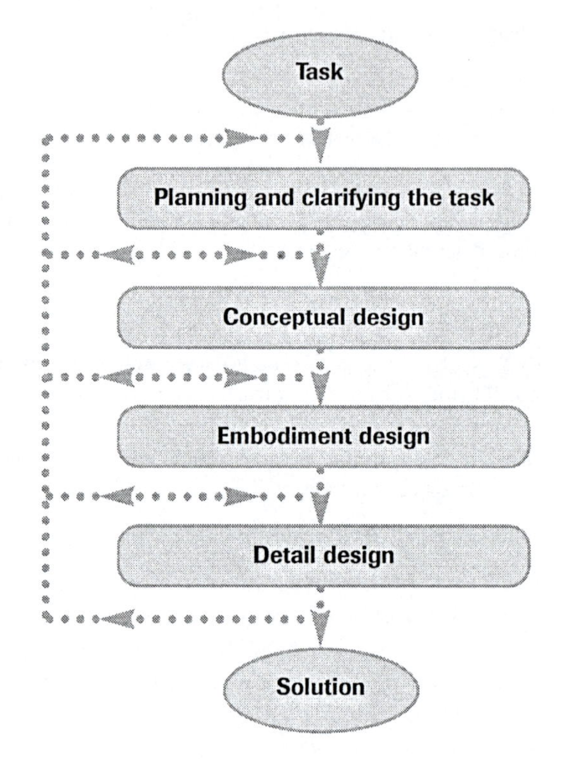

Figure 2: THE PRODUCT PLANNING AND DESIGN PROCESS

Source: Based on Pahl and Beitz 1996

Planning and clarifying the task

The step for planning and clarifying the task includes a formulation of a product proposal. This proposal is based on product ideas that are found and selected after an analysis of the market and company situation. The task is then clarified with the elaboration of a requirements list (Pahl and Beitz 1996).

The first three steps in the **backcasting** method, described in the preceding section, can facilitate the incorporation of sustainability aspects when generating product ideas. The fourth step can then, to some extent, be realised with the development of products that are helpful or even critical in the transition towards sustainability.

It is important that the suggested products can lead in the right direction towards sustainability and be flexible platforms for forthcoming products heading towards sustainability. As mentioned before, this is also one of the most important characteristics of this approach; while many existing tools for 'design for environment' have the capability to direct new investments in directions to decrease environmental impacts, few of them consider if the investments are good platforms for further investments in the right direction.

The generation of product ideas can result in a description of a service or a set of services that the new products should aim to deliver. Examples of questions that can be asked when evaluating services are: 'What human needs are fulfilled by this service?' and 'Which restrictions and possibilities can be found if this product is incorporated in a sustainable society with ten billion people?'

When a product idea has been generated, it has to be clarified. This can result in a list of requirements that must be fulfilled and a list of desirable qualities. Examples of aspects that can be considered are economic, reliability and safety, as well as physical surroundings such as geography and weather. Ambitions concerning sustainability aspects also have to be specified at this stage. If the ambition is to develop a product that works as a bridge towards products that could fit into a sustainable society, how far should this product reach in this direction and in which ways? The four principles for sustainability can be used as guidance for the requirements list.

The planning and clarifying step results in a product proposal that is based on conditions for a future sustainable society besides additional aspects that are traditionally considered in product development, and a requirements list that includes sustainability criteria.

◻ *Conceptual design*

The conceptual design step implies a development of the principle solution. Essential problems are identified and function structures are established. Working principles and structures are found and combined and firmed up into concept variants. The variants are then evaluated against technical and economic criteria (Pahl and Beitz 1996).

Sustainability criteria can also work as a basis for the evaluation of different concept variants. Principle solutions must be able to fulfil the overall function of the product and the requirements on the specified list from the clarification of the task. Desirable qualities can be weighted according to their importance and the evaluation of concept variants can be based on them. Sustainability aspects can be considered as part of the requirements list or as desirable qualities.

Sustainability aspects are relevant on a systems and sub-systems level. The services that should be delivered by the product give an overall function of it. This overall function can be divided into sub-functions. A variety of possible solutions can be found for separate sub-functions as well as for the combination of them. Different concept variants for the overall function could be developed into different products that more or less head in the direction towards sustainability. For example, if we look at some already developed technology—a conventional steam-electric coal power plant can be considered a dead-end technology, while a coal integrated gasifier/combined-cycle power plant may be a transitional technology towards sustainability. The reason is that much of what has been learned in developing this technology is readily transferable to the biomass integrated gasifier/combined cycle power plant which has better qualifications to fit into a sustainable society. Different possible solutions on a sub-functional level can be, for example, different kinds of systems for the power supply in a car, such as electric motors that may substitute for combustion engines for cars in urban traffic. Different options on a sub-system and system level are further elaborated, in more detail, in the embodiment design step (see below).

The conceptual design step results in a principle solution. The choice of principle solution is made from an evaluation, on a system and sub-system level, partly based on sustainability criteria.

☐ *Embodiment design*

The embodiment design includes a development and a definition of the construction structure. The development of the construction structure starts with a preliminary form design and material selection. A refinement and improvement of layouts are done from a selection of the best preliminary layouts. The layouts are evaluated against technical and economic criteria. The definition of the construction structure includes the elimination of weak points. Errors, disturbing influences and minimum costs are identified. Further, the preliminary parts list and production and assembly documents are prepared (Pahl and Beitz 1996).

The choice of construction structure influences the environmental impact during the consumption and the waste-handling phases of the product. Environmental impacts in earlier phases of the life-cycle of the product and its components and materials, such as extraction of resources and various production processes, are also influenced by the choices made in this step.

In our approach, the evaluation of the layouts can, besides other criteria, be based on sustainability criteria. The criteria can be based on options for **transmaterialisation** and **dematerialisation**, and such options can be based on the four principles for sustainability. Transmaterialisation is a strategy to decrease the environmental impact per amount of materials while dematerialisation can decrease the amount of materials needed to maintain a specific service level.

The **transmaterialisation** strategy is an option when different materials can substitute for other materials without changing the qualities of the product more than is allowed by the set of requirements. The principles can be used as guidance when evaluating choices of materials. Information about hazardous materials with known environmental effects can, of course, be used in the evaluation. But, during the process of substitution, it is essential to also apply principles 1 and 2 to identify, and avoid, materials with a high probability of leading to increased concentrations in the ecosphere later on—if used. Such potentially hazardous materials are, for example, metals that are scarcely occurring in nature—they have a larger potential to increase systematically if emitted to the ecosphere compared to more abundantly occurring metals. This also goes for persistent substances that are foreign to nature, for the same reason. Abundant metals and easily degradable substances could substitute for such substances.

Besides on a material level, substitution is relevant on a raw material and a component level. The same material may be obtained from different raw materials: for example, methanol can be produced either from fossil fuels or from biomass. Different types of component can give the same function: for example, the choice of a battery can be made from various types. In the product development process, the selection of raw materials, materials and components are further elaborated in the detail design step.

Dematerialisation can be accomplished in two principle ways: reducing the flow, i.e. less material/energy flow to achieve a certain service; or closing the flow, i.e. increased recycling of materials. This strategy is strongly related to the fourth principle of sustainability. **Reducing the flow** implies a more efficient use of a given material for a given function. This can be achieved, for example, by miniaturisation of products or components and by increased quality or reparability to extend the product's lifetime. The copper wire in power transmission is one example of dematerialisation. By raising the transmission voltage it has been possible to reduce the amount of copper needed to transmit a given amount of electricity.

Closing the flow of materials within society implies that the same material is used again and again. The flow can be closed on different levels: by re-using products, by re-using parts of products or by re-using materials or raw materials. Cycles can be closed within the production process or in the exchange between producers and consumers. An important condition for successful recycling of materials is that flows are sufficiently pure or separable. Unnecessary **mixing** of different kinds of materials can make separation more difficult, and should therefore be avoided. 'Design for disassembly' (see e.g. Luttropp 1997) has evolved as a strategy specifically for facilitating the recycling of products. The effects of the inevitable loss of quality in materials and energy can be minimised through the implementation of **cascading use,** where each step involves a drop in quality requirements. After each recycling step of a certain material, it should be used in such a way that the quality can be kept on the highest possible level. There are mainly three qualities that are interesting—**purity**, **structure** and **exergy**[3] (Holmberg *et al.* 1996). For instance, a special steel should not be used as reinforcing iron after only one cycle if one wants to save **purity**. The bulk **structure** of wood, for instance, can be utilised if wood is first used as a construction material before the fibre structure is used in paper of stepwise declining quality and, finally, the chemical structure is utilised in the chemical industry or fuel production and combustion. In the energy sector, one can also improve the **exergy efficiency** through cascading use of energy, where each step involves a drop in temperature. In this context, Ayres and Ayres (1996) have discussed **waste mining** as a strategy that utilises waste-streams from (currently) irreplaceable resources: for example, recovering elemental sulphur from natural gas and petroleum refineries. This strategy reduces: the environmental damage due to the primary waste-stream; the rate of exhaustion of the secondary resource; and the environmental damage due to mining the secondary resource.

---

3 The exergy of a certain system is the amount of perfectly ordered energy, such as mechanical work, that can, in principle, be extracted from the system in an ideal reversible process (Kotas 1985).

A **dissipative** use of materials, such as chemicals that are emitted to the ecosphere during their consumption, is the opposite to closing the flow strategy. The dematerialisation strategy can be used to decrease the amount of emitted materials and the transmaterialisation strategy can be used to avoid using hazardous (or potentially hazardous) materials in a dissipative way.

The embodiment design results in a construction structure which has been chosen after an evaluation of layouts. The evaluation has, besides other criteria, been based on sustainability criteria.

Detail design

The step of the detail design implies that production and operating documents are prepared. Detail drawings and parts lists are elaborated. Complete production, assembly, transport and operating instructions are concluded and all documents are checked. This final step should result in product documentation (Pahl and Beitz 1996).

So far, we have mainly discussed the **design** of the product, which is connected to other activities such as the development of the production process and the organisation of transports. Sometimes, the design of the product can influence other activities in a positive way. Hitachi, for example, designed a model of washing machine containing just six screws to facilitate the recycling of old machines. The design also had the consequence that the production process was facilitated and manufacturing time cut by 33% (Esty and Porter 1998). Although other activities are dependent on the product design, they should also be considered from a sustainability perspective by themselves.

Different activities can be connected to different roles of a firm. With some of these roles, the firm has the potential to influence other actors in their use of resources. A firm can be divided into four different roles, by which it directly or indirectly influences the ecosphere. By the choice of purchased goods or services, the firm in the role of **purchaser**, influences the supplier's interaction with the ecosphere. In the production and delivery of goods and services the firm acts as a **resource converter** and influences the ecosphere directly. As a **supplier** the firm delivers goods or services which, due to their properties, influence the customer's interaction with the ecosphere. In the role of **communicator**, the firm takes part in personnel education, advertising and lobbying and it sends product information to customers and specifications to suppliers, thereby influencing a whole range of actors (Holmberg 1998).

Discussion and conclusions

We have described the structure for the incorporation of sustainability aspects within the process of product development. The main advantage with this approach is that progress towards sustainability is emphasised and facilitated through the use of four principles for sustainability. Based on sustainability criteria, it is easier to avoid the development of products that lead to dead ends. New products should at least be able to work as sufficiently flexible platforms for other products in progress towards sustainability. Further, a structure for sustainability makes it easier not to miss considering relevant aspects of sustainability.

The structure we have presented in this text is a first step towards a method for sustainable product development. In our continuing work with this project we will develop this method in more detail with the objective of achieving a method that can become a useful tool for planning and product design in SMEs.

Our ambition is to develop a tool that has the form of a set of **hierarchically structured questions**. The questions will be sorted under the steps of the product development process and based on the four principles for sustainability. The questions will be sorted in a hierarchic structure from general to more specific. Most questions will have a qualitative approach while others can result in quantitative measurements where relevant. There are many advantages with using a semi-qualitative approach:

1. The framework and the principles make more sense and will be easier to grasp.
2. It is easier to identify which aspects should be quantitatively analysed.
3. It is easier to identify which aspects are not quantitatively analysed.
4. It is easier to communicate the results, due to the logical structure.
5. It is possible to find solutions with less effort, since it is easier to avoid unnecessary quantitative analysis.
6. It is possible also to cover aspects that are difficult to analyse quantitatively.

Principles 1, 2 and 3 make it possible to focus upstream on fundamentally different mechanisms in the societal impact in the ecosphere:

- A systematic net increase in concentration of matter that is introduced into the ecosphere from outside the system (the lithosphere)
- A systematic net increase in concentration of matter that is produced in society
- A systematic physical deterioration through harvesting and manipulation

This means that the questions can start from these mechanisms. For instance, the first principle raises the following question: will the activity or investment cause emissions (directly or indirectly) of substances extracted from the lithosphere? And will they be a part of a flow that leads to systematic increased concentrations in the ecosphere of elements extracted from the lithosphere? (Maybe some of the actual elements are already known to cause harmful effects?) If this is the case, the next question is then to figure out how this can be avoided. For different activities and investments different actors find different solutions: substitute an abundant chemical element for a scarce/toxic element; avoid dissipative uses; increase recycling; reduce downcycling or fulfil the same service with less of the material. The other principles can be used in the same way.

There are three reasons why we believe that the presented integration of frameworks for sustainable development and product design, respectively, will be particularly suitable for SMEs:

1. Experience tells us that quantitative measurements of environmental impact, for example LCA, take a lot of effort. This can make it expensive to find sufficient data for the measurements. Our qualitative approach to include sustainability aspects in product development can therefore be of special usefulness for SMEs which do not have the same economic possibilities as larger enterprises.

2. The suggested planning model takes even short-term economic rationales into consideration ('low-hanging fruits' that are technically flexible platforms), which means that it can be implemented relatively quickly in business activities.

3. It is easier to establish transparent flows of information between different sectors of activities in SMEs than in large firms. The suggested integration of a sustainability perspective into a model for product development is simple enough to apply as a shared mental model for dialogue—stimulating business creativity—by virtually all employees in an SME.

13
Environmental innovation and small and medium-sized enterprises

Sandra Meredith

Innovation, we are constantly told, is the key to business survival. The current rate of technological change is high as companies strive for competitive advantage through constant new product development or by superior product or process performance. From the start, it must be emphasised that innovation is not simply coming up with new ideas—that is invention. Innovation is a three-stage process that consists of the original idea, the development phase, and exploitation of the developed idea. It is only when practical applications have been achieved, making them work technically and gaining commercial benefits, that innovation can be said to have taken place.

Much discussion around innovation concerns the drivers for change. In particular, the debate often focuses on technology push versus the market pull. The work of Rothwell and Zegveld in the 1980s did much to clarify the issues when they concluded that 'the process of innovation represents the confluence of technological capabilities and market needs within the framework of the innovating firm' (Rothwell and Zegveld 1985). In other words, no one factor can be said to drive innovation, and the conditions within the company need to be 'right' for successful innovation to occur.

The arrival of environmental concerns into the business context can be said to have added another dimension to the need to innovate. Not only must industrially active organisations fight their corner in the innovation game, but they now need to do so within an increasingly stringent regulatory regime. This means that the development of a new product or process now entails a consideration of environmental impact, as well as economic feasibility. History shows that to follow a narrow economic view, giving inadequate attention to environmental or social repercussions, can be disastrous to society and costly for the organisation or industry concerned. The hole in the ozone layer, the consequences of the use of certain pesticides, and the catastrophic results of the inadequately tested drug thalidomide all bear witness to this.

Small and medium-sized enterprises (SMEs) have to be aware that environmental legislation can be instrumental both in opening up new pathways of innovation and in closing others down. In these circumstances, it can be said that the demands of the environment are just as much a trigger for innovation as are the traditional economic and market drivers. To make environmental innovations of economic benefit to the company the technological innovation must carry with it some added value. That is, it should at least comply with regulatory demands while accruing competitive advantage through increased efficiency in production, the development of a new product, or through improved company image. In this chapter we focus on the question of how SMEs approach innovation, and examine the nature of the innovation process, in particular identifying practices and behaviours that support innovation that is specifically environmentally oriented.

The innovation process

Traditionally, technological innovations have sprung from the hands of large and multinational companies rather than SMEs, since it is within these large organisations that resources reside—financial investment in research and development, technological expertise, human resources and the space in which to take risks. Innovation always involves some element of risk and, by their nature, SMEs are risk-averse and so are less likely to innovate. However, smaller firms do have advantages over larger ones. They benefit from a lack of bureaucracy; they are able to respond rapidly to change; their internal communications are efficient; and they are often characterised by a strong entrepreneurial spirit. It is these characteristics that SMEs must exploit.

Despite the fact that most research in the field has been directed at the problems of managing innovation in large, complex organisations, a generic process has emerged that can assist a company no matter its size. Bessant *et al.* (1996) offer the following four phases that make up the core of the innovation process:

1. **Identify market signals**. This first stage is the organisation's recognition and management of signals from the business environment about market needs, competitor behaviour and regulatory (environmental) requirements. This phase sets targets for the direction of innovation. The organisation needs to adapt to these forces and develop new ways of meeting these changes. At the same time, it must also process signals about technological developments, the emergence of new opportunities arising out of research, out of competitor behaviour, and out of equipment supply markets. These represent possible threats if they are not taken up.

2. **Strategic concept**. Having taken in the 'market signals', the organisation needs a focused strategy that sets out where, why and when it will commit its precious resources to change the way it does things.

3. **Search and selection**. This stage entails research into, and identification of, alternative paths of innovation, ensuring that problems are recognised and dealt with. This stage is about getting agreed choices of 'what' the innovation will entail and 'how' it will be done. The output of this stage is a chosen solution or set of solutions.

4. **Implementation**. This phase involves the management of change along several dimensions in parallel. Not only must the innovation itself work, but the context in which it is placed must accept and absorb it effectively. The more radical the change, the more critical this process of change management becomes. If this is a 'process' innovation rather than 'product', totally successful implementation will involve a major element of internal marketing.

A fifth element can be added to the above—that of learning. An SME taking a proactive stance towards its environmental responsibilities will take time to review the previous phases (both in terms of success and failure factors). This ensures the capture of the relevant knowledge in order to learn from the process. Of course, as well as following the four-phase process, all organisations must understand their own situation and the contingencies likely to influence them. Companies are, invariably, at different levels of technological and organisational development, so environmental issues will affect them in different ways. Hence, it can be said that innovation is, to some extent, company-specific. Finding the most appropriate innovation process for your organisation, at the heart of which is placed the basic sequence of activities outlined above, means that successful management of the innovation is more likely to take place.

Innovation in the environmental context

It is within certain sectors that environmental innovation needs to be particularly developed and supported. For example, new technology-based firms with their reliance on software, information technology and knowledge, have few environmental demands made of them. They are often valued as being important sources of new ideas and of 'high' technologies which can be exploited by larger organisations, and are perceived to pose little threat to the environment. This chapter refers to those SMEs that operate in mature manufacturing sectors such as textiles, printing, metal-working and coatings, which make up one in five of SMEs in the UK (Hillary 1995). Important as the new technology-based firms are, the mature companies must also be considered as important by reason of their association with industrial regeneration through their industrial production. Their importance is social as well as economic. The distribution of economic power through a system of small firms leads to a more favourable distribution of power in society in general. In addition, SMEs may provide a stabilising effect during times of sharp fluctuation in employment (Rothwell and Zegveld 1982).

However, if mature industrial SMEs are perceived as an important component of the UK's total industrial production, it may be logically assumed that they have been responsible for a significant part of the damaging environmental impact which, in the not so distant past (the last 20 years), has been largely ignored. Until the late 1980s and early 1990s the UK had a rather ad hoc way of dealing with environmental impact control. Responsibility for protecting the environment was broken down into different media and placed in the hands of local authorities, the then water boards and various other agencies. It was not until the introduction of the Environmental Protection Act 1990 that environmental protection took on a unitary and comprehensive enforcement framework which now works within the stringent guidelines of EU environmental regulation. Formerly, it was the large companies that took the brunt of the blame for environmental pollution. The new regime has meant that smaller companies, which once fell through the regulatory net, are now being held accountable and asked to comply with environmental legislation.

Environmental regulation is now a reality that affects all types of organisation. Whatever size your company, at some point in time you will be confronted with environmental issues that will demand solutions. Gouldson and Murphy (1998) point out that, historically, a mutually antagonistic relationship has existed between the interests of economic development and environmental protection. The impact of the Industrial Revolution bears witness to this and in the 18th and 19th centuries it was the environment that lost out. Now, meeting environmental responsibilities and remaining economically competitive is a challenge for *all* companies. Instead of viewing the rise of environmentalism as an obstacle to technical change, it is increasingly becoming an aid to innovation. This chapter will point to overt environmental innovation—as opposed to minimal defensive strategy—as a way of meeting the challenges that face SMEs.

The power of innovation

In mainstream business literature, innovation or 'the ability to manage change—in products, technologies, services and processes—in the face of a turbulent and often uncertain business environment' is cited as a key success factor (Rothwell 1992). Indeed, one the most enduring of management commentators—Drucker (1995)—sees innovation as the one essential core competence that every organisation needs. Generic examples can be found of companies working in intensely competitive markets which have survived where others have gone under, simply because they did not wait for the economic climate to change, but took control of their own future by committing themselves to innovation. Good examples of this are the innovative designs of the Flymo lawnmower company and of Richardson of Sheffield which developed new products in the cutlery industry, saving it from the fate of extinction faced by the majority of other companies in this sector in the UK (Tidd *et al.* 1997).

In a similar sense, the politicisation of environmental issues and the demands made on companies by environmental regulations place SMEs in just such a turbulent and unpredictable situation. Environmental innovation is, however, subtly different. Whereas innovation *per se* is triggered by a combination of market pull or technology push, environmental innovation is usually triggered by regulation and the responsibility placed on the SME to minimise its environmental impacts. The company itself must help sensitise market demand for environmentally sound products, as well as develop or adopt technological solutions to improve production processes. Case study research indicates that a strategy built on innovation is best and that the first rung to survival is to take a proactive stance (Taylor 1992; Meredith and Wolters 1996).

Whether large, small or medium-sized, how firms think about the environment and their approach to the problem will almost inevitably dictate outcomes. Simply ignoring pressing environmental issues can have a catastrophic effect on small companies. Sitting in a room listening to a telephone conversation concerning a small paint shop which was about to be closed down by the UK Environment Agency because its owners had failed to respond to a new regulation brought home the perils of ignoring demands for environmental innovation. And, again, taking minimal action—for example, fixing end-of-pipe temporary technology as opposed to adopting more radical innovation of integrated processes or redesigned products—only delays the inevitable and can prove more costly in the long run.

How, then, can SMEs deal with a fairly major new piece of environmental legislation *and* make technological choices? As with many situations, the end result depends on how they view and approach the issues. Research within industrially active SMEs stresses the importance of **perception** of the environmental challenge.

Importance of perception of the environmental challenge

From the outset, it is accepted that environmental management, particularly meeting legislative requirements, can be difficult. The difficulty is exacerbated by the fact that environmentalism in the late 1990s has become highly politicised and is characterised by four main features: the growth of public concern for the environment; the growth of green consumerism; a general diffusion of ecological values; and the intensification of regulation (McGrew 1993). All these factors can have major repercussions for all firms and make the issues impossible to ignore, despite the fact that many companies perceive the environmental challenge purely in terms of cost.

However, the question of whether environmental regulation enhances or erodes economic competitiveness almost becomes irrelevant. When confronted with specific regulation an SME must react in some way, and it is the nature of the response that is important. Research within manufacturing SMEs (Meredith and Wolters 1996) categorised the nature of this response to environmental responsibility and linked its innovative outcomes. Over time, SMEs' perception of environmental issues has been influenced by the rise in environmental awareness and reflects itself in differing levels of strategic behaviour towards the environment.[1] Perception of the issues plays a major role in shaping company behaviour and appears to strongly influence strategic outcomes and the pursuance—or not—of environmentally oriented innovation.

1 See Chapter 4.

Meredith and Wolters (1996) have outlined three possible lines of strategy distinguished through research carried out within SME manufacturers in the UK and in the Netherlands which can be said to provide evidence of types of strategy followed by individual companies (see Fig. 1). Companies perceive the environment in different ways and adopt different strategies accordingly:

- **Defensive strategies**. SMEs exhibiting defensive strategies are those that perceive environmental responsibilities in purely negative terms and assume there are no market opportunities to compensate for environmental costs. Company strategic orientation fails to include environmental issues and it may even lobby against specific regulation. No innovation takes place.
- **Offensive strategies**. The company assumes that market demand is sensitive to the environmental impact of its existing processes and products. Perception of the environment is not *entirely* proactive, but the company accepts the need to improve existing products and processes. Some innovation takes place.
- **Innovative strategies**. Companies following an innovative approach have a perception of the environment that takes it for granted that market opportunities exist in the development, production, use and/or marketing of innovative environmentally oriented technologies.

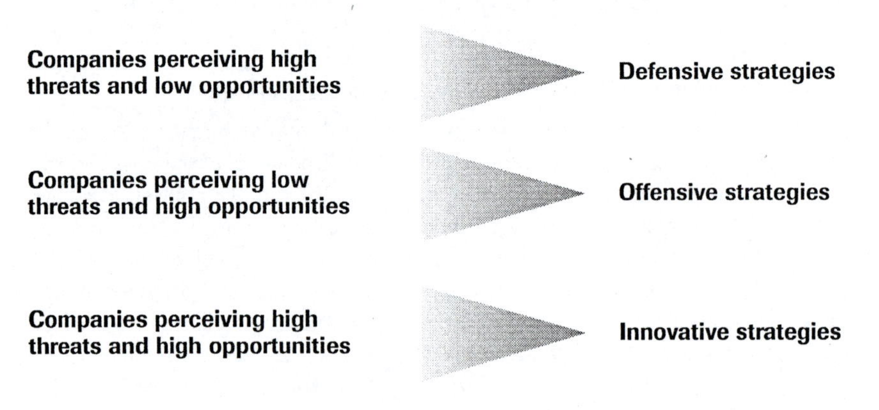

Figure 1: COMPANY PERCEPTIONS AND STRATEGIC OUTCOMES

Source: Meredith and Wolters 1996

Further research undertaken among a wide variety of SMEs (van Dijken *et al.* 1998) within four environmentally sensitive sectors supports the argument that a proactive approach can be most profitable. This often involves tackling a technological problem head-on and demands that the SME examines its ability to be innovative. The section below, based on this research, identifies and discusses those elements identified as essential to the environmental innovation process.

Conditions necessary for environmental innovation within SMEs

Whether large or small, an environmentally oriented innovation does not happen in isolation. Once an SME has understood the implications that environmental regulation may have and aligned its perception of environmental demands towards an innovative rather than a defensive strategy (as detailed above), the path is clear for the firm to behave in an environmentally innovative way. How that is best pursued, and the elements necessary to succeed, are identified in the 'dynamic triangle'.

Again, we call on the results of a research study (van Dijken *et al.* 1998) which was specifically designed to identify key strategies for successful innovation, in both environmental and business terms. Data indicated that, for successful environmentally oriented innovation to occur, three crucial elements must be present within the SME. Together, these are termed the 'dynamic triangle' (see Fig. 2). The elements of this triangle define and bring together the determinants of SME innovation, namely:

- The strategic orientation
- Competences and capabilities
- Networks

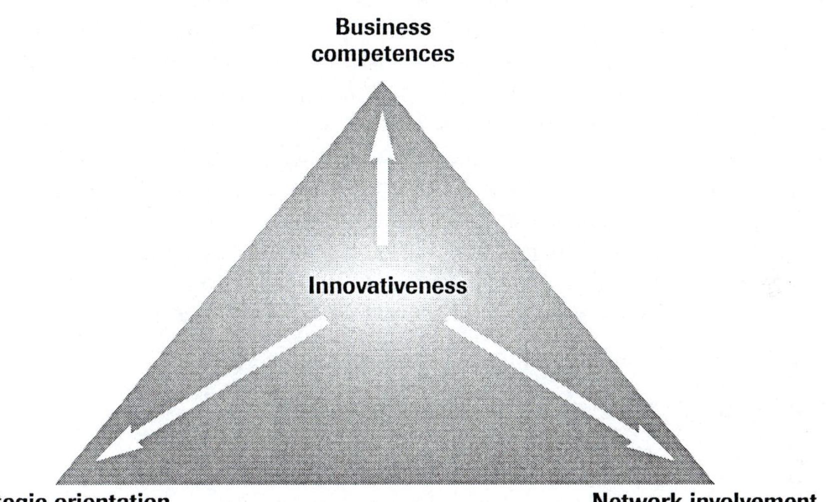

Figure 2: THE 'DYNAMIC TRIANGLE'

Source: van Dijken *et al.* 1998

In the process of innovation, the evidence gained from companies suggests that there is a dynamic interaction between these three dimensions. They are variables which are partially interrelated. For example, a company's internal competences and strength of network relationships define the limitations of its strategic options, while the strategic orientation influences the kinds of competences and external networks that are developed. Within each company it is the dynamic interplay between these three dimensions that supports the innovation processes. If any of the elements is missing or under-developed then the firm's environmental innovativeness is weakened or becomes non-existent. Each dimension is discussed below.

Strategic orientation

Within the context of this discussion, strategic orientation includes the level at which environmental issues are integrated into the company's business strategy. One of the most consistent findings of the research was that the decision to innovate was strongly related to the overall business strategy of the enterprise. An enterprise's strategy is, by its nature, built on its values, motives and goals, the fulfilment of which rely on level of resources, the business scope of the firm, how it perceives its present and future markets and, of course, its perception of the 'environment'.

The way environmental issues are communicated by a specific SME into the business network is very important. In environmentally sensitive sectors, such as electroplating firms, research indicates that customers demand certain environmental practices, which lead to raised environmental awareness throughout the sector. In one specific case, a customer's demand for both state-of-the-art technology *and* environmental soundness led to the development of a highly innovative piece of closed-loop equipment which meant that the manufacturing process no longer posed an environmental threat. This innovation solved a pressing environmental problem, had a payback time of only 15 months and, thereafter, saved the company in excess of £100,000 a year through major reductions in water consumption and other environmental costs; but this did not happen overnight. This company had to review its strategic thinking, identify its technological options, develop its internal competences and capabilities and, of course, gain support from a variety of external sources (Meredith and Biondi 1997). All these elements link together in a way that turned what at first seemed a major environmental cost into a major economic gain.

The 'strategic orientation' interlinks with the 'competences' element because it:

- Determines which competences are to be developed
- Determines the nature of core competences
- Determines viable strategies for environment and business
- Determines the level of market search and identifies 'signals'
- Assists technological selection and development
- Determines the capability to engage in and implement innovations

Competences–capabilities

Competences refer to the capacity to carry out internal innovation processes and to develop relationships with external resources; they relate to the technical system, knowledge, skills, values and organisational routines that are of special advantage to the firm (Leonard-Barton 1995). The skills of an SME's employees constitute the backbone of such capacity, together with the value system and 'routines'—the way the company does things to achieve specific goals. Hence, human and organisational resources rank high.

The competences and capabilities of the SME are of great importance when selection between different possible courses of innovation has to be made, for the level of environmental knowledge and management systems may define the technological innovativeness of the company.

The competences element, in turn, interacts with the functioning of the network, determining:

- The configuration of the network
- Levels of communication
- Level of resources, activities and people used to expand competences
- Scope of the interactive learning
- Which environmental competences need to be developed

The innovation networks

The third variable in the 'dynamic triangle' is that of networks. Key writers on innovation (Rothwell and Zegveld 1982) have stressed the importance of external links or networks in the development of innovation, and this has been supported by research looking at environmentally oriented innovation (van Dijken *et al.* 1998). The importance of networks to the innovation process is the different streams of information and support they provide for the company.

Three very specific networks, with which the innovating companies had relationships, have been identified. They are identified in Figure 3 as the 'business network', the 'regulatory network' and the 'knowledge network'. The research indicated that, without a strong link into each of these three networks, companies were likely to gain only a partial understanding of, or support for, the environmental innovation process. For example, linking into the technical knowledge part without adequate liaison with the regulatory network may result in products or processes that fail to take the innovation far enough in terms of environmental compliance. And, similarly, making strong links with both the regulatory and knowledge networks without adequately consulting the customer or end-user may result in products that do not sell.

Distinguishing between these three networks allows each company to identify where it may find information and technical and financial support. However, while the individual networks themselves may function independently of each other, the company needs to marry up the three to gain a balance of ideas, information, technological options and financial power to support its innovation strategy. Networks can be seen to be more important for SMEs, as owners and managers in small firms are less likely than their larger counterparts to have undergone formal management training and may be less familiar with the analytical techniques associated with the strategy formulation and decision-making that underpin innovation.

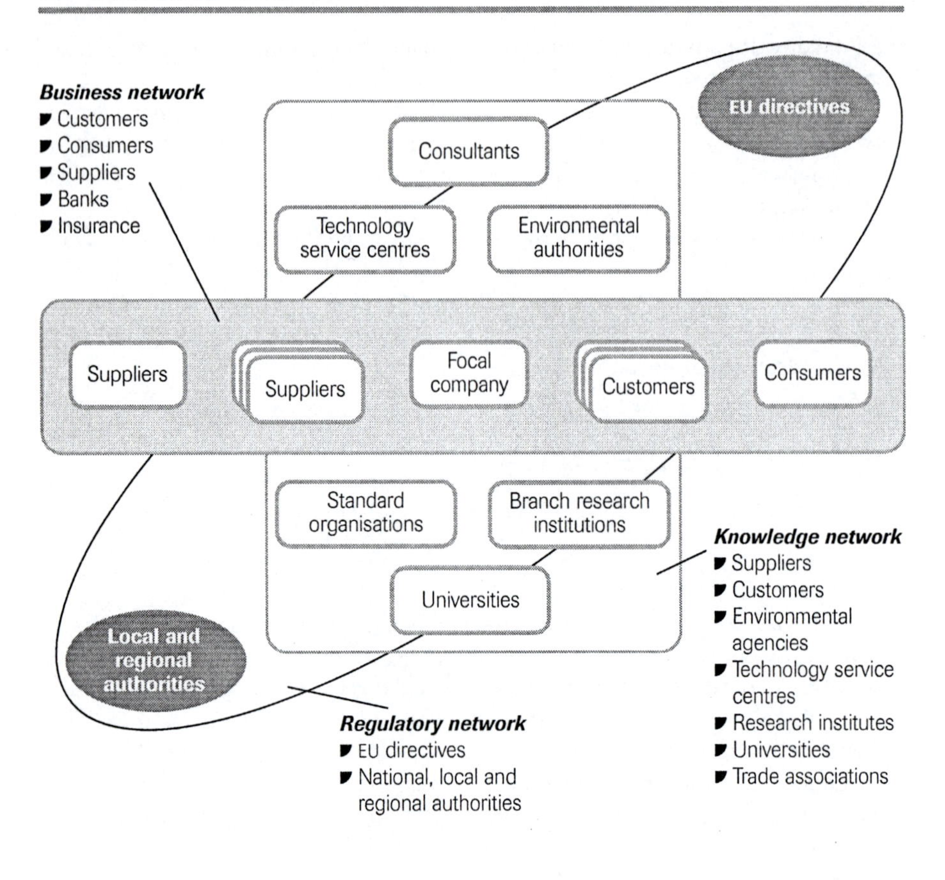

Figure 3: THE DIFFERENT NETWORKS IMPORTANT IN ENHANCING AN SME'S ABILITY TO ACHIEVE ENVIRONMENTAL INNOVATION

Source: Hansen *et al.* 1998

Where SMEs suffer from a lack of resources they are able to participate in very few network activities.[2] This means that their contact with networks may be through one important customer or supplier only. This situation can constrain a company's capability to innovate, as a variety of sources of information is required to make up an environmental innovation network.

To understand the character and workings of the network, it is important to differentiate between differing elements, and the type of support they offer to SMEs. The quality and strength of each network relationship has implications for SMEs in terms of the level of success they may achieve at every stage of the environmental innovation. Such relationships are important at the beginning of the process when the company must gather information on possible solutions. The success of this stage

2 See Hunt's discussion of networks in Chapter 15.

relies on the company's links with both their *knowledge* and *regulatory* networks. At the development and testing stage, reliance is placed on the *business* and *knowledge* networks. Finally, at the implementation stage, when engaging other organisations' technological and project management expertise is pivotal to success, all three elements (*knowledge*, *regulatory* and *business*) can be of equal importance. Competences and network relations are interrelated, since the competences of SMEs are extended by fully developing external resources. The converse can be stated in that weak internal competences may explain why external network resources remain remote and unavailable, and the route to environmental innovation remains blocked.

Having identified the three very specific elements of the 'dynamic triangle', it may seem that the innovation process can be seen from each of the three dimensions. When viewed at a deeper level, however, it can be seen that they are all inextricably linked. From the strategic perspective, the capabilities are the firm's strengths and weaknesses and the network presents both opportunities and threats; the internal capabilities limit the environmental strategic orientation and, thus, decide the final configuration of the network; and the network conditions the strategic orientation, the interactive learning that can take place, and the development of core competences in the firm (Hansen *et al.* 1998). In taking an environmentally innovative stance SMEs must ensure that they understand the complexity of the process and, importantly, that all three elements of the 'dynamic triangle' are included.

Towards an environmentally innovative company

The aim of this chapter has been to outline the benefits of following an innovative strategy at both an environmental and a business level. It has endeavoured to provide guidelines for pursuing such a strategy. The core of the message has been to stress that the nature of a company's perception of the environmental challenge is of paramount importance to that company's wellbeing. How an organisation manages its environmental responsibilities can mean the difference between success and failure.

A major part of the difficulty posed for an SME is that it is often unable to estimate the environmental costs and economic benefits in pursuing a specific environmental innovation. It is only through understanding the issue and making it part of the enterprise's business strategy that the opportunities or threats become apparent. What is clear is that, with increasingly stringent regulation coming from Europe, an innovative strategy that takes the company well beyond compliance is called for, rather than a short-term defensive approach.

Emphasis has been placed on the importance of SMEs realising their strengths as innovators in a generic sense. Some characteristics of the small company give them an advantage over large organisations: for example, they have a flexibility and speed of response that is rarely replicated in their larger rivals, and there is usually a closeness with customers that allows the development of products or technologies that meet both business and environmental goals.

The importance for SMEs to link into networks has been stressed. Research (van Dijken *et al.* 1998) has identified three specific networks from which companies can draw information and support to compensate for lack of resources. However, networks are only one element of the whole and it is the 'dynamic interaction' between the three dimensions that determines the innovativeness of a company. But environmentally oriented innovativeness does not happen overnight: these determinants of innovation clearly evolve as companies learn to see the opportunities as well as the costs in the environmental challenge.

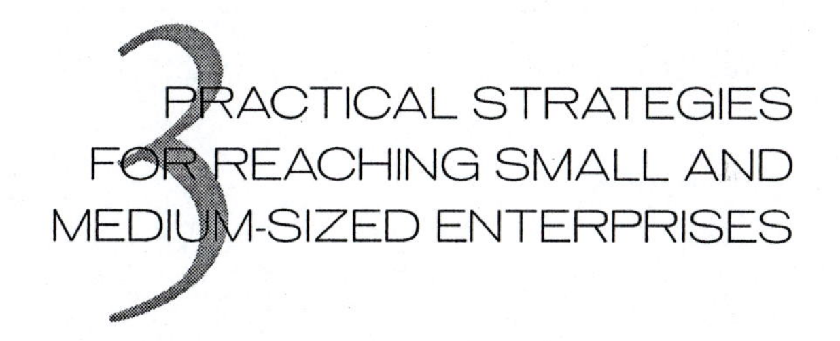

3 PRACTICAL STRATEGIES FOR REACHING SMALL AND MEDIUM-SIZED ENTERPRISES

14
Developing better systems for communications

Environmental best practice in small business*

Garrette Clark

The mission of the United Nations Environment Programme's Division of Technology, Industry and Economics (UNEP TIE) is to encourage decision-makers in government and industry to develop and adopt policies, strategies and practices that are cleaner and safer, and use natural resources more efficiently, as well as to reduce pollution risks to human beings and the environment. The Production and Consumption Unit specifically aims to reduce the environmental consequences of industrial development and the pollution that arises from the ever-greater consumption of goods and services. A focus needs to be kept on enhancing the resource efficiency of industry, adopting cleaner and safer methods of production, and improving the design and life-cycle of products.

As part of its work on promoting cleaner and safer methods of production, with the financial support of European Commission Directorate General III: Industry (DG III) and in conjunction with organisations such as the World Business Council for Sustainable Development (WBCSD), UNEP TIE carried out a two-phase project on how to improve systems for communicating environmental best practice in industry. This chapter presents the results of that project.

* UNEP TIE would like to thank European Commission DG III for its support of this project. For any further information please contact: UNEP TIE Production and Consumption Unit, 39–49 Quai André Citroën, 75739 Paris Cedex, France; tel: 331 44 37 14 50; fax: 331 44 37 1474; e-mail: *unepie@unep.fr*; website: *http://www.unepie.org*

Background

The need to provide industry, including small and medium-sized enterprises (SMEs), with information on how best to incorporate environmental management strategies is widely recognised. Numerous organisations have initiated information dissemination activities targeted at the local, national, regional and global levels. These activities consist of awareness raising, technical training and assistance, and they disseminate information via hard-copy, one-on-one exposure and electronic means such as databases and the Internet. But there is a consensus that improvements need to be made and that more needs to be done.

Industry has both positive and negative impacts on the environment. Over the last 30 years, industry's response to environmental issues has evolved from ignoring problems to recognising that environmental and economic performance are not mutually exclusive. On the contrary, combining the strategies of doing more with less, wasting less, recycling more and reducing inputs brings more economic benefits to society, as well as direct benefits to industry. Although this approach has been adopted by a number of industries, it is not the case for all industry, particularly SMEs. Issues associated with environmental management in industry are very dependent on the size of the individual enterprise.

Communicating environmental information to industry in order to get it to incorporate environmental concerns into management decisions faces a number of obstacles. Larger companies, especially multinationals, tend to have the motivation and the resources necessary to identify relevant environmental concerns and to translate them into action. International trends such as environmental management systems (particularly various international standards), environmental reporting, and other forms of voluntary codes of conduct, are known. Although action has been taken on this level, much remains to be done.

The problems posed and faced by SMEs can be taken to be a good example of the problems faced by industry in general. The study focused on providing answers to the following questions:

- What information do enterprises need?
- What information is available? (What additional information should be provided?)
- How is information delivered? (How could delivery be improved?)

Methodology

The project was carried out in two phases. First, a meeting of experts specialising in communicating environmental information to enterprises (particularly SMEs) was convened. They discussed how to improve industrial environmental performance through improved access to environmental information. Discussions centred on presentations by each participant of successful and unsuccessful examples of information dissemination efforts. From these, suggestions on how best to communicate with enterprises evolved, as well as general recommendations to other organisations involved in the dissemination of environmental information.

The second phase, from April to August 1997, involved a survey of information providers familiar with promoting environmental issues to industry. The survey findings were used to supplement the outcomes of the expert meeting. Respondents came from cleaner production centres and other information intermediaries such as public- and private-sector-sponsored information centres and industry associations.

One of the basic issues that arose dealt with vocabulary. Various terms are currently in use to describe preventative environmental management strategies. For example, the concept and actions implied by the term 'environmental best practice' are similar to other terms such as 'cleaner production', 'eco-efficiency' or 'pollution prevention'. Each term promotes a hierarchy of environmental management practices in which preventative actions are preferable to treating waste after it has been generated. Therefore, experts and information sources were contacted if they were involved in activities that could be described using this vocabulary.

In addition, it was acknowledged that information itself does not guarantee improvements. First, environmental best-practice information for an enterprise varies depending on the industry sector, geographical situation, and the regulatory, environmental and management commitment to implement change. Second, industry needs to be able to process the information and to make changes. An information campaign should enable industry 'to ask the right questions'. This would, in turn, create a demand for information and thereby create incentives for local-level 'information gatekeepers' or intermediaries to develop and deliver more information.

This is highlighted by the fact that information is available but much of it provides 'solutions' to problems of which companies are not aware. In fact there was a consensus that companies receive too much information—little of it useful.

The information collection effort resulted in several kinds of information that are summarised below: issues specific to SMEs; success factors for reaching SMEs; what type of information do SMEs need; and recommendations for information providers.

Issues specific to SMEs

What is an SME?

Definitions of what is meant by SME depend on where the enterprise is located and on what standard of measurement is applied. SMEs have a 'small' number of employees, although 'small' in China may mean 150 people and in Denmark it may mean five. The European Commission officially defines an SME as having below 250 employees. Another way to define an SME is by its annual financial returns.

Globally, environmental problems faced and posed by SMEs are similar and they seem to be exacerbated in developing countries. SMEs cumulatively account for a large percentage of economic activity and, hence, have a major environmental impact. But their small size and isolated nature makes influencing their behaviour difficult. A small enterprise usually has limited access to necessary information on environmental issues, and a limited infrastructure to handle them. Also, they tend not to have the resources to evaluate and implement environmental strategies.

In the EU, for example, 70% of economic activity is carried out by SMEs (DG XIII: CEC 1996c), but a variety of factors, including difficulties in accessing information, keep a majority of SMEs from meeting applicable environmental regulations. In developing countries, the percentages of SME economic activity are similarly high, but further compounded by a less well-established regulatory structure. Furthermore, all the experts concurred that the most effective way to reach SMEs is through personal contact.[1] Hence, the potential workload for reaching SMEs is considerable. Experts noted that, although it is not possible to reach all SMEs, it is important to make priorities and start taking action. Recently, there has been a global recognition of the important role SMEs play and of the need to address them. A variety of international, national and local organisations have begun to target SMEs. The time is ripe to evaluate these activities and to make recommendations for improvement.

Disseminating information: getting wired globally

There were mixed responses and feedback regarding electronic information routes. Although few companies may currently use them, most experts agreed that they have a huge potential to reach industry. Associated factors include the raising of environmental awareness in industry in general, the advancement in electronic communication technologies and a reduction in costs, and the overall increase in environmental activities globally.

Regarding the role of the Internet, it is worthwhile to note that, globally, the Internet is changing the way communications operate. The potential to narrow the information gap is there, but it is still dependent on technology that is relatively expensive and often inaccessible in developing countries. Internet accounts are doubling in the industrialised world, but take-up rates are substantially lower in developing countries. For example, Finland and the US have more than one Internet host per 100 people, but 80% of the world still lacks telecommunication access (Woods 1996).

Use of communication technologies is growing in developed countries. In the EU, an estimated six in ten jobs depend on information and communication technology. These include transport, retailing and manufacturing as well as service industries such as banking, insurance and tourism. This has resulted in availability of the necessary electronic equipment even though it may not currently be used to access the Internet. One survey showed that 50% of industries in four EU countries have a computer, 75% in companies with over 26 employees (90% of these companies have a fax machine). Company size, rather than industry sector, appears to be the most important determinant. The phone and fax are now considered basic business tools. Taking account of the increasing rate at which people adopt new technologies, combined

1 See Chapter 1.

with the reduction in price of computer equipment and the cost of long-distance communications, it is predicted that electronic communication will probably play a greater role in information transfer in the future (DG XIII: CEC 1996).

Environmental information for industry (including SMEs) is already on the Internet. Numerous other information sources, including publications, training and technical assistance, and centres of capacity-building also exist. Currently, however, most companies, particularly SMEs, do not consult the Internet for information.

Developed as well as developing countries still rely on hard-copy and personally delivered information. Videos (especially targeted training materials) and information diskettes are commonplace and have been shown to be effective.

How companies currently receive information

The expert meeting compiled the model illustrated in Figure 1 to explain how companies currently receive information. Information efforts that do not take into account this flow are less likely to be effective in reaching or influencing their behaviour (UNEP TIE 1997).

Following this model, information campaigns are most effective in reaching SMEs when they use existing information routes through local-level information sources such as rotary clubs and local chambers of commerce. What was also apparent and worthy of note was that information campaigns from the international governmental level directed at SMEs are *not* effective. However, information campaigns on the part of international industry via supply chain management were more effective.

Information also flows among organisations at each level. In general, industry, regardless of size, trusts information coming from other industry representatives over that coming from governmental bodies. However, the exchange of information at the SME level is weak.

Survey responses confirmed this model (UNEP TIE 1997). Although experts noted that it is difficult to generalise about what works with companies in all cases, survey responses indicated, for example, that SMEs use or may use information from local business associations (84% of survey respondents), suppliers (78%), consultants (75%), and other industry/SME contacts (80%)—all sources are in direct contact with SMEs. Less effective in reaching industry were government agencies (international and national), banks, insurance companies, universities and mass media.

Experts compiled a general list of barriers that exist to reaching SMEs and changing their behaviour. These barriers include:

- The competing priorities within an industry often mean that preventative environmental issues are not at the forefront of industrial managers' thoughts.
- Companies will only seek information when they make the decision that they need it.
- Currently, there is a lack of personnel responsible for environment.

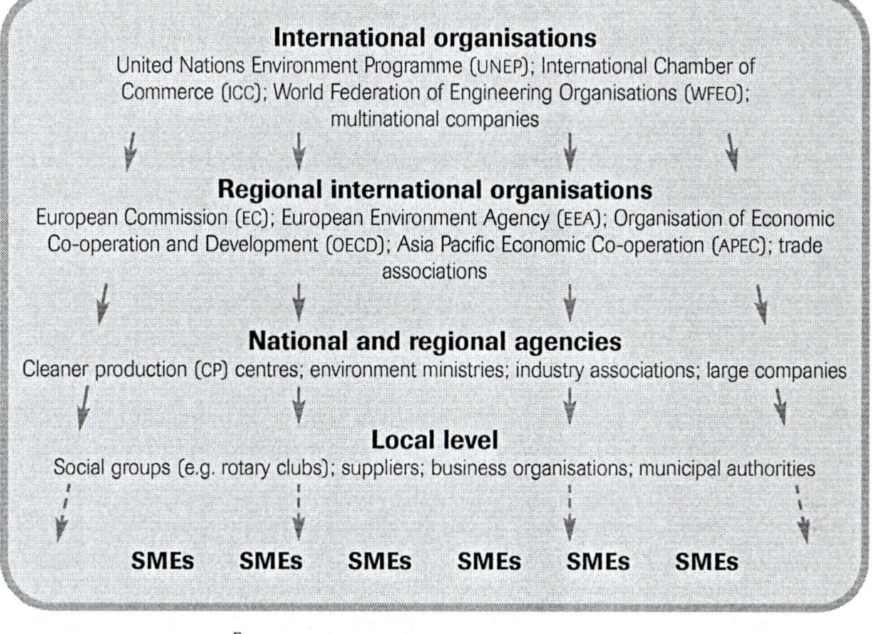

Figure 1: CURRENT INFORMATION PATHWAYS

- It is impossible to reach all companies, especially SMEs. Some information needs are very specific and may be too resource-intensive for an information provider to put together. Efforts made to reach SMEs should prioritise whom they wish to reach and commence action.

Success factors for reaching SMEs

Experts' conclusions on how to reach SMEs can be summarised as: 'Communicate the right message with clear next steps, personally, to top company managers and provide ongoing support while they are taking action.' In other words, the message must be formulated in the right language—using industry terms and motivations via the current routes through which SMEs directly receive information.

There was strong consensus that the only way to effectively reach and change the behaviour of SMEs is through personal contact. Personal contact made through existing, trusted routes of information dissemination appears to be the best path to follow. But the routes of dissemination or sources of information will differ depending on the local culture and nature of the geographical location and industry sector. In addition, consensus was clear that a need or demand within industry for the information must be established or it will be impossible to get and maintain interest. This commitment will define the success of the effort.

The challenge is to identify the 'agent of change' that has access to industry (and if they do not exist, to create them). These mechanisms can be used to deliver a simple, clear, message that highlights economic benefits, easy-to-do actions and to indicate where to go for more assistance. These 'agents' will vary greatly and therefore the role of local/regional national and international information intermediaries will need to be flexible enough to identify these agents and tailor a specific programme.

Additional points to consider include:

- The major concern of SMEs is the short-term financial bottom line. Selling the idea that environmental management can save money, reduce costs and increase efficiency is more favourably received.
- SMEs are compliance-driven and reactive. Reference to regulations and compliance could be effective, depending on the previous effectiveness of comparable regulatory schemes.
- Keep it simple. The message itself should be simple and clear about the problem, its solution and where to go for more information. The message needs to be general enough to pique the interest of a broad variety of companies and the proposed solution or suggested next action must be 'do-able'.
- The information should be perceived as 'reliable' and 'competent'. Information received directly from governmental authorities or 'unknown' organisations, such as international governmental bodies, will not be as effective.
- Start by promoting 'easy' changes that can be quickly implemented and show a result, and work up to more complicated, costly efforts. The kind of information that works best with companies is examples of how other companies have applied the principles, both in terms of the technical application and the management process.
- The clear, well-focused, practical message should be delivered through as many media as possible to reinforce its validity. Media routes include radio, mailings, TV and local publications. In addition, the more interactive the learning, the more effective it will be.
- Charge the 'appropriate' price for information. It is not possible to charge for general awareness information. However, as information becomes more specific and or technical, companies value information for which they pay more highly.
- Target company owners. Commitment is necessary from the top.
- Contact through government. Such routes either work or do not work depending on existing relationships. As implied earlier, some responses indicate that changing industry behaviour depends on receiving information

from a source that does not play a policing function. Many comments directly stated that information received from governmental bodies would be ignored. Other comments stated that the role of public authorities was important. Companies will 'take notice' and change their behaviour only when the failure to comply with legal requirements becomes costly to them.

- Respondents indicated that the least effective means for reaching industry is via general information with no follow-up such as general direct mailings or form letters. SMEs get too much information and 'time is money'. In addition, documents not in the language used or courses that take up too much time have little impact.

What types of information do SMEs need?

In general, experts agreed on the types of information needed. All the survey respondents felt that companies need at least some sector-specific information on how to improve environmental performance; 95% felt they need at least some information on where to go for more assistance; 90% felt they need at least some information on specific environmental issues. Companies require information tailored to their own situation as well.

In considering the appropriate format for information, experts felt that checklists, case studies and technical guides were useful. There were mixed results about the usefulness of electronic sources of information. Although the potential is great, there was no clear agreement that they reach companies. Other ways of effectively delivering information include: information centres, phone-in 'hot-lines', and government-funded training and technical assistance and demonstration projects.

Experts and survey respondents provided a number of additional comments regarding how to facilitate increased use of information by companies.

Help companies process information

- SMEs need help thinking through the information they receive.
- Companies should be given a snapshot of the 'big picture' of what their potential benefits might be, the environmental impacts in general and relevant environmental regulation.

What specific types of information are useful?

- Offer clear direction on what technologies or methods are the most relevant, available funding sources and help accessing them (for example in preparing proposals), the benefits that would accrue, and on suppliers.
- Offer company-specific case studies (benchmarking) highlighting the commercial benefits/opportunities for environmental management, the

potential for cleaner production in a given enterprise (e.g. specific inputs and energy consumption levels, product quality levels).

- Assist in writing policies and setting up environmental management programmes.

Recommendations for information providers

In addition to the recommendations outlined above, experts and the survey respondents provided recommendations for the different levels of information providers on how best to approach the SME issue.

General recommendations to all organisations

- Assess current activities in light of the study recommendations. The conclusions and recommendations in the study, both general and specific, provide guidance for how to reach industry, particularly SMEs. Organisations should analyse their current activities in light of the findings to determine where modifications are needed. The recommendations should also be used in defining new communication efforts.
- Monitor information dissemination mechanisms and modify activities accordingly. The Internet, as well as other emerging electronic means of communication, is changing the way the world exchanges and uses information. Although they have a high potential to reach and effect change, they are not currently used extensively. This, however, should be monitored and studied so that, when new forms of information exchange become widespread, they can be incorporated into information dissemination activities.

International organisations

International organisations such as the European Commission, UNEP, multinational companies or associations such as the International Chamber of Commerce (ICC) should:

- Focus on existing information intermediaries (or create new ones in the long term). For example, depending on information routes such as the Internet to reach SMEs is not effective because SMEs do not use the Internet to search for information. They prefer to go to local, established information sources, such as local chambers of commerce. Hence, if the Internet is used, it should be targeted at information intermediaries who, in turn, could communicate with SMEs. For example, Internet web pages focused on trade associations that outline the benefits of environmental best practices have greater potential for success. Supporting the association's ability to access and understand the information is an important element as well.

- Identify global priorities. Assistance can take the form of providing general policy guidance, information, and financial support. The information should be easily modifiable for national/local situations. In addition to general awareness raising, industry-specific information should be made available.
- Network among other international organisations and provide support to national and local initiatives.

National organisations

National organisations from both the public and private sectors, including ministries, industry and industry associations, information centres and non-governmental organisations, should:

- Increase efforts to adapt international policies and strategies to the national situation, transmit relevant information, ensure integration of international policies in national environmental policies.
- Network among other national organisations and provide support to national and local initiatives.

Local organisations

Local-level organisations, including chambers of commerce and local authorities, should:

- Deploy resources at the local level in order to ensure change in SMEs. Regional or local organisations should be flexible and creative to adapt to various situations and provide training and seminars, information, ongoing support and facilitate networking among SMEs and with industry associations.

Conclusions

UNEP TIE carried out this expert consultation to identify how to improve systems for communicating environmental best practices to SMEs. The results highlight that activities under way and those being designed should take into account how SMEs actually receive information and what kind of information works best to motivate behaviour change. Note should also be taken of the rapidly changing role of electronic information sources. UNEP TIE has made the study publicly available and it has provided the baseline of subsequent UNEP TIE SME work. For example, in September 1998 at the Fifth High-Level Cleaner Production Seminar, held in Phoenix Park, Korea, the report was the background paper for the parallel session held on improving the implementation of cleaner production in SMEs.

15
Environment, information and networks

How does information reach small and medium-sized enterprises?

Jane Hunt

The importance of disseminating appropriate information as a primary strategy for encouraging the greening of SMEs is virtually unchallenged. However, it is also widely recognised that getting the right information to the right person at the right time is a problematic task. This chapter discusses the dissemination of information to SMEs in relation to the networks in which they participate.

The idea that businesses require information to facilitate greening comes from an essentially economic perspective. First, it reproduces the assumption that greening is economically beneficial and therefore something that businesses should pursue. The question, then, becomes one of why businesses, as rational economic actors, do not take up the opportunities that are open to them; the answer is that they are prey to imperfect information or to an information deficit. The solution is to provide the missing information.

This model of business activity, however, leaves out many important factors such as the way in which information is perceived by the users, particularly with respect to the authority and motivation of the information provider, the activities that a business sees as appropriate or as priorities (which may have nothing to do with economic rationality but be a result of more cultural influences), and whether there is a perceived need for information.

This chapter explores some of these other factors, drawing on a study of business responses to environmental change carried out at Leeds University, UK, from

1996–98[1] (see Hunt *et al.* 1997; Purvis *et al.* 1999). This study involved semi-structured interviews with 60 managers with environmental responsibilities in the refrigeration, printing and baking sectors. The majority of firms were located in the UK, with a quarter being in France or Germany. Small, medium and large companies were equally represented. Unlike many other studies (particularly questionnaire-based research where response rates are notoriously low), very few managers refused to participate after being approached first by a letter outlining the study, followed up by a telephone call to arrange a time for interview; this suggests that perhaps business managers are more willing to talk than to write! More seriously, it emphasises the point that paper communications are often ignored by busy managers.

Information help-lines and services

Although the economic perspective provides a rather narrow understanding of the multiple factors that influence the choices and decisions made by SMEs, it nonetheless underlies and justifies the resourcing of substantial information campaigns. For example, the UK Environment Minister, Michael Meacher, has recently announced a new environment and energy telephone help-line, which provides information and advice specifically for SMEs. But, as SMEs themselves will recognise, there is rather more to the successful provision of information than simply making it available, although this is of course important. The Leeds research showed that, in line with other studies in this area, SMEs suffer from information overload and discount much that 'comes across the desk' without further consideration:

> It's a question of time . . . I just haven't got around to it . . . there are lots of things that absorb your time, but I'm one of those people that, really, you come in on the day, you see what there is to do, you get done what you have to and then the bloody time goes (factory general manager, medium-sized baker's).

> The problem is, from our point of view, we're that busy running the business; we tend to sort of let these things go (part owner, small printers).

Indeed, of the 60 interviewees in this study, only two had even heard of the long-standing Environmental Technology Best Practice Programme (ETBPP) help-line, and none had used it. The Small Company Environmental and Energy Management Assistance Scheme (SCEEMAS) is a government-funded system of grants which assists small businesses in establishing environmental management systems and pays part of the verification costs for EMAS registration; while many of our interviewees identified the provision of government support in the form of grants as being useful, none had

1 This chapter draws on an ESRC-GEC funded project award number L320253204 entitled 'Global Environmental Change and European Business: Global Atmospheric Change, Reactions and Responsibilities'. I would like to acknowledge my colleagues on the project Martin Purvis, Deborah Millard and Frances Drake.

heard of SCEEMAS.[2] Notwithstanding the excellent quality of the information provided by the ETBPP, both by telephone and in the form of free leaflets and brochures, if this is reaching only a small (though growing) minority of users, the question of how best to provide information to SMEs remains unanswered.

Just as importantly as lack of knowledge of the ETBPP programme, the Leeds study found that most information claiming to offer a service that would improve business was dismissed by SMEs as being driven by the provider's commercial interests, and even free services were seen as loss-leaders which would have consequent charges attached. This suggests that widespread and untargeted information campaigns are likely to have only very limited impacts. SMEs stated that they are more likely to be influenced by information for which they have a direct and immediate need, and that trust in the information provider is highly important. Trust in information providers, and relevance to their immediate needs as defined by SMEs, are thus two crucial factors in the successful dissemination of information.

Green business clubs

Partly in recognition of this, in 1992 the Advisory Committee on Business and the Environment (ACBE) recommended establishing a number of green business clubs, and 11 of these were subsequently set up, jointly funded by the UK Department of Trade and Industry (DTI) and the then UK Department of Environment (DoE). The rationale was that these clubs should operate as a forum for information exchange and mutual support for businesses interested in becoming more environmentally friendly. However, it has proved difficult to enrol businesses into these clubs and many of their original functions have now been subsumed into other agencies, such as Business Link. The one activity businesses in Leeds were enthusiastic about was attending seminars on imminent regulation.

This indicates the difficulty of establishing new networks of businesses and enrolling them into generalised, rather than specifically focused, activities. What it suggests is the necessity to consider existing business networks and how they can be utilised for the dissemination of information. The rest of this chapter goes on to do this.

Supply chains

The possibility of using supply chains to create pressure on businesses, for instance by buyers demanding environmental statements or environmental management system certification from suppliers, has recently come to the fore as a strategy for greening businesses. However, supply chains are also channels through which information flows, although they have received less consideration in this light. Moreover, supply chain pressure is generally thought of as coming from the customer—often a large company keen to establish its own green credentials—rather

2 SCEEMAS was reviewed and removed due to the poor uptake by SMEs.

than from the supplier. But it is apparent, both from our own work and from that of others, that suppliers are actively engaged in selling and directing customers towards specific purchases. Following from this, it is suppliers who are providing significant information sources as well as being the direct providers of new technological options:

> If the supplier came along and says, look, there's a free sample of ink here, give it a go, this is what they cost, then we'd probably try it . . . it comes from them, rather than us saying 'have you got . . .' (small printer).

It is supply networks that are used between businesses to carry information about products; they can be similarly used to carry environmental information, especially when this is attached to products.

Wholesalers are particularly important in this respect, as they will offer a range of products rather than promoting sales of 'own brands', and are a locus for information about product information and availability. Our interviewees confirmed this, telling us that it was wholesalers who would suggest new products and would advise on the best purchase for a particular use. Wholesalers are a crucial node in the extensive network of production and supply in which all businesses are embedded. Smaller businesses are much more likely to use wholesalers rather than buying direct from producers, although the structure of the supply chain varies in different sectors. As ideas such as Integrated Product Policy (IPP) begin to take hold, the role of wholesalers, as well as suppliers more generally, is likely to become more significant in the promotion of environmentally friendlier purchasing and the dissemination of information. However, this presupposes that wholesalers are well informed, and makes them an appropriate focus for information campaigns. A recent study of energy labelling on white goods showed that training for retailers on the meaning of the energy labels, and how to promote them, had a substantial impact on overall sales of the most energy-efficient appliances (Boardman *et al.* 1997).

Regulation and regulators

Regulation—and regulators—are a significant channel of information provision and one that is preferred or most often utilised by SMEs (Groundwork 1998).[3] Regulators provide direct information on compliance requirements and, in some cases, more detailed advice on the mechanisms that need to be instituted to ensure compliance. Regulators are also more likely to be aware of where and how regulation impacts on different sectors. However, there are also instances of regulators providing incorrect information and, in our study, of failing to encourage enquirers who showed an interest in compliance-plus, perhaps by redirecting them to the ETBPP. Nonetheless, our study also showed that, particularly in larger SMEs, there was a relatively close relationship between regulators and businesses, and the regulatory authorities were used as an information source both to achieve compliance and in planning future developments.

3 See the full survey results in Chapter 1.

This contrasts with some other studies which have shown that SMEs are unaware of the relevant regulation (e.g. Baylis *et al.* 1997), but we suggest that the form of the question asked may explain this difference—SMEs may not know the correct name for a piece of regulation, but they know (more or less) what they have to do to comply.

In line with the finding that business club activities were most successful when focused on imminent regulation, we found that SMEs were largely ignorant of forthcoming regulation. The only exception to this was when there was a development of existing regulation which directly impacted their business activities, such as the ongoing discussions at a European level of further restrictions on refrigerant gases, particularly HCFCs or the expected increase in landfill charges. This is partly a product of the different time-horizons that operate in business and regulatory domains: SMEs generally work at the most with a two-year plan, while regulation can take more than a decade to develop and implement. However, many capital commitments have longer lifetimes. Buildings, particularly in relation to energy consumption, provide an especially acute example where businesses could find themselves with high outgoings if they do not take account of likely future increases in energy prices when refurbishing or taking on new premises. The immediacy of much business activity, especially in smaller business without the managerial and administrative infrastructure to facilitate longer-term planning, pre-empts their adopting a longer time-horizon.

Coupled to this is the commonly voiced cry from businesses that they want to know about forthcoming regulations to reduce their uncertainties when planning for the future. Again, this is most acute when a sector has some experience with a particular regulation and expects it to develop further. This places regulators in a difficult position, given that regulation is negotiated over time between a number of parties, and that outcomes are hard to predict. Keeping abreast of ongoing discussions, however, was something that we found was the preserve only of managers with a direct environmental responsibility, and a budget to subscribe to relevant publications or information sources. Again, the immediacy of other concerns squeezes 'active' searching for information and keeping up to date out of the picture for most small businesses.

Perhaps most importantly, we found that regulation itself defined what was considered to be an environmental impact.[4] For example, in commercial refrigeration environmental issues were defined by all our respondents as to do with ozone depletion and climate change:

> There's two issues, obviously. There's the CFC issue, the ozone, and there's the greenhouse effect, the CO_2 emissions, so it's coming from both angles. So it's the amount of power we're consuming on-site, the amount of refrigerant we're pushing up into the atmosphere (medium-sized refrigeration company).

The logic seems to be that, if something is problematic, then the government will regulate it, and thus that if something is not regulated then it is not problematic. This thinking excludes the possibility of compliance-plus, not because businesses feel no

4 See Chapter 3.

responsibility for their environmental impacts (all our interviewees stressed that, at least in principle, they were concerned for the environment and believed businesses 'should do their bit'), but because they are simply unaware of the impacts of those of their activities that are unregulated, or at best consider them inconsequential—because, if they were a problem, they would be banned, wouldn't they? This reliance on the government to define what constitutes an environmental problem was coupled to a lack of any sort of systems thinking of the sort that underlies practices such as life-cycle assessment. While many of our interviewees made a connection between discharges to water and local river quality—a simple connection in a field that has been regulated for decades—they did not see a connection between their own energy consumption and climate change, and none identified transport as an environmental impact of their business. So, the answer to the question of what environmental impacts a business thinks it has is: 'Those things that are regulated'.

If regulation is the predominant way in which government informs businesses about environmental impacts, this presents a strong argument for increased regulation, given the comparative lack of effect of other attempts at information dissemination. Our respondents favoured regulation over voluntary agreements as it was seen to provide a level playing field; complying with regulation was seen as a form of citizenship similar to, for example, not breaking the law regarding theft, especially when the reason for the regulation was clear (such as not polluting local rivers).[5] Arguments for voluntary agreements tend to be made by large companies and trade associations (where small businesses are likely to be under-represented). For the SME, there is a good case to be made for more and stronger regulation both to provide a clear signal on what their environmental impacts are, and to promote equity. The same arguments can be applied to environmental taxation.

Trade journals and organisations

One information network that seems to have been under-used by those concerned with disseminating environmental information is the trade journal. These are widely read by businesses, small and large.[6] To keep up with what is happening in the industry, subscription to at least one trade magazine is essential. We examined the environmental coverage in trade journals of the three sectors studied and found that it mirrored businesses' own identification of environmental issues, similarly covering the environment as mainly defined in terms of regulation, which is the one environmental area consistently covered by the trades:

> We generally get to know [about environmental issues] from what we call *QPMA* [an equipment manufacturers' trade association journal] . . . they keep us informed technically about what's happening in the industry, what regulation is coming up for ovens . . . (medium-sized baking oven manufacturer).

5 Petts's research in Chapter 3 supports this assertion.
6 See Chapter 15.

> *Interviewer*: If you wanted to pursue some of the environmental issues in a more detailed way, where would you go for information about that?
>
> *Respondent*: BPIF [printing trade association]. That would be the only place I could think of going really, in the first instance. And then I'd expect them to pass me on to the relevant departments or whatever (small printer).

Baking trade journals had virtually no mention of environmental issues, reflecting the common perception of baking as a 'clean' industry and the lack of environmental regulation specific to the baking industry (food hygiene, in contrast, was a subject of great concern). The refrigeration journal showed a clear progression through the issue of CFCs and their substitutes to concerns with energy efficiency. Printing trade journals discussed the issues around recycled paper, sustainable forestry and the control of hazardous substances.

The trade press contributes to the way in which businesses define themselves and their concerns and, as such, could be highly influential. Certainly, it appears that advertising and the contribution of articles to trade journals by agencies such as the Environmental Technology Support Unit (ETSU) is likely to be a productive means of information dissemination.

Trade associations also organise a variety of local activities and networks, often incorporating social activities. These pre-existing networks seem to offer an underused resource to spread environmental messages and information. There has been some success using environmentally committed business people as speakers at events organised by trade associations: business speakers are generally trusted more than 'outsiders' and are familiar with the detailed practices of their sector. They can thus provide solid information on how to go about making environmental improvements and, in many cases, the cost savings that have accrued. They are also more likely to be believed than glossy pamphlets or 'exhortation literature':

> He's a well-respected sort of bloke that's been in business a lot of years, so whatever he's talking about I'll go and listen (part-owner, small printer).

> We did [establish waste minimisation] and so we explain it to other people; its what we do . . . the trouble is it's a bloody marketing exercise: I hate it when the marketing guys grab hold of something and want to say it's something that it's not (medium-sized baker).

As well as sectorally based trade associations, many SMEs are members of local traders' associations. These tend to have a stronger social component and a stronger awareness of specifically local issues, such as planning development. They thus provide a forum for providing information on local environmental issues and for developing initiatives that can be best tackled at a local level, such as waste management.

Sectoral networks

Entwined with, and facilitated by, some of the networks discussed above are sectoral networks. SMEs tend to identify strongly with other businesses in the same industry, more so perhaps than larger businesses involved in several different industries. The trade press and trade bodies are organised sectorally; regulation often requires different practices in different sectors, and purchasing and supply networks are structured in relation to their sectoral context. The social structure of sectoral networks is organised through trade bodies and the supply chain. As noted above, different sectors tend to have different 'environmental identities', that is, they perceive their environmental impacts as to do with sectorally specific activities. What is especially relevant when considering the provision of environmental information on a sectoral basis is to recognise the different factors applying to different industries; one of the strengths of the ETBPP is its sectoral expertise. Information on generic environmentally friendlier practices such as energy efficiency and waste minimisation—and the as yet largely invisible business transport issue—cannot go very far before it needs to engage with the specific needs of different sectors. Tailored information for different industries, disseminated through trade groups and the supply chain, is more likely to be useful than generic information distributed through across the board mailshots.

Finding the right information

Coupled to this is the difficulty of finding the relevant information to facilitate environmentally friendlier action, the importance of being in touch with the relevant networks, and in accessing information at the right time. One interviewee, with reference to finding a waste recycling contractor, commented:

> [We've] torn our hair out a few times, yes, and it's been a lot of talking to people and, I know a man who does and, you know, networking type thing to get to know. [We've] had a bit of time doing that but that's been the frustrating part and the time-consuming part is getting to know. Now once you know then it's easy . . . Once you're in, you're in, you are in the club (medium-sized baker).

With reference to energy consumption in a new building, another interviewee observed:

> We employed an energy consultant to come in and give us advice . . . he gave me some useful information but unfortunately we'd already done most of it . . . because we're printing, we had to make sure the lighting was spot on, and he came in after we had put all the lighting in, and he recommended these like foil sort of things for the tops to make sure the light went down (small printer).

These examples emphasise the need to integrate information provision with the immediate needs of business—in these examples, in relation to new regulation, and in relation to new premises. These are points at which businesses are actively seeking information, and thus offer an opportunity for information providers.

From information to action

Implicit in information provision is the assumption that, once information is received, action will follow. Research in the public sphere has shown that it is extremely difficult to separate the impact of information on behaviour from other influences; what is clear is that there is no simple relationship between information provision and action (Tietenberg 1997).[7] From the point of view of the SME, the most effective information is relevant, accessible, and provided at a time when it is needed. This presents difficulties for the information provider, who can only meet these SME requirements if a specific request is initiated by the SME. Nonetheless, the finding that regulation on the brink of implementation generates an information need in business offers clues as to how best to organise information provision. Simply put, if a need for action, along with the associated need for information, is generated—through new regulation or through changing taxation—and then that need is met through dissemination of sectorally specific information via the networks in which businesses are already engaged, then the chances of that information being used are relatively high. A carbon or energy tax, for example, coupled to information provision in the trade press about how to increase energy efficiency in that industry, is very likely to produce larger energy savings than generic exhortations and examples that may seem irrelevant.

Conclusion

The simple economic assumption that information leads to economically rational behaviour fails to take account of the real experience of SMEs. Studying this experience raises issues about trust in the information providers, the relevance of the information provided at the time, and the way in which it supports or challenges business identity and perceptions of the environment. The networks in which every business is involved—supply chains, trade and sectoral associations, regulatory compliance—are key routes for the dissemination of information. What is encouraging is that many SMEs do acknowledge an environmental responsibility; what is discouraging is the limited changes to business practice that result from this generalised responsibility. Providing relevant information to interested businesses at the appropriate time is a challenge for all information providers. This chapter offers some clues as to how that might be done.

7 Results in Chapter 15 confirm this point.

16
Local authorities in dialogue with small and medium-sized enterprises

Charlotte Pedersen

The aim of this chapter is to describe the possibilities of getting local authorities to support SMEs' environmental improvements. The experience outlined in the chapter is based on Danish industry structure, environmental legislation and specific pilot projects. It shows that co-operation between local authorities and SMEs on cleaner production and environmental management creates incentives and conditions for environmental improvements. By using different kinds of tools, the local authority develops a more advisory and supportive role towards companies.

The chapter is structured in four main parts: the industry sector in Denmark; the main features of Danish policy and environmental legislation; Danish case studies/projects that have been carried out with the local authority as a co-operation partner for the SMEs; and the local authority as a dialogue partner.

The industry sector in Denmark

The industry sector in Denmark supports a total of 500,000 jobs, 25% of which are in trades and 75% in industries. The main sectors in terms of gross domestic product (GDP) are: iron and metal works (35%); food, drink and tobacco (19%); chemicals (14%); paper and printing (10%); and wood and furniture (5%). Danish enterprises are small by international standards, 80% of jobs being in firms with fewer than 500 employees. Fewer than 300 enterprises have more than 500 employees. The three largest sectors (in terms of GDP) also encompass the largest polluters, most of the problems being attributable to the discharge of trade effluent. There is little heavy industry in Denmark and, hence, no pollution from associated mines and blast furnaces (Moe 1995).

The relative contribution of SMEs[1] to total industrial environmental impacts is unknown, but is likely to be considerable given their contribution to total production and their dominance in some sectors.

Making environmental improvements takes time, and, if SMEs are to obtain an environmental performance beyond what is required by law, it will require both resources and knowledge to which many do not have direct access.

An analysis of 200 SMEs within the industrial production, construction and service professions by Håndværksrådet (the Federation of Crafts and Smaller Industries) in Denmark (SMV-analyse 1998) has been undertaken to determine the general attitude of SMEs to environmental demands, their handling of these demands, and their attitude to green taxes and duties.[2]

The analysis shows:

- SMEs are very aware of environmental matters with 40% thinking that environmental matters will continue to play a growing role as a competition parameter. Three out of four enterprises confronted with legislative environmental demands find them reasonable. It is a widespread observation by the surveyed SMEs that the business sector should participate in protecting the environment. Every third enterprise has, during the past three years, invested in environmental or energy-saving measures.
- A quarter of the SMEs are familiar with environmental management and approximately 15% have adopted environmental management practices; 20% of the enterprises without environmental management have more informally defined environmental targets. But the analysis also shows that those enterprises with more than four employees are familiar with or have introduced environmental management. Many enterprises with fewer than five employees think they know how to handle environmental matters without any specific environmental management tools.
- Two out of three enterprises with environmental management make environmental demands of their suppliers, thereby assisting in the spread of environmental improvements throughout the business sector. A further 25% of the enterprises without environmental management make environmental demands of their suppliers.

A conclusion of the survey is that there is a need for work to focus on assisting SMEs in achieving environmental improvements. The enterprises have expressed an interest in this, but information, tools and guidance need to be supplied for success.

1 An SME is understood as a company with fewer than 500 employees.

2 See Chapter 1 for an analysis of 300 UK SMEs.

Danish policy and legislation

The first Danish Environmental Protection Act in 1973 was primarily aimed at regulating the local environmental effects resulting from industries. An administrative apparatus was developed to implement and administrate this Act. The Danish Environmental Protection Agency (EPA) was established as the national administrative element and the 14 counties and 275 municipalities were chosen to be the implementing elements. The counties are responsible for the 10,000 companies believed to be the largest polluters while local authorities are responsible for approximately 120,000 companies.

With the amendment of the Danish Environmental Protection Act in 1991, it became possible to introduce a number of new instruments to promote the reduction of environmental impacts from industries. For instance, there are environmental agreements, eco-label schemes, voluntary consultancy arrangements, and an extension of the use of financial management tools—including refund fee/discount arrangements and waste disposal fees. In addition to this comes the establishment of the voluntary EU Eco-Management and Audit Scheme (EMAS) for industrial enterprises, and the introduction of a number of environmental duties, fees and contribution arrangements.

Subsequent legal initiatives, concerning green accounts and the duty to inform the authorities when exporting second-hand production plants from special environmentally significant enterprises, have focused attention on the involvement of the population through increased public insight into the enterprises' environmental conditions and arrangements.

Approval scheme

The approval scheme has been valid since 1974 and comprises enterprises which, according to today's standards, are considered potentially very polluting as well as enterprises with very complex environmental problems that require special expertise. The scheme implies that this type of enterprise may not start, expand or change its production in a way that increases pollution until this has been approved. Approvals for enterprises established before 1974 are being gradually phased in so that, by 2002, all will have an environmental approval. A total of 7,000 enterprises will then be covered by the approval scheme.

Notification scheme

Most enterprises are covered by the Environmental Protection Act's notification scheme. This requires around 14,000 enterprises to notify the environmental authorities before establishment or material extensions of processes. The notification enterprises are only met with specific environmental demands in cases where the environmental authorities (after a short assessment) find that the enterprise poses a pollution risk. The same goes for the kind of enterprises that are included in neither the approval nor the notification schemes.

Cleaner production

Denmark has a clear policy and strategy for cleaner production, and many initiatives have been started within this strategy.[3] The preventative strategy was introduced in 1983 and appeared for the first time in legislation in 1984. From 1992, the Environmental Protection Act introduced the principle of cleaner production and committed local authorities to engage actively in promoting cleaner production in any environmental activity involving private companies (Ministry of Environment, Denmark 1997).

The cleaner production policy has been implemented through a strategy with various financial programmes and action plans. In 1999, the EPA launched a programme for product development with cleaner production as one element. The change in strategy is based on the assumption that significant and increasing environmental impacts are caused by the production, use and disposal of products.

Local authority dialogue with SMEs

The Environmental Protection Act emphasises the development of tools to prioritise environmental protection work 'and the establishment of a basis on which the environmental authorities can co-operate with companies and their organisations to a far greater degree than before'.

In respect of this new law, it was recognised as an essential component to have the local authorities create new roles and design a range of instruments *vis-à-vis* companies so that local authorities may meet legislative requirements and co-operate more extensively with companies than in the past. However, the Act does not propose any concrete tools or methods to be used for this co-operation.

There could be differences in the form of co-operation and the methods and tools that could be used for small and large companies and for environmentally positive and negative companies. This difference is illustrated in Table 1 which shows the relationship between companies' attitudes towards the environment and the management tools that the environmental authorities could employ.

Not all companies should be dealt with in the same way. In other words, the management and tools will vary from company to company in reaching the same target (cleaner production and environmental improvements). Companies with a positive attitude to the environment are implementing environmental management systems anyway and can be left to get on with it. Those with a negative attitude will need to be confronted with rules, regulations and intensive supervision. This results in more differentiated and more effective local authority environmental supervision (see Table 1).

In order to enable the participating local authorities to effectively stimulate cleaner production in their city or region it is important to know more about the dialogue around companies' environmental performance and the various players involved. Figure 1 shows some important variables in this dialogue (Pedersen 1994).

3 See Hobbs's discussion on cleaner production initiatives in Chapter 11.

	Company with a positive attitude to the environment	Company that merely complies with legislation	Company with a negative attitude to the environment
Development of rules and requirements	• Dialogue on specific issues • Framework permits	• Dialogue on rules at industry level	• Dictation of rules and regulations
Enforcement and supervision	• Environmental management • Spot checks	• Standard supervision • Environmental management and spot checks	• Intensive supervision and any necessary use of powers

Table 1: COMPANIES' ATTITUDES TOWARDS THE ENVIRONMENT IN RELATION TO THEIR USE OF MANAGEMENT TOOLS

Both the company and the local authority are described as organisations characterised by four variables:

- The tasks the organisation has to perform
- The structure of the organisation—expertise, management hierarchy, procedures, etc.
- The technology or methods used to perform the tasks (tools)
- The staff involved and their roles, skills, etc.

These four variables are all interdependent which means that a change in one may well lead to compensation or counteraction in one or more of the others. For example, when the organisation's tasks are changed, a change in structure, technology and staff skills should also be considered. In this organisational concept, the stimulation of environmental improvements and cleaner production is seen as the task of the organisation. The tools will therefore be the technology and the procedures used by the organisation. In addition, the staff will also require the necessary skills in, for example, environmental expertise, project management experience and knowledge of business management/economics.

In this field, counties and local authorities have the possibility of supplying the enterprises with resources and guidance especially concerning the parameters dealing with technology/methods and staff. Information and knowledge about legislation, cleaner production alternatives, contact persons, methods, etc. can be disseminated in many ways and training courses for a company's staff is an obvious way.

When working with companies, it is important to recognise that the prerequisites for success include goodwill, expertise and knowledge within the organisation of both the local authority and the company. Both organisations are characterised partly by:

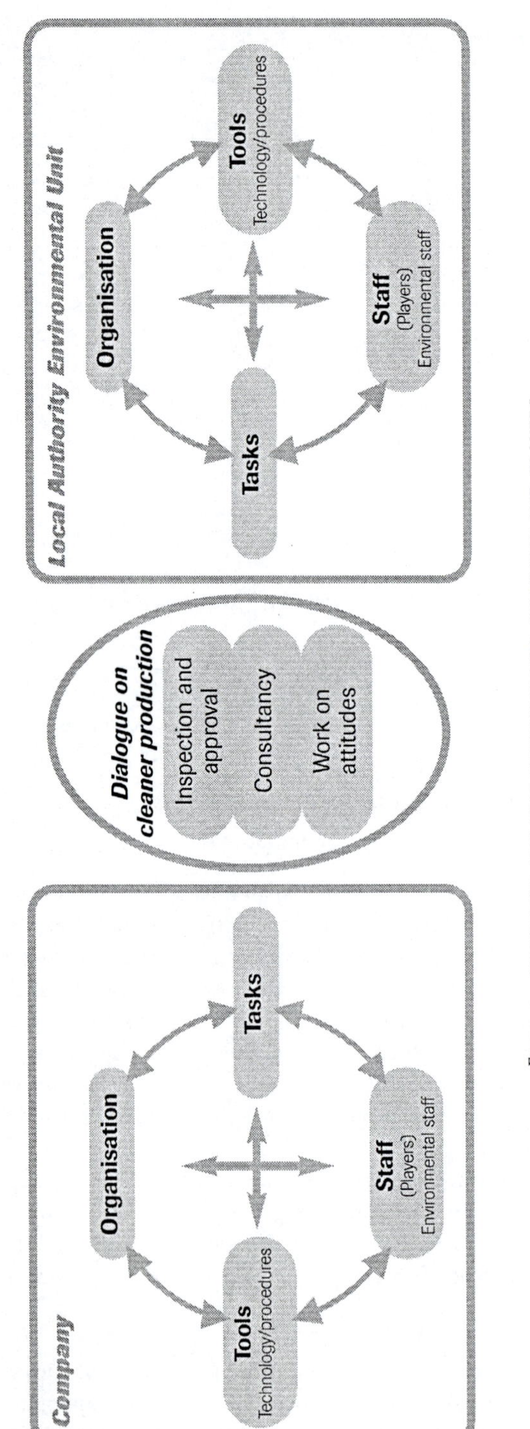

Figure 1: COMPANIES AND THE ENVIRONMENT: DIALOGUE AND PLAYERS

- The organisation that has been tailored to the tasks that have been performed up to now
- Staff with different skills
- The organisation of work in the form of a management and expertise structure, and finally
- The tools used to perform the tasks

The preceding organisational parameters together define the organisation's culture.

The objective is the same for both organisations: namely environmental improvements and the introduction of cleaner production in the company. At the same time it is important to realise that both organisations have tasks and roles in areas other than cleaner production which may at times appear to be in conflict with it. This is one of the reasons for involving more players or intermediaries in this dialogue: for example, customers, suppliers, local authorities/counties, authorities in health and safety and management and labour force.

The local authority in Denmark has turned out to be a suitable partner for SMEs. Experience shows that some of the advantages of using the local authority as the initiator and catalyst for environmental improvements are:

- The local authority knows the local business community.
- The permitting department of the local authority is responsible for the local environmental policy and the issuing of permits to the companies.
- The permitting department of the local authority has knowledge of environmental legislation, cleaner production alternatives, experience from other companies, etc. that could be disseminated to SMEs.
- The department of economic affairs has an important role in promoting a healthy local economic climate in which the companies can flourish. In addition, this department is responsible for attracting new companies to its city.
- The local authority will have to agree with the reconstruction and building activities of companies and is, therefore, in many ways the logical partner of the local business community.
- The local authority is not only familiar with the local business community, but also with other local intermediaries that could help promote the proposed environmental activities, such as banks, insurance companies, chambers of commerce, etc. (BECO Environmental Management and Consultancy and Deloitte & Touche Environmental Services 1998).

Several projects have been completed in Denmark on how local authorities and counties can help SMEs implement cleaner production projects and introduce environmental management by adopting a more guiding role. Some examples are listed in Boxes 1–4.

Participants/ initiative	There are approximately 30 known environmental networks or green networks around the country. Initiative is typically taken by the local authority or the county, often with a large company in the municipality. Other participants: ▸ Companies ▸ Business clubs ▸ Industry organisations ▸ Environmental organisations One network, ELM (Erhvervslivets Ledelsesforum Miljøfremme [The Business Society Management Forum for Environmental Improvements]) is only for businesses.
Role of the local authority	Initiating the network, often hosting the secretariat and planning the meetings. Sometimes co-financing.
Objectives	The objectives of the different networks are to disseminate information on: legislation; experiences with cleaner production and environmental management; and to initiate work on cleaner production in more companies.
Organisation and structure	The local authority or county is often secretary for the network, while in some networks there is a formal board with private as well as public representatives. The first network was founded at the beginning of the 1990s.
Finance	Some networks are solely public (local or county) financed while more and more are being established as limited liability companies where different kinds of companies, organisations, municipalities, etc. can buy shares.
Project description and methodology	The network typically holds a meeting after work for two to three hours, approximately four to six times a year. The meetings are often combined with company visits to one of the members. Presentations are given by the members (private/public) and by external experts. The activities of the network can vary between: ▸ Meetings with presentation and discussion on specific subjects ▸ Company visits ▸ Newsletters ▸ Organising waste exchange ▸ Certification to a specific network standard ▸ Training ▸ Advising on EMAS or ISO 14001
Instruments	Instruments used to get the network going are: ▸ Voluntary membership ▸ Low financial cost for participating companies ▸ Knowledge dissemination ▸ External inspiration through guest speakers ▸ Building a network between companies and other players

Box 1: ENVIRONMENTAL NETWORKS

(continued opposite)

Experiences/ anecdotes	The idea of the environmental business network is growing around the country. It is very popular among the authorities. So far, there has not been a systematic evaluation of the experiences, environmental results, etc. of the networks. One network (Green Network) has been evaluated and one conclusion was that the companies want closer contact with the authorities. Recently, the Danish EPA has co-financed a secretariat for all networks in Denmark (Miljøforum Danmark). The EPA has financed 50% and Green Network has provided the remaining 50%. The purpose of this secretariat is to disseminate information to and from the local networks and initiate co-operation and new networks. Miljøforum Danmark will also act as the focal point for contacts with international networks such as, for example, the European Roundtable on Clearer Production.
Results	Participating in the network is voluntary and the local authority/county can use the network as an instrument for awareness raising, training (both for companies and authorities), improving business contacts, etc. So far, the network concept seems a success judging by their rapid expansion.

Box 1: ENVIRONMENTAL NETWORKS
(continued from previous page)

Participants/ initiative	This is an initiative of NALAD (The National Association of Local Authorities in Denmark). Other participants are: ▪ The environmental departments of nine local authorities (Frederikshavn, Gladsaxe, Hvidovre, Herning, Lyngby Taarbaek, Noerre Rangstrup, Nexoe, Nakskov and Ry) ▪ Selected companies and industries of between one and 115 employees, except for one company with 580 employees ▪ Consultants
Role of the local authority	The nine local authorities were project managers of their own project, conducting the role of environmental authority in the dialogue with the company in a new way.
Objectives	The aim of the project has primarily been to create new roles and design new instruments in order to introduce cleaner production in private companies by local authorities. Also, various potential co-operation partners, which local authorities may successfully enlist in the dialogue, have been tested. Co-operation efforts targeted other public authorities and inter-municipal companies, counties, the Danish Environment Service, the Occupational Health Service, and private consultants.
Organisation and structure	The project was carried out in 1993–94 and the total project was organised in a steering committee (NALAD, EPA, local authority representatives, industry organisation, etc.) as well as a project group (NALAD and consultants). A working group was created with members from the environmental department of the authority and the production unit of the company. The consultants were connected to the working groups as advisers.
Finance	The project was financed by the Danish EPA and the local authorities. The companies contributed their time. The total budget was €426,000.
Project description and methodology	The project was carried out through the following steps: ▪ Defining the different roles of the local environmental authorities. The project ended up with three role categories: inspection, advice and information. All three derive from the legislative role which remains the sole legitimate means of access to the companies in general, and company environmental performance in particular—this means that the authority role forms the basis of the other roles. ▪ Each authority defined its own role and selected the company(ies) for the project. ▪ Consultants developed different instruments to be used (tested) in the dialogue between local authorities and companies. ▪ Training sessions for the local authority staff before the dialogue. ▪ Testing different instruments (having the dialogue) in the local authorities/working groups coached by the consultants. ▪ Project evaluation and reporting.

Box 2: KURS* PROJECT
(continued opposite)

* KURS means, in Danish, the work of developing cleaner production strategies in the local authorities.

Instruments	• Participating in the project was voluntary for both companies and local authorities and the only cost was the time put in. • The project consultants were available for free for a certain number of hours. • There were possibilities to create networks and disseminate experiences. Specific instruments for the dialogue were developed during the project and tested as well. Such instruments may be industry-targeted supervisory campaigns, information exchange on cleaner production methods, environmental courses for company leaders and employees, preparation of specific industry-oriented information, or advice on cleaner production.
Experiences/ anecdotes	The experiences indicate that it is essential that local authorities are conscious about the conflicts of interest inherent in any co-operation; these conflicts are unlikely to disappear. Seen in this light, it is important that the mutual relations and interests of the co-operating partners are clearly described. This will enable them to establish unambiguous limits to the level of expectations.
Results	The conclusion reached in this project is that it is necessary for each local authority to define its scope of advice, level of competence, and knowledge, and to identify to what extent external competence is required. The project has clearly demonstrated that some of the local authority strong points include being the dynamic and leading partner in this process, as well as being the co-ordinating partner in charge of establishing co-operation with other relevant players. In testing the new roles, the participating local authorities have succeeded in expanding the traditional inspection role and in establishing closer ties with companies. None of the local authorities has had a pure role with respect to the companies, e.g. those who have maintained their inspection role have also engaged in advice and information. The experience of local authorities indicates primarily that one instrument rarely works on its own. Also, available experience indicates that the instruments should be applicable in an existing organisation on a technical administration level.

Box 2: KURS PROJECT
(continued from previous page)

Participants/ initiative	This is an initiative of NALAD. Other participants are: ▪ Central Union of Local Authorities of Greece (KEDKE) ▪ London Borough of Sutton, UK ▪ Consultants ▪ Local authority associations, local authorities and SMEs (all more than 50 employees) from Denmark, Greece and the UK
Role of the local authority	Facilitating the process towards EMAS, together with the companies and testing the guiding role in this process. At the same time, the local authorities took part in the training in EMAS and environmental management systems.
Objectives	To explore the scope for co-operation between local authorities in Europe and local businesses in terms of environmental management, specifically testing the applicability of EMAS in the process. LACE stands for 'Local Authorities helping Companies implementing EMAS'.
Organisation and structure	The project was carried out from 1995 to 1997 under the project leadership of NALAD. The total project was organised in a European steering committee and in national groups with representatives from the local authorities and the participating companies.
Finance	The project was funded by the European Commission under the LIFE programme and by the Danish EPA. The participating parties invested time. The total budget was €973,000. Funding from the LIFE programme was €485,000.
Project description and methodology	The project contained the development and testing of a manual for companies on how to implement EMAS and a guide for local authorities on how to participate and help the companies in this process. The project was divided into steps, parallel to the seven steps in EMAS. The main ingredients of the project were several European and national workshops with introduction and training for both local authorities and companies to the different steps in EMAS. Between the workshops EMAS was implemented at the companies and the local authority's role was tested in this process. The project results have been reported in: ▪ A guide for those local authorities wanting to promote EMAS to SMEs. ▪ A promotional brochure designed for those SMEs that have yet to make a decision on whether to use EMAS. ▪ A manual written for SMEs that have already taken the decision to use EMAS and which need practical assistance. ▪ One international conference in Brussels, two Danish workshops, one Greek and one UK workshop.

Box 3: LACE PROJECT
(continued opposite)

Instruments	Instruments for carrying out the project were: ▪ Funding ▪ Voluntary participation by both companies and local authorities ▪ Support from consultants ▪ Development of a manual ▪ Development of a guide for local authorities ▪ Training sessions ▪ Dissemination of experiences both at national and European level
Experiences/ anecdotes	▪ The local authorities have a wide range of existing environmental expertise which can be of practical use to companies wishing to improve their environmental performance. ▪ It is important to find the right mixture of roles and responsibilities to encourage the maximum number of companies to improve their environmental performance. ▪ In order to advise on EMAS, local authorities need to develop an understanding of the spirit, requirements and technicalities of the scheme—this takes time. ▪ Keep it simple when communicating environmental management principles and techniques to companies. ▪ Local authorities need to understand the real opportunities and barriers confronting SMEs in order to be able to advise effectively on the implementation of EMAS.
Results	The project showed that there are many ways in which local authorities can encourage the use of environmental management systems to improve the performance of local companies. In the project, several local authority inputs on the seven steps in EMAS were identified, e.g. information about legislation as the background for the environmental review, comments on draft versions of the environmental statement, examples of other company environmental policies and examples of good practice. The methodology and tools in the manual were developed by taking the difference in national legislation into consideration and are, therefore, applicable in Greece, UK and Denmark, as well as in other European member states.

Box 3: LACE PROJECT
(continued from previous page)

Participants/ initiative	This is an initiative of the Aalborg local authority, the permitting department. Other participants are: ▸ The local authority department on energy supply and other utilities ▸ Companies (all SMEs with 10–50 employees) There are no external consultants.
Role of the local authority	Initiator of the project and catalyst or guide in the implementation of EMS. At the same time, the local authority still worked as the inspecting authority, although inspections were carried out by others than those involved in the project. For some companies the local authority functioned as the chair for meetings, etc. The manual that developed as a tool in the project is designed for the companies own implementation, with not too much involvement from the local authority. This is due to the authority's limited resources.
Objectives	The objective of the project is to create environmental improvements through the implementation of environmental management systems (EMSs) and energy management in companies in the municipality of Aalborg. Besides this, the local authority wants to: ▸ Collect knowledge on EMS ▸ Change the order of priorities in the inspection work—more preventative work towards environmentally positive companies—and less inspection ▸ Reach the SMEs with guidance on energy ▸ Test new ways for co-operation with the local companies
Organisation and structure	The project was initiated in 1993 with three pilot projects. It is organised by a steering committee with representatives from the two departments in the local authority.
Finance	Annually, the local authority has spent 0.7 person-years on the work in the first few years of the project, in addition to the resources spent on the inspection work. Besides that, the local business council has sponsored some parts of the project. The local authority has spent 0.3 person-years on the project in 1998.
Project description and methodology	A pilot project was initiated in 1993. This resulted in the development of company agreements. The agreements are voluntary and imply that the companies should work after a specific concept on implementation of EMS and strengthened co-operation between company and local authority—with less of an inspection role on the part of the authority. Based on the pilot project, the concept was improved and developed as a manual on EMS in SMEs. In 1996 the manual was tested and adjusted in co-operation with six companies. The manual is now being offered to companies interested in environmental improvements with the intention of increasing the numbers of companies involved.

Box 4: COMPANY AGREEMENTS, AALBORG LOCAL AUTHORITY
(continued opposite)

Instruments	Through the project the local authority has used several instruments: • Voluntary agreements which commit the company to implement EMS, involve the local authority in the work and produce a yearly statement on its environmental status. • No cost for the company besides time. • Manual on EMS for SMEs. • Revised inspection procedures. • Advice from local authority staff on significant aspects and the environmental programme in the implementation of EMS. • Information and update on environmental legislation. • Yearly meeting on significant aspects and the environmental programme. • One responsible project manager and contact person at the local authority.
Experiences/ anecdotes	• The project has in several cases identified non-compliance at the companies. This has not, however, caused any problems between company and local authority because all the companies are interested in solving the problems. • The project depends on participation of environmentally positive companies. • There have not been any problems with the role of the local authority as both the authority and the catalyst. • Preconditions for the project/co-operation are mutual trust and openness. • The local authority had several ideas for improvements at companies.
Results	The local authority evaluates that the objective is reached. It now has a manual and a concept for voluntary agreements for further work with EMS in SMEs. The local authority has tested the catalyst role and gained some very positive experiences. Through the project, the local authority has gained improved knowledge of the companies and their environmental impacts and production—knowledge that had not been collected by traditional inspection work.

Box 4: COMPANY AGREEMENTS, AALBORG LOCAL AUTHORITY
(continued from previous page)

Conclusion

Experience from Denmark shows that most SMEs are aware that environmental issues have a significant importance for their future situation in the market. In that sense they should be motivated to work towards a proactive environmental profile.

Many SMEs will derive significant advantage by forming a partnership with the local authority. The authority can support the SME with a wide range of environmental expertise, e.g. by providing examples of cleaner production alternatives, experiences from other companies and contact with knowledge centres.

In order to ensure success, it is necessary for each local authority to define its scope of advice, level of competence and knowledge, and identify to what extent external competence is required. In testing the new roles, Danish local authorities have succeeded in expanding their traditional inspection role and have established closer ties with companies.

The limited resources at the authorities disposal makes it important to find the right mixture of roles and responsibilities to encourage the maximum number of companies to improve their environmental performance.

Local authorities need to understand the real opportunities and barriers confronting SMEs in order to be able to advise effectively on the implementation of EMAS—and when communicating environmental management principles and techniques to companies it needs to be kept simple.

17
The mentoring of small and medium-sized enterprises

Insights from experiences in the United States

Walter W. Tunnessen III

Businesses of all sizes face a variety of challenges while working to succeed in the marketplace and as environmental stewards. Companies seeking, at a minimum, to meet government environmental regulations often find that the pathways to compliance are not always well marked. Other companies, looking to exceed mandated standards and maximise the economic benefits of more efficient, less wasteful production processes, frequently have dozens of questions about where to begin. Pursuing environmental excellence can require significant amounts of information and resources that some companies—especially smaller ones—simply may not have. Mentoring programmes can help bridge this resource gap and pave the way for small and medium-sized companies (SMEs) to become environmental leaders.

Mentoring is not a new concept in business training. Historically, most trade and business skills were cultivated through the mentoring relationship established between the master and his apprentice. This tradition has been continued in the US and elsewhere by business development and assistance programmes that recruit older, more experienced professionals to engage less experienced professionals as protégés and provide them with expert guidance and counsel.[1] Under this mentoring arrangement, the protégés benefit from the veterans' advice while the mentor receives recognition and gains the satisfaction of having provided a valuable service.

1 The Service Corp Of Retired Executives (SCORE) is a non-governmental organisation that works closely with the US Small Business Administration to provide such services. SCORE has 389 chapters in the US and Puerto Rico, with over 12,000 volunteers.

Formal and informal networks for peer-to-peer assistance, also a form of mentoring, have long been used by business to exchange information and practices. With the advent of environmental regulatory systems over the past 30 years, larger companies and industrial sectors interested in sharing best management practices have formed numerous peer-mentoring programmes.[2] The use of mentoring to engage SMEs in initiatives aimed at improving environmental performance is, however, a relatively recent development.[3]

Increased concern about the environmental impacts of SMEs by both governments and corporations is driving interest in non-regulatory approaches towards improving the performance of companies either outside the current regulatory arena or within the corporate supply chain.[4] The appeal of mentoring is that its can provide a non-threatening, low-risk, low-cost and effective means of introducing small companies to concepts and strategies for achieving better environmental results.

Over the past few years, innovators in business, government and non-governmental organisations (NGOs) have developed a range of approaches towards mentoring in the context of environmental management. In addition to the classic one-on-one model of mentoring and peer-mentoring, there is now a creative mix of programmes led by companies, trade associations, government agencies, public/private partnerships, and NGOs that are reaching 'mentees' in specific regions, industries, through supply chains and other networks.

This chapter provides an overview of environmental mentoring. It describes basic types of mentoring programme, reviews the rationale for mentoring SMEs, highlights the benefits of mentoring, offers some elements for success, and provides a few examples of specific programmes. The chapter draws primarily from mentoring experiences in the US; however, the information is applicable in other parts of the world.

What is environmental mentoring?

Environmental mentoring is simply the use of expertise to help another entity improve its environmental management and performance. The basic goal of environmental mentoring is to provide help that enables the mentee to achieve and maintain

2 The Global Environmental Management Initiative (GEMI) is an example of a peer-to-peer assistance programme established by large corporations from different industries. The Chemical Manufacturers' Association's Responsible Care initiative reflects an industrial-sector approach to peer counselling and benchmarking.

3 This is the case in North America (USA, Canada and Mexico). The information in this chapter primarily draws on mentoring initiatives in the US.

4 In the US, there is not a standard definition for the size and scope of a small business. In general, the federal Small Business Administration (SBA) considers a company with 500 employees or fewer an SME. However, it is important to note that the SBA has established specific definitions for SMEs by industry sector based on either number of employees or revenues. These definitions can be found in Title 13 of the Code Federal Regulations at Part 121 [13 CFR 121]. For US environmental regulations, SMEs are defined in specific legislation. For example, the Federal Clean Air Act uses both number of employees (100 or fewer) and emission levels. The Resource Conservation and Recovery Act uses a definition based on waste generation.

compliance (with legal requirements) or go beyond legal compliance and establish best management practices. Such help can range from being highly technical to merely increasing awareness of potential options. The role of the mentor is to facilitate intentional learning that enables the development of the mentee's skills through instructing, coaching, modelling and advising (Kaye and Jacobson 1996). Mentoring also differs from consulting in that mentoring is nearly always provided for free.

A secondary goal of mentoring is to facilitate a change in the mentee organisation's cultural perceptions of environmental management and its relationship to economic performance. Many SMEs frequently view programmes aimed at improving environmental performance as creating additional costs or threatening future growth. Mentoring, especially if it is done by another business, can help change these perceptions by providing a form of cultural benchmarking. SMEs are more likely to be persuaded by another company about the value and importance of having a good environmental management programme.

Mentoring programmes or initiatives can be structured in a variety of ways. Typically, the form of a mentoring programme is shaped by the type of function and activities it performs. Most mentoring activities can be roughly categorised into three general types:

- Information sharing
- Compliance assistance
- Environmental management assistance

Information sharing primarily involves developing newsletters, workshops, guidebooks, websites, listservs and other information-based resources capable of reaching a broad audience. **Compliance assistance** usually requires direct interactive activities such as site visits, audits and reviews, telephone counselling, training, and assistance with corrective action that specifically focuses on compliance issues. **Environmental management assistance** also usually requires some degree of direct interaction in order to assist in creating management systems, and identifying pollution prevention and process change opportunities, etc.

The form of a mentoring programme reflects the relationship between the mentor and mentee.[5] Basic mentoring forms can be described as:

- **One-to-one.** The mentor works directly with one mentee company.
- **One-to-several.** The mentor works with a small number of companies, often in the same business sector or industry. Sometimes referred as 'cluster mentoring'.
- **One-to-many.** The mentor provides assistance to a wide range companies.
- **Customer-to-supplier.** The customer company mentors its suppliers to improve their performance, frequently on a set of specific issues.

5 Note, a 'mentor' can be an individual, a group of experts, or even an environment, health and safety department. The term is used to describe the entity that embodies the expertise.

- **Supplier-to-customer.** The supplier mentors its customers on the best management of the mentor's products.
- **Peer-to-peer.** Individuals or companies share information and counsel each other.

Mentoring programmes can be further distinguished by sponsorship. Sponsorship describes who organises, supports and provides the mentors with materials and direction for the programme. It is the 'brand' or label of a mentoring programme. The sponsorship of a mentoring programme influences its legitimacy, shapes how the programme is perceived by potential mentees, and affects the marketability of the programme. For example, in the US, many small businesses are suspicious of mentoring initiatives and technical assistance programmes sponsored by government agencies with regulatory and enforcement authority. These companies fear that involvement in such programmes might invite inspections and enforcement action.

Common types of mentoring sponsorship include:

- **Business-to-business.** Business mentoring other businesses either outside or inside their industry sector.
- **Association/industry group.** Experts or companies within a trade or industry association mentor other members within the same group.
- **Non-governmental organisation (NGO).** Mentors come from an NGO, academic or research institution.
- **Government.** Mentors are from a government agency.[6]
- **Public–private partnership.** Mentors come from a variety of sources.

Mentoring programmes can be and are structured in a variety of ways. How a programme is designed depends on its goals, resources and the commitment level of the mentors. Additionally, mentoring programmes can be targeted around a specific issue, such as improving the quality of a watershed[7] or meeting a specific set of regulations.[8] Table 1 provides some examples of the different structures of actual mentoring programmes.

There are trade-offs associated with each form of mentoring. For example, one-on-one mentoring is the most effective since it addresses the specific needs of the mentee through hands-on counselling. However, it is also the most time- and resource-intensive. Mentoring focused on information sharing is usually the least expensive kind of mentoring, but its effectiveness is difficult to measure and it is dependent on

6 In the US its is important to distinguish which level of government (federal, state, local) is the sponsor.

7 For example, Businesses for the Bay is a business-to-business, one-to-many, multi-state mentoring programme focused on reducing environmental impacts from business on the Chesapeake Bay watershed.

8 For example, states' governments are required by section 507 of the Federal Clean Air Act to establish Small Business Assistance Centres to help SMEs comply with federal air regulations. Many of these centres offer mentoring programmes.

Function	Form	Sponsor	Description
Information sharing	Peer-to-peer	Chemical Manufacturers' Association (CMA) (trade association)	The CMA's 'mutual assistance network' sponsors meetings and listservs where members can mentor and provide advice to one another.
Compliance assistance	Supplier-to-customer	3M and Akzo Nobel (business)	The sponsors developed a powerful electronic environmental guide for the furniture industry, an important customer base.
Compliance and environmental management assistance	One-to-several	John Roberts Company (business)	Works directly with small companies within the printing industry to help them establish EMSs for compliance and beyond.
Compliance and environmental management assistance	One-to-many	Texas Natural Conservation Commission (state government)	The Texas EnviroMentor programme provides on-site assistance to regulated small business. The State Commission matches volunteer mentors from industry with SME mentees.
Environmental management assistance	One-to-one	WasteCap of Maine (NGO)	Offers mentoring and technical assistance focused on waste management.
Environmental management assistance	One-to-several	US EPA's Environmental Leadership Programme (federal government)	Facilitated mentoring of SMEs by large companies within the same industry sector.
Environmental management assistance	Customer-to-suppliers	Volvo Cars NA (business)	Helps key suppliers develop advanced EMS.

Table 1: EXAMPLES OF THE DIFFERENT STRUCTURES OF MENTORING PROGRAMMES

Source: Adapted from Makower 1998

the distribution vehicle for the information. Cluster (one-to-several) mentoring can provide valuable information and training to a wide number companies in the same industry or supply chain in a cost-effective manner. The downside is that it is difficult to address specific issues of relevance to a particular company.

As noted earlier, sponsorship is also an important factor for effectiveness since it affects the legitimacy of the programme. For example, government and university-based mentors might be seen by some businesses as lacking an adequate commercial experience. Yet such mentors may be able to dedicate more resources towards addressing technical issues. Trade and industry association-based mentoring programmes are highly effective for benchmarking and disseminating best management practices. However, these programmes and their resources are usually limited to the fee-paying membership.

Business-to-business mentoring between a corporate environmental leader and small company can be a powerful form of mentoring. Under a business-to-business mentoring relationship, smaller companies have the opportunity to benefit from larger companies' greater access to resources and experience in creating environmental management strategies. Additionally, business-to-business mentoring is important for helping to dispel negative perceptions that the small business might have about environmental initiatives and programmes. However, some larger companies may not see the advantages of mentoring for them and some small business might be sceptical about whether large corporate programmes can be scaled down to their facilities.

Why mentor SMEs?

Mentoring can provide a non-threatening, low-risk, low-cost and effective means of introducing and engaging small companies with concepts and strategies for achieving better environmental results. While there are numerous reasons why many small businesses are not addressing their environmental impacts or have failed to reach compliance, lack of information is usually an important factor.[9] Indeed, most SMEs do not have an environmental manager or staff person who focuses solely on environmental performance. Frequently, SMEs are failing to comply because they are unaware or misinformed about industry-specific regulations and do not want to appear ignorant when asking for help. For companies that are not regulated, most are either unaware or sceptical that improved environmental performance can have the potential to improve their economic efficiency as well, while other companies may have no understanding of the environmental aspects of their operations. Since managers in small companies tend to be very busy, they prefer to maintain the status quo and do not take the time to learn about regulations or to become active in trade associations that offer regulatory assistance information or pollution prevention strategies.[10]

9 See Chapter 14.
10 See Chapter 2.

In the US, as elsewhere, small companies typically look for help only when they become aware of a potential penalty or regulation that applies to them. Sometimes, they will ask for help from regulators if they have not previously had bad experiences, or seek out training if it is close by and inexpensive. More often, they will look for help from their suppliers or from other business owners.

Like any business concerned with compliance and environmental performance, SMEs have numerous questions, such as: 'What are my technical requirements?'; 'How will I pay for the expertise needed to meet compliance?'; 'Are there ways I can avoid being covered by regulations?'; and 'What are my options?' However, SMEs often lack the knowledge and resources to answer these questions efficiently and effectively.

Mentors, with technical and managerial expertise, can help SMEs find the answers. Mentors can help small companies to get started on the road to compliance and beyond by working with mentees in areas such as:

- How to effectively map inputs and outputs into the business
- Tips for creating a 'compliance drawer' organised by permits/applications, licences/fees, compliance reporting, compliance plans, and other environmental management documentation
- How to make a compliance schedule
- Strategies for creating performance plans
- Ways to integrate training into employees' schedules
- Strategies for organising a small safety and environmental team to help identify pollution prevention opportunities

Benefits to the SME mentee

For an SME that participates in a mentoring programme, the benefits are apparent. Mentoring by an experienced professional provides the opportunity to gain access to expertise for free and catalyse environmental improvement. Mentoring can help demystify technical issues while transferring insights into how to manage environmental issues, purchase control technology, make process changes, use consultants effectively, manage paperwork for reporting, and so on. Mentoring also provides a source for advice and suggestions. Furthermore, operational changes that a mentee might implement as a result of mentoring, can frequently save the company money. For companies that are mentored by their customers, the experience represents a chance to build stronger relations with that customer and ensure inclusion in the supply chain.

Benefits to the mentor

The benefits to the mentor depend, in part, on who they are. For companies that act as mentors to companies outside their supply or customer chains, benefits may include:

- Recognition as a corporate environmental leader by the public and regulators
- Improved community trust by demonstrating that the company's commitment to environmental excellence extends beyond its facility walls
- Improved employee satisfaction
- Increased knowledge about environmental management systems

This last point should not be discounted. As one mentor from the WasteCap of Maine mentoring programme observed:

> I've been on several [site] visits and in some ways I feel guilty because I think I take away more ideas than I bring to the company (Hess and Bishophric 1995: 30).

Jeffrey Adrian, a mentor and Environmental Director of the John Roberts Company in Minnesota, has observed that involving staff in mentoring small business both enabled his staff to discover important changes for their own facilities while also helping to invigorate their own company's environment, health and safety programme[11] (see Box 1).

Companies concerned about the environmental performance of their supply chain may find mentoring an effective way of ensuring that their suppliers meet certain standards, such as ISO 14001 or internal corporate performance standards.[12] Mentoring suppliers also enables corporate mentors to develop better relationships with their supply chain, gain a better understanding of their suppliers' capabilities, as well as assess which suppliers are more adaptable to change. Additionally, some companies, such as Volvo, believe that supplier education and mentoring programmes are important for strategic positioning in the development of new, more environmentally friendly products such as alternative-fuel vehicles (ICEM 1998).[13]

Mentoring customers can have strategic benefits. For companies whose products present potential liabilities for their customers, mentoring on proper use can help reduce risk for the customer while strengthening the supplier–customer relationship and the brand's name. When asked why 3M has mentored its customers, Tom Zosel, former manager for environmental initiatives, stated that mentoring increases the company's competitive advantage by adding more value to 3M's products. 'You risk less by doing it than by doing nothing,' Zosel commented. '3M did not have to justify the cost because it is part of the business plan; it's a responsibility' (ICEM 1998: 15).

---

11 From conversations with Jeffrey Adrian, who is also a member of the Institute for Corporate Environmental Mentoring at the National Environmental Education and Training Foundation, of which the author is Director.

12 It is worth noting that, with the increased use of outsourcing, many large corporations are concerned about becoming liable for their suppliers' poor performance. Companies are using a variety of methods to reduce potential risks from the supply chain, such as performance standards, auditing and reporting, etc.

13 *http://www.neetf.org*.

The John Roberts Company (JRC), a commercial printing company in Minneapolis with over 300 employees, is a recognised environmental leader in the printing industry. As such, the EPA selected it for participation in the Agency's Environmental Leadership Pilot Project in 1995. As part of this programme, JRC was asked to mentor smaller printers.

From spring 1995 to autumn 1996, environmental staff from JRC mentored four small printing companies—Bromley Printing, Dorholt Printing, Hoppe Printing and Reindl Printing—to help them develop their own environmental management systems (EMSs). The programme's mentoring activities included: site visits, hands-on practical guidance, assistance in developing an environmental management plan, and technical support over the phone.

JRC developed a variety of technical and administrative tools for their mentees, such as:

- A tab system for organising documents in a compliance document file drawer
- Combined training notification and record of training forms
- Materials for all basic required employee training
- A model reporting schedule to facilitate timely reporting for state and federal licensing requirements

As a result of the mentoring programme, Bromley Printing established an effective EMS, Dorholt Printing addressed its licensing and annual reporting issues, Hoppe Printing was able to gain a competitive advantages available through its EMS, and Reindl Printing was able to refine its documentation and compliance plans.

As the mentor, JRC benefited as well. The company's participation in a mentoring relationship resulted in new ideas for the mentor's own operation. According to Jeff Adrian, Environmental Director of JRC, 'We do not have all of the good ideas. By working with other printers, we see new opportunities which we had not previously realised, because we were too close to the operations in our own facility.'

The mentoring process also reinvigorated the company's own environmental efforts. The ongoing discussions between the mentor and mentee have helped John Roberts's employees sharpen their own skills in environmental management. Furthermore, the mentoring programme provided opportunities for effective networking.

In addition to these benefits, JRC was able to 'give something back to the community.' This is an important value of the company's culture, which the company believes includes more than just monetary donations. Mentoring has now become an additional way in which the company contributes to its community. Today JRC continues to mentor other printers and has provided counselling to smaller printers across many parts of the country.

Box 1: MENTORING CASE EXAMPLE

For governments, the benefits of mentoring include providing low-cost, low-risk and effective ways of ensuring compliance and improved environmental performance.[14] Additionally, in areas where the relationship between the government and business tends to be adversarial, mentoring can help build trust.[15] Indeed, one of the stated

14 For example, a US EPA-sponsored mentoring programme for waste-water treatment facilities with poor non-compliance records was able to increase the compliance rate of these facilities by 89%. In another EPA mentoring initiative for dry-cleaners, mentee compliance rates increased by 20% over that of non-participants.

15 Governments that work in partnership with other businesses as mentors can also help to build better trust and understanding of each others' issues.

objectives of the US Environmental Protection Agency (EPA)'s Environmental Leadership Program was to 'foster constructive and open relationship between agencies, the regulated community, and the public' (EPA 1996). Some local and state governments have also used mentoring as a way of both maintaining and luring companies to their jurisdictions. In addition, government personnel can gain a greater understanding of the issues that small businesses face in managing their environmental issues through mentoring. This process, in return, can help inform policy, rule-making and enforcement decisions.

Elements for successful environmental mentoring

Environmental mentoring, as noted earlier, is the use of expertise to help another entity improve its environmental management and performance. Therefore, only individuals or teams with expertise that relates to the function and goals of the mentoring programme should be mentors. Generally, businesses that are either in the same industry sector or use similar processes frequently form the most successful mentoring relationships. Regardless of background, all mentors must be willing to share knowledge, have strong communications skills and, if a company, must be in compliance with applicable environmental regulations.

Before beginning a mentoring relationship, mentors and mentees must define and understand their roles. The degree to which this needs to be done depends on the type of mentoring. For a formal mentoring relationship, the mentor should prepare a letter of commitment outlining the roles of both parties in the mentoring relationship. Such a letter should include specific project objectives to be addressed during the mentoring programme so that the outcome can be measured. Any security, legal and liability issues that might affect the relationship between the mentor and mentee must be addressed, understood and agreed on by all parties.[16] With regard to liability, in most cases the mentee must agree not to hold the mentor liable if a problem occurs. Nevertheless, mentors must be careful about the advice they give their mentees. Additionally, in some cases, a confidentiality agreement may be required if the mentee is concerned about trade secrets.

The baseline knowledge of the mentee should also be evaluated before work begins. This enables the mentor to match 'skill sets' with the mentee's staff, plan for a self-directed project, and give the mentee a choice of what to do. Conducting a baseline assessment also allows the mentor to develop realistic exceptions regarding what to expect on the first visit. Other elements that are important for enabling a successful mentoring relationship include the following.

16 Legal and liability issues vary with jurisdiction. Therefore, it is important to review relevant legal requirements regarding responsibilities for such things as making a finding of non-compliance with the law and time-period required for taking a corrective action (fixing the problem.) In the US, the EPA has issued policies regarding small business environmental compliance issues, as have many states. Such documents should be reviewed and understood.

- **Commitment.** Both companies in a mentoring relationship—particularly the mentor company—must have a strong commitment to seeing the process through. Uncertainty on the part of either party may yield disappointing results for one or both participants.
- **Goals.** Clear, realistic, goals that include time-lines for meeting established benchmarks should be set.
- **Criteria.** Appropriate goal-setting is dependent on the criteria that both the mentor and mentee have set for themselves and their partners. By establishing criteria, the mentor and mentee can engage in a screening process that is more likely to yield a successful partnership. Criteria might include the distance between facilities, the willingness on the part of the mentee to implement recommendations, and the existence of a successful and replicable programme on the part of the mentor.
- **Obstacle identification.** Identify in advance any obstacles that would prevent the implementation of any proposed recommendations.
- **Needs assessment.** Mentees, with help from the mentor, should complete a needs assessment survey to determine their interests, capabilities and resources.
- **Similar business interests.** Businesses that are either in the same industry sector or use similar processes are more likely to form successful mentoring relationships. Also, it is important to match 'skill sets' between individuals from the mentor company to those in the mentee's staff.
- **Plan a 'self-directed' project.** Look for opportunities for the mentee to put what it has learned into place.
- **Flexibility.** Mentors should be flexible in their approach and adjust to the needs of the mentee.

In addition, mentors should remember the following:

- **Keep it simple.** Mentors cannot assume that the mentee has the same level of experience, education and understanding of environmental issues as they do. Therefore, mentors should be able to translate complex concepts into plain language and avoid using technical jargon.
- **Focus on the practical.** Mentees from small business respond best to 'how to' information they can put to work right away. Abstract discussions about materials flows and industrial ecology are not particularly helpful to the small business person who is just learning the basics of environmental management.
- **Understand time constraints.** Most small businesses are spread thin. Time is money so, consequently, SMEs operate under tight time constraints.
- **Eliminate preconceived notions of what the SME needs.** Mentors will be more successful if they listen to the needs of the mentor as opposed to telling them what they need.

- **Use business language.** Frame advantages in terms of cost savings, efficiency and profits.
- **Seek industry affirmation.** If possible, check with contacts in industries similar to the mentee's to identify benchmarks and general suggestions to make sure you are going in the right direction.

Mentoring can face potential obstacles; however, with proper preparation, the pitfalls can be avoided. Some of the common problems that mentoring programmes face, and ways to avoid them, are outlined below:

- **Poor recruitment process.** An effective recruitment process with clearly spelled-out criteria and goals is critical to recruiting mentees that can successfully complete the programme.
- **Failing to establish safeguards regarding liability and competition.** Clarifying the roles and responsibilities of each participant, and establishing guidelines on protecting proprietary information, will help ensure that both parties feel adequately protected.
- **Over-commitment.** Mentoring can be time consuming. Be realistic about how many companies can be mentored simultaneously and do not assign personnel who are already over-committed.
- **Fear of regulatory oversight.** Small companies may worry that participation in a mentoring programme, particularly one associated with a government agency, will invite government scrutiny. In the US this fear is generally unfounded since there is no evidence that mentoring increases government enforcement or regulation. However, it is still important to review compliance and enforcement policies. In the US the federal Environmental Protection Agency has established a special small business compliance and enforcement policy that provides flexibility for fixing self-discovered problems.
- **Lack of goals and deadlines.** Regular contact is essential to ensure that the stated goals of the mentoring relationship are met.

Conclusion

In January 1998, the Institute for Corporate Environmental Mentoring (ICEM),[17] which is a project of the National Environmental Education and Training Foundation (NEETF)

17 The ICEM is a project of the National Environmental Education and Training Foundation (NEETF), an NGO based in Washington, DC. The Institute is currently developing mentoring resources and exploring the strategic use of mentoring for regulatory compliance and in supply chain management. Additionally, it is developing an environmental management and industrial ecology curriculum programme for small business. The NEETF was charted in 1990 by the US Congress to develop programmes that address the nation's critical environmental issues through

and the White House Council for Environmental Quality, sponsored a forum to examine the role of environmental mentoring in the US. This meeting found that environmental mentoring is increasingly being used as a means of engaging small businesses in environmental initiatives (ICEM 1998). In fact, ICEM has identified and surveyed over 250 active mentoring programmes in the US alone (see Box 2).[18]

The variety of these mentoring programmes reveals a need for education-based programmes to address the diverse environmental management issues facing business and society today. The existence of these programmes also reflects at least two general trends in contemporary environmental management. The first, as noted earlier, is the search by government at all levels, and others, for more flexible, cost-effective, and innovative ways of encouraging superior environmental performance by all companies. In the US, this search has led government agencies to tap the willingness of some companies to go beyond requirements on a voluntary basis and share their knowledge and experience with other companies seeking similar improvements.[19] For these companies, mentoring is one way for companies to demonstrate their commitment to environmental results beyond the facility gate.

The second trend is the increased focus by corporations and stakeholders on the environmental performance of supply chains rather than just individual companies. With increased reliance on outsourcing, companies are now seeking ways to manage risks within their supply chains. Related to this trend are efforts by some companies, whose customer base is comprised of smaller businesses, to improve the environmental performance of the customer base. Consequently, supplier performance and customer education programmes are becoming more common, both domestically and overseas.[20] Companies involved in these forms of mentoring recognise that good environmental performance reflects good business practices needed for success. And success is obviously important for their business as customers, suppliers and neighbours. As Brad Allenby, Vice-President of Environmental Health and Safety at AT&T, observes: 'Large companies that have committed the resources to improve their own environmental performance understand that, by helping existing suppliers and customers achieve environmental excellence, they can improve their performance and the overall environmental health of the community' (ICEM 1998).

---

education and training. The Institute was established in 1997 to encourage and support these trends. Its steering committee includes representatives from business, government, academia and NGOs. Corporate members include companies interested in mentoring, such as AT&T, BP Amoco, Compaq Computer, R.R. Donnelly & Sons, Mobil, John Roberts Co., 3M and Volvo. The Institute is currently developing mentoring resources, exploring strategic use of mentoring for regulatory compliance and in change management, as well as developing an environmental management and industrial ecology-focused curriculum for community colleges and post-secondary vocational institutes.

18 ICEM has created a database that allows companies interested in mentoring to search for mentors. Along with other resources, this is available at the Institute's Mentor Centre website at *http://www.mentor-center.org*.

19 The US EPA's Environmental Leadership Program mentoring project is an example of this approach.

20 For example, United Technologies Corporation has established a mentoring programme for some of its Asian suppliers. In Guadalajara, Mexico, the World Bank, along with 11 large Mexican companies, has funded an ISO 14001-oriented mentoring programme for SMEs that are suppliers to Mexican subsidiaries of transnational corporations.

3M and Akzo Nobel: The American Furniture Manufacturing Association

3M and Akzo Nobel entered into a mentoring relationship with the trade association of one of their principal customers, the American Furniture Manufacturers' Association (AFMA). The AFMA has over 350 member companies in more than 1,000 local facilities, 60% of which are small businesses. The AFMA wanted information on the environmental compliance requirements specific to their industry. 3M and Akzo Nobel developed a plain-English, 1,000-page environmental compliance guidebook that could be used, not just by the facilities' environmental managers, but by the person on the shop floor. While some members were initially sceptical of the value of the guidebook, it has been tremendously popular with AFMA members. The guidebook is now available on the Internet and as a CD-ROM and has been distributed to all AFMA members. It has also become the basis for a series of environmental management training seminars on topics such as environmental auditing which are given to AFMA members around the country.

Santa Clara County Pollution Prevention Program: Proto Engineering

The Santa Clara County Pollution Prevention Program (SCCPPP) in California is a county government-sponsored peer-mentoring programme that facilitates the exchange of pollution prevention expertise. The programme brings together companies in the metal finishing and printed circuit board industries to discuss ways to reduce pollution. The SCCPPP organises workshops on technical subjects using experts from local companies. The workshops use business language and are designed to help small and medium-sized companies improve process efficiency, reduce chemical purchases, save money, and ease the regulatory burden. For Proto Engineering (a printed circuit manufacturer), the direct benefits from participation in SCCPPP's peer-to-peer mentoring programme have been a reduction in hazardous waste generation and waste-water discharges, saving US$97,000 annually and enabling production to nearly double.

The Voluntary Protection Programs Participants' Association

Although not focused on environmental performance, the Voluntary Protection Programs Participants' Association (VPPPA) mentoring programme is one of the most established mentoring initiatives the in US. The programme was launched in the spring of 1994 to expand the use of mentoring to promote worker safety and health programmes offered by the US Occupational Safety and Health Administration's (OSHA) Voluntary Protection Programs (VPPs). Companies that have established a VPP share their safety and health expertise with facilities that are interested in and/or are pursuing application to OSHA's VPP. Mentors work with an applicant site to ensure that its application is properly documented, well organised, and contains all the proper elements. The mentor may perform benchmarking on safety and health programmes as requested.

Surveys conducted by the VPPPA show that nearly 90% of mentor programme participants credit it for easing their VPP application process. The most effective mentoring techniques have been one-on-one relationships, telephone consultations, needs assessment surveys, and self-guided study materials. Small workshops have been slightly more effective than large ones. Website programmes and videos have not been effective (but the website programmes have not been completed). VPPPA has recently begun pursuing 'cluster' mentoring programmes, in which several companies of similar size and focus are teamed with a matching mentor company. Efforts are made to match companies of similar size, but comfortable relationships have developed between programmes of differing sizes. Mentees generally prefer to be matched with a company in a similar business and geographical area. Non-managers appear to be as involved in the mentoring programme as management. Graduates of the programme now mentor other companies.

Box 2: MENTORING SNAPSHOTS

18
Greening the supply chain

Are formal environmental management systems appropriate for the needs of small and medium-sized enterprises?

Alan Powell

It was the scholar Gredler who said: 'Those that focus on winning can lose sight of the objective, which is: to solve the problem.' In the context of this publication, the objective is clearly to improve the environmental performance of business, so the question is: 'Do formal management systems help achieve such objectives?'

While this chapter does not set out to prove one way or another the merits of formal management systems, it is centred on the need to demonstrate organisational competence in a supply chain context. This can be a major problem for a small and medium-sized enterprise (SME), as it balances the demands of fulfilling a formalised system alongside the potential environmental and business benefits. Fortunately, this problem is understood by decision-makers in government and the problem is being left to industry to resolve.

The acknowledgement that environmental management is a strategic business issue is a very recent phenomenon, and the rules for the true integration of economic and environmental performance are still being written. As the debate continues it becomes clearer that primary environmental management requires significant co-operation between upstream (suppliers) and downstream (customers).

This chapter sets out to explore, in some detail, the whole essence of the business environmental agenda and how it is being influenced by the expectations and demands of large organisations. For any business that supplies, or aspires to supply, blue-chip businesses or any type of governmental organisation, knowledge of such organisations' current and future thinking is essential. Like it or not, supply chain management is here to stay!

But the idea of supply chain management is not new. It is, for example, a standard and regular practice in the automotive and component industries and there are famous examples in the retail trade where it has been practised by companies such as Marks & Spencer for over 50 years. The supplier standards introduced by the UK Ministry of Defence in the 1950s became the forerunners of international quality standards, and the requirement for suppliers to provide evidence of quality performance through certification to one of the standards in the ISO 9000 series is now common practice worldwide.

Historically, supply chain management has evolved from a desire to control product quality, price and, latterly, logistics, where the customer–supplier relationship is significant enough to warrant such attention. But is there a wind of change occurring?

Any and *every* business has an environmental impact! Many are now fully appreciating the consequences of this. Key among such consequences is the fact that an organisation is ultimately responsible, and may be held to account, for its 'environmental footprint'.[1] In many cases, the environmental impact of an organisation's procurement programme, or the use of its products, may far exceed the actual direct environmental impact of its own operational activity. It is for such reasons that the international environmental management system standard ISO 14001 requires participants to address such issues within their management system/ programme. In addition, 'name and shame' exercises such as that undertaken by the UK-based Business in the Environment in its *Index of Corporate Environmental Engagement* series, identify key areas of failure (and success) in the environmental management programmes of the top 350 corporate businesses in the UK—with the express purpose of driving improvement.

In addition, the real economic impact of poor environmental performance is now well understood. Costs associated with discharging legal obligations, inefficiencies resulting from poor operational performance and lack of management controls are all effectively passed on to customers. The financial risks associated with an environmental failure in a supply chain cover not only the threat of fines, but those associated with combating negative PR and defending against litigation. They might also include: capital costs in having to change a process or a procedure; downtime costs associated while finding a new source of supply for a component or ingredient; and, of course, the cost of potential lost orders. Appreciation of these factors is leading some to radically consider the way they do business. How many suppliers do they really want, given the potential risk, exposure or vulnerability of a supply chain containing, in some cases, tens of thousands of suppliers when, following Pareto's law, 80% of their business is placed with 20% of their suppliers.

However, such awareness creates as many problems as it solves for the corporate organisation. Experience shows that trying to impose complex or demanding

---

1 The total environmental impact caused by an organisation resulting from: the production, processing and supply of its resources and raw materials; the use of its facilities and operational activity; the distribution and use of products by customers; and the ultimate disposal of its product at the end of its useful life.

requirements across the board rarely works as it is not only difficult, complex and expensive to police, but can place significant burdens on suppliers which can cause them to take their eye off their core purpose and may, in extreme cases, threaten their viability.

In practice, there are several significant hurdles to the general application of such an approach when it comes to managing a supply chain's environmental performance, including:

- A general reluctance to demand full compliance with international standards by customers
  - The effect this may have on supplier choice
  - Concerns about the knock-on effects on general business performance
  - The appropriateness of such standards, related to the significance of the customer–supplier relationship
  - The effort and resources required to achieve co-operation
- Historic relationships and working practices
  - The location of power in the chain and the ability to demand and police expectations outside of the normal product/service specification
  - The ability to extend influence beyond first-tier suppliers
  - Personality, the customer–buyer partnership, trust
- Relative priority
  - Related to other management objectives

Supply chain management

There is relatively little generally available information that specifically covers the 'greening' of the supply chain and, indeed, analytical information on the issue of 'management' of the supply chain generally is, for the most part, relatively recent. This is despite the fact that Marks & Spencer could be said to have started the process back in the 1930s when it started to get its suppliers to produce to detailed specifications for sale in its stores.

Of, perhaps, the greatest relevance to this current project is the work being done by the Global Procurement and Supply Chain Benchmarking Initiative at Michigan State University in the US whose Director, Professor Robert M. Monczka, has conducted research and consulted with 150 of the *Fortune 500* top world companies. Monczka (1997) notes that, even after a decade of existence:

> Supply chain management continues to be a poorly understood, badly explained and wretchedly implemented concept.

He identifies a number of factors responsible for this, including:

- Fragmentation in the way supply chain management is understood and applied
- Failure of companies to develop true integration of the processes used to achieve supply chain management
- Organisational resistance to the concept
- Lack of 'buy-in' by many top corporate managers
- Lack, and/or slow development, of needed measurement systems
- Lack of good and sufficient information, including integrated information systems and electronic commerce linking firms in the supply chain
- Failure of supply management thinking to push beyond the bounds of individual companies

Monczka goes on to define supply chain management, concluding that it is built on the assumption that a supply chain is a resource to be exploited for market position where a number of factors come into play, based on the relationships of the companies forming the chain of supply to the final users.

This position presupposes that companies will do a number of things to maximise the strategic use of such a resource. These include the following:

- Gain a closer understanding of their customers' and future customers' needs.
- Understand their internal and supplier core competences in meeting customer needs in such areas as technology, production capabilities, marketing agility and organisational competence.
- Determine where redundancies and inefficiencies lie in relation to current and future competitive needs.
- Develop relationships and alliances with suppliers who have key competences that strengthen, supplement and enhance internal core competences.
- Form closer, cross-functional relationships with outside suppliers, inside suppliers and internal customers that deal with many and varied uses of sourcing as a competitive advantage.

Monczka further claims that there seems to be increasing recognition that, in the future, it will be the **whole supply chain** that will compete, not just individual firms. It is very much in this context, therefore, that this project seeks to define further the supply chain relationships prevalent in major organisations based in the UK, with particular respect to environmental issues, but also looking beyond these to other areas where external education and training may be relevant and helpful.

The Global Procurement and Supply Chain Benchmarking Initiative has provided a modified model of supply chain linkages, shown in Figure 1.

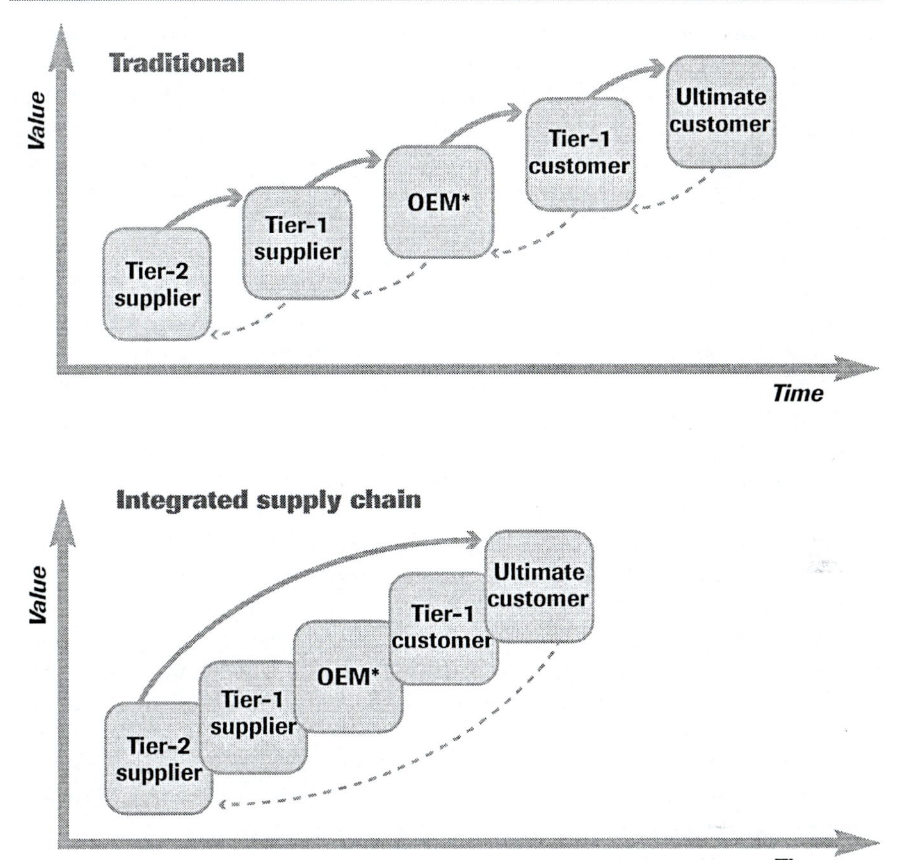

Figure 1: SUPPLY CHAIN MODELS

Supply chain modelling

Historically, businesses in the supply chain have operated relatively independently of one another to create value for an ultimate customer and profits for themselves. Independence was maintained by buffers of material, capacity and lead times. This is represented in the 'traditional' model shown in Figure 1. Market and competitive demands are now, however, compressing lead times and businesses are reducing inventories and excess capacity. Linkages between businesses in the supply chain must therefore become much tighter. This new condition is shown in the 'integrated supply chain' model in Figure 1.

It is clear that interdependency is increased in the latter model, giving more weight to environmental imperatives which may otherwise cause problems for the whole of the supply chain. In either of these models, if the various 'actors' in the supply chain feel there is some advantage to be gained from an even closer relationship, it has always been possible, given appropriate commercial conditions, for some kind of vertical integration to take place. This, however, is not a part of the consideration in this chapter.

What may be assumed is that over time the general status of the issues considered in the overall supply chain remains fairly similar. In other words, the financial and business objectives of all the players in any supply chain are driven by the desire to maximise value-adding processes and eliminate or minimise costs and waste factors.

These objectives have, over recent years, been responsible for the development of many business planning systems and models which have been elevated to 'buzzword' status, appealing more or less, at different times, to professionals in the operations and procurement fields—Just-in-Time management and logistics (JIT), Total Quality Management (TQM), Enterprise Resource Planning (ERP)—all of this now enhanced (obscured?) by information technology (IT) and electronic data interchange (EDI). All of these models have been applied at different levels and in differing degrees to manufacturing, service and retail industries.

As one industry source put it: 'Business plans are based on forecasts—and forecasts are *always* wrong.'[2] The key to improved planning is to minimise the margins of error in the forecasts. From an environmental point of view, such improved planning is important in that it can reduce waste of resources by a significant degree. It seems likely and logical that closer co-operation in the supply chain should be capable of bringing better information to the planning process and this may well be a factor in making supply chain management appeal to many more organisations.

A study in the US chemical industry by Kearney (1997) showed that 90% of respondents in the industry were planning supply chain initiatives. The study also highlighted, however, that only 18% of companies believe that their IT organisations effectively support their supply chain activities. Also, 'the most serious problems companies face are the continuing internal functional focus, a failure to align their IT systems and organisations with supply chain needs and the traditional nature of their relations with external supplies and customers'.

The same study also cited success stories of companies who had used supply chain management to good effect. Zeneca Ag Products cut its company-owned inventory by US$50 million in year one of a two-year re-engineering project 'just by focusing on business processes'. Procter & Gamble was also reported to have cut its cost of goods sold by 4.4% through initiatives such as developing continuous supply programmes with major retailers. This set of initiatives also saved an estimated 25% of other supply chain costs, according to the study.

It should be borne in mind, however, that inventory savings may not *always* lead to environmental savings if, for example, increased numbers of journeys are required

2 Richard Sherman, Senior Vice-President for Strategic Research at Numetrix, reported in *Chemical Week* 1997.

to ensure JIT delivery of goods, with all the extra pollution caused by increased mileage and fuel use entailed.

Power in the supply chain

Another paper by Ogbonna and Wilkinson (1996) referred to the changing power relationships in the UK grocery trade. They pointed out that:

> A whole range of factors, more or less subject to the control or influence of actors within the chain, determine the location of power. Within the grocery distribution chain, retailer concentration, retailer control over distribution and market access, and the emergence of own-label goods have combined with great effect to shift power.

This is an issue we feel it is important to address in the present study on supply chain relationships, although to do it full justice would warrant a full-scale research project on its own.

Eventually, from an environmental point of view, either legislation or customer influence, and probably both, will be seen as the key drivers for change in the marketplace and it will be interesting to get a view as to how far and how quickly such change might be effected in the retail area as a result of consumer pressure.[3] There is already evidence of changes in retail behaviour as a result of pressure group activity—for example B&Q's stance with respect to forest management and the sale of hardwoods. The experience of B&Q also suggests that supply chain pressure is being exerted some distance down the line, and this 'distance' needs to be checked with this research approach.

The Ogbonna and Wilkinson article quotes other authors (Pfeffer and Salancik 1978: 271):

> Organisations are controlled by an external source to the extent they depend on that source for a large proportion of input or output.

Ogbonna and Wilkinson go on to add:

> Power capacity depends on the pervasiveness and immediacy of the impact of withholding the resource and the substitutability of the resource.

In the context of this research, which is targeted at very large organisations, the power of these organisations is of some considerable significance. It may be interesting to gauge the degree of impact this potential for power has, or is perceived to have, on the supply chains involved and in particular on SMEs.

We should at this point differentiate between an SME and a small supplier. The former relates to the overall size of the business in terms of turnover and/or number of employees, whereas the latter is simply an organisation whose sales value to any one customer is relatively minor. In some supply chains—the aerospace industry for example—SMEs are frequently critical suppliers to some key processes, but in terms

3 See Chapters 1 and 10 for details on the degree of customer pressure.

of the business relationship the customer may not be that important commercially to the SME supplier. This raises real problems relating to supply chain influence and power.

Supply chain integration and information

Monczka and other proponents of the importance of the supply chain claim that there is clear movement toward supply chain integration, but that it will occur most effectively if approached both from the top, at chief executive officer level, and from the point of view of the purchasing professionals in whose day-to-day control the key issues fall. For instance Morgan (1997) states:

> Opportunities start to be seen in terms of shortening product development time, reducing cost, taking time out of the system, responding more flexibly and quickly to customer wants.

To do all of this effectively means managing the supply chain up to three or four tiers either side of the particular company being studied. One study, conducted by the Food Marketing Institute in Washington, DC (Fox 1996), estimated that 42 days could be removed from the typical supply chain in the food industry, freeing up US$30 billion in current carrying costs and reducing inventories by 41%.

It is interesting to note that in none of the comments above do environmental issues appear to be of any particular consequence. This appears to be the case in most of the articles and papers uncovered in the background information search for this project. It will be again useful to discover, in the primary research, whether the respondents see environmental issues associated with supply chain management at all and, if so, to what extent.

According to Monczka (1997), perhaps the most significant enabler of supply chain management is information system technology, an area of recent and rapid change for many global companies. It is also an area of some considerable complexity for such companies when they take over different organisations with different, if not incompatible, computer/software systems. White (1996) claims that:

> Effective information sharing, the fundamental building block of supply chain management is not supported sufficiently . . . with traditional technology.

He proposes a solution to this problem gap, using the Internet and dedicated intranet systems, giving 'real-time' information for decision-making purposes across many suppliers and customers on a worldwide basis.

Similarly, Andel (1997) stresses that:

> Your enterprise will succeed or fail on how well you use the information flowing through it.

He goes on to quote Peter Weis, Vice-President of Information Services with American Consolidation Services:

> Too much effort going into redundant software and not enough into co-operation. To date, supply chain partners have had pretty fragmented goals and there hasn't been a lot of co-operation for the greater good of the customer . . . The power of technology has always outstripped the ability for people to implement it.

Andel concludes that the goals in setting up your information supply chain should be rooted in the following elements:

- Operation strategy
- Performance measurement
- Information
- Systems

Co-operation may take many forms, but one that seems to be attractive, at least to the technical people in supplier organisations, is the kind of event run by Black & Decker, termed 'techno-share'. Supply managers at Black & Decker invite 60 or so top design engineers from their suppliers to attend a full day of seminars (grouped loosely around a focus topic) given by roughly 20 suppliers, all with new technology under development. This allows 'relationships to build and product innovations to occur'. A Black & Decker executive suggests that these events generate around 120 concrete ideas and, by his estimation, the last event generated US$1 million of cost savings.

One of the key points to shine through most of the literature already referred to is that supply chain management appears to be most developed in original equipment manufacturers (OEMs), and the leading exponents are seen to be automobile manufacturers, where the benefits appear to be most palpable.

Enviro-Mark™

Before the supply chain can be used as an effective means of driving environmental improvement, it is clear that an understanding of how supply chains interact and how they are likely to develop in the future is essential.[4]

It is known that many large businesses are independently charging suppliers with a requirement to demonstrate environmental credentials or a policy towards adopting environmental management procedures. Such a process, if widely adopted, would be highly inefficient, requiring suppliers to respond to, and accommodate, many different agendas and administrative procedures. Illustrative of this point was one small company in the electronics field which received 15 questionnaires from the different operating divisions of a large group that it supplied.

In addition, and from the customer's perspective, there is a significant resource and administrative burden in formulating, gathering, analysing and auditing data

4 Much of the background research relating to this was undertaken under commission (Business Environment Association) by Keith Melton of Nottingham Business School.

received back from such enquiries. Many organisations recognise this dilemma, but no such solution is readily available until such time, perhaps, when all businesses will be required by law or commerce to hold certification to an appropriate internationally recognised management standard. Most acknowledge that a solution needs to be found in a much shorter time-scale than this road might offer. In addition, many acknowledge that full environmental certification is simply not necessary or required for many SMEs or businesses with low environmental impacts.

In 1996–97 a strategy to address such issues through the development of an environmental benchmark scheme that would measure an organisation's environmental performance against a predetermined set of progressive criteria was developed—Enviro-Mark™.[5] The five stages[6] of the programme lead from compliance with legislation, through the development of policy, establishment of a management programme, development of procedures and manuals to a fully established, working and auditable environmental management system meeting the full requirements of the ISO 14001 standard.

Access to the programme is through a simple on-line package[7] that helps an organisation self-determine its current level against the five benchmarks. It is a simple process thereafter to determine where the organisation aspires to be, given the pressures or requirements placed on it, and for a programme based on the advice given as part of the process to be put in place.

In September 1999, over 40 of the UK's top 500 businesses have participated in trials and discussions to evaluate and prove the value of the process, involving hundreds of their regular suppliers. Enviro-Mark™ is particularly appropriate for SMEs as it allows for a pragmatic 'hands-on' approach and places rather less emphasis on systems bureaucracy at the earlier stages.

Buying into the environment

The UK-based Business in the Environment produced a guideline document in 1995 called *Buying into the Environment* which establishes seven key principles for businesses to apply when dealing with purchasing from an environmental point of view. These are:

- Understand the business reasons
- Know your environment
- Understand your supply chain
- Adopt a partnership style
- Collect only information needed

---

5 The trademark of Enviro-Mark Systems Ltd and the name of the copyrighted five-stage environmental benchmark process.

6 1. Enviro-Mark™ Bronze; 2. Enviro-Mark™ Silver; 3. Enviro-Mark™ Gold; 4. Enviro-Mark™ Platinum; 5. Enviro-Mark™ Diamond.

7 Enviro-Mark Planner™ at *http://www.enviro-mark.com*

- Validate suppliers' environmental performance
- Set a timetable for performance improvement

The report was seen at the time as part of an international move to place environmental issues higher up the business agenda. Expert advisers on the Business in the Environment group included executives from two of the organisations that form the small sample of in-depth interviews for this research project—Boots and BT. The formulation of the research instruments for this research has taken into account the need to test the extent to which the principles inherent in the Business in the Environment report have been incorporated into normal business activity more than two years later.

Conclusion

Much of what happens (or does not happen) in a supply chain appears to depend, in part at least, on the power relationships in the chain. This is most apparent in work on key retailing companies and the automotive industry, but it may be that major companies, such as those studied, are not making most use of what power they do possess.

Most of the work on supply chains deals with value analysis in a purely financial sense, with the emphasis on inventory reduction and savings available on time and resource use. Indeed, one piece of recent research identifies that, while 88% of respondents buying MRO (maintenance, repair and operations) items had made strategic changes in the way they buy these items for their plants, 82% of respondents taking action were doing so to 'control costs'.

Environmental issues are more noticeable by their absence than by their presence. We make the point, for example, that JIT deliveries may actually be unfavourable to the environment, increasing fuel use overall.

If the purpose of supply chain management is, as was determined from the perspective of Gredler in the opening paragraph, 'to improve the environmental performance of business', then ultimately it will be necessary to understand the consequences of such relationships. Arguably, therefore, environmental performance can only be managed by the systematic measurement, collection, collation and analysis of actual information pertaining to organisational and product performance. However, such information is only valuable if it can be directly related to other business performance indicators, such as sales value, capital employed, number of employees, output, etc.

It seems clear that the supply chain is a phenomenon of influence and potential power in a number of dimensions that have yet to be fully realised. Its influence as a value chain is partially perceived and acted on, in the sense that first-tier suppliers are partially 'managed' by many large organisations for cost benefit. Much more may be possible, however, particularly if we are able to determine appropriate indicators that provide a proper basis for real comparative decision-making.

19
The interrelationship between environmental regulators, small and medium-sized enterprises and environmental help organisations

Tim Fanshawe

When embarking on any journey, one needs to know the starting and finishing points. The starting point may be a particular 'problem', such as non-compliance with environmental legislation resulting in pollution. A 'journey' must be undertaken in order to attain the 'end-point' of full compliance and pollution prevention. A number of mechanisms or 'good ideas', such as guidance on regulations, may be offered in order to facilitate such a journey. Unless these 'good ideas' fully address the specific needs of those undertaking the 'journey', the end-point (of full compliance and pollution prevention) may never be reached.

In the same way, small and medium-sized enterprises (SMEs)[1] have specific needs compared to large businesses. When considering the 'problem' of environmental legislation, it can be seen that there is a need for SMEs to embark on a 'journey' to understand how to comply with environmental legislation, in order to prevent pollution incidents and avoid fines, clean-up costs and even prison. A well-meaning regulator or environmental help organisation may have a 'good idea' for a road map for this journey, for example by listing all the legislation relevant to SMEs. It may even include a detailed interpretation of each piece of legislation. But does that actually address the needs of individual SMEs? In many cases it does not. SMEs simply want

1 SMEs in this chapter are defined as those businesses that have fewer than 250 employees.

to know 'What do I have to do?' They remain hungry for the right information but both regulator and environmental help organisation have thus far failed to provide that information in such a targeted way.[2]

Of course, the situation is more complicated than this because many SMEs do not realise that they are affected by environmental legislation, or that they need guidance to help them comply with it (Hillary 1995). This means that the interrelationships between SMEs, regulators and environmental help organisations can be far more complex than at first anticipated. The legal standpoint that 'ignorance is no excuse' may be a flawed philosophy to follow if the result of such an attitude is businesses being forced into liquidation due to non-compliance with the law and environmental damage requiring remedy from the public purse. In the age of the soundbite and instant global communications, it is, perhaps, more pertinent than ever to reflect on both the importance of understanding the specific needs of SMEs as well as the role of marketing to address them.

In this chapter I shall explore some aspects of these interrelationships with regard to raising the awareness of SMEs to their legal responsibilities. Using a simple model and focusing on the Environment Agency for England and Wales, I hope to clarify the key needs of the various actors involved in bringing about compliance. The 'problem' of SMEs remaining non-compliant must be looked at very carefully, otherwise the 'journey' towards pollution prevention will remain a very tortuous path.

Moving from inaction to good business practice: a system model

Before the interrelationships between the various actors involved in bringing about compliance can be discussed, we need to look at the stage on which they are playing. To do this it is useful to break down the whole picture into more manageable components. Figure 1 presents a simplified 'cycle of action', depicting the interaction between SMEs (shown as the active stakeholder) and the regulators, help groups, trade associations, consultants and other intermediaries who have various roles in assisting SMEs on environmental issues. This picture should be familiar as a positive feedback loop of continuous improvement. However, the feedback loop arrows are shown as heavy double lines to indicate that, although this is the ideal, in reality such continuous improvement does not always occur.

Business-support organisations often produce guidance that SMEs can use on their journey of enlightenment. The main philosophy behind such a journey is that businesses will realise that good environmental management is exactly the same as good business practice. For example, if fewer resources are used, if any waste produced is limited, and if pollution incidents are prevented, the result will be greater efficiency for the business leading to cost savings. However, while the environment is still thought of as a separate issue, it will always be lower down the priorities of

2 See Chapters 14 and 15.

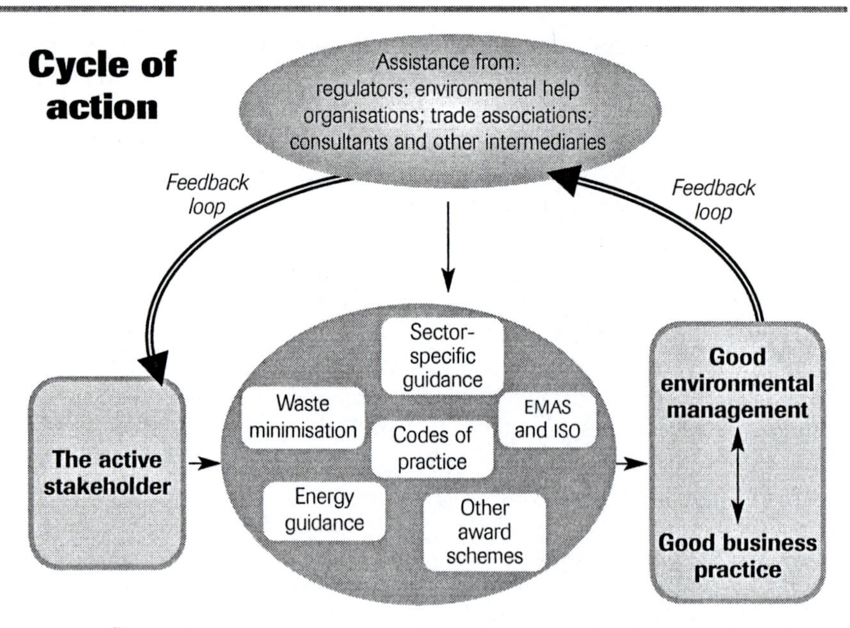

Figure 1: THE CYCLE OF ACTION AND CONTINUOUS IMPROVEMENT

SME managers than, for example, finance. However, once environmental issues are accepted by SMEs as just another aspect of business efficiency, it may be said that the journey towards continuous improvement has truly begun. As a result, the SMEs will move closer to complying with legislation. Consequently, the needs of the intermediaries to prevent pollution will start to be satisfied.

The central circle in Figure 1 depicts some of the ways an SME can move towards a more environmentally friendly way of functioning. It is not meant to be exhaustive, but does suggest that there is the potential for information overload and confusion, which can have a negative impact on SMEs. For example, how do SMEs, new to the concept of environmentally friendly business practice, cope with the plethora of environmental terms? With environmental help organisations offering products under headings such as energy efficiency, waste minimisation, recycling and sustainability, can we first be sure that there is enough understanding of the overarching term 'environment'? Anecdotal evidence suggests that very often there is little understanding.[3] The potential for confusion only emphasises the need for good co-ordination between the various intermediaries.

3 Personal communication with Nicky Chambers, Best Foot Forward, 1999.

Figure 1 represents the ideal situation. The reality has been alluded to earlier: namely, that the vast majority of SMEs remain in ignorance of either the fact that there will be some environmental legislation that affects them or if they are aware of any legislation they generally don't fully understand what the practical implications are. Figure 2 depicts a 'Catch 22' 'cycle of inaction', whereby SMEs remain in a negative feedback loop of inaction towards environmental management and good business practice. There are many reasons why a business may remain stuck in such a cycle of inaction. Most fall under the four categories shown in the circle in Figure 2: fear of reprisal (by a regulator), (confused by) mixed messages, conflicting pressures and genuine disinterest. The situation is further complicated by the fact that a single SME may be affected by a combination of some or all of the above. The general consensus seems to be that the bulk of SMEs (perhaps 70%–80%) may fall into this cycle of inaction.

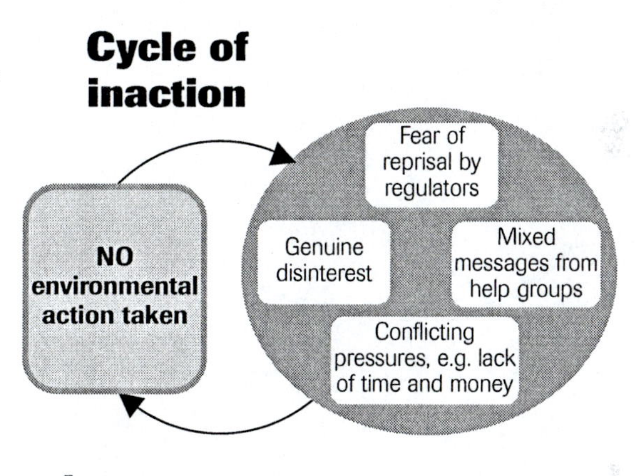

Figure 2: CATCH 22 AND THE CYCLE OF INACTION

Having identified the two key components of the model, the next step is to establish the link between the two. This takes the form of what is termed the 'action barrier', shown as a heavily dotted line in Figure 3. Along the bottom of the figure is a simple representation of movement through time from one extreme, the point of inaction, to the other, the point of enlightenment. This is to show that the model is not static, but rather a simple depiction of an extremely dynamic and complex set of interactions.

It should be clear that the main problem revolves around the fact that most of the guidance offered by the regulators and environmental help organisations is utilised by only 20%–30% of SMEs, i.e. those who have already started to take action of some sort.[4] The key question must be to work out how to break the 'action barrier' and engage the main bulk of the business audience caught in the 'cycle of inaction'.

4 See Hunt's discussion in Chapter 15.

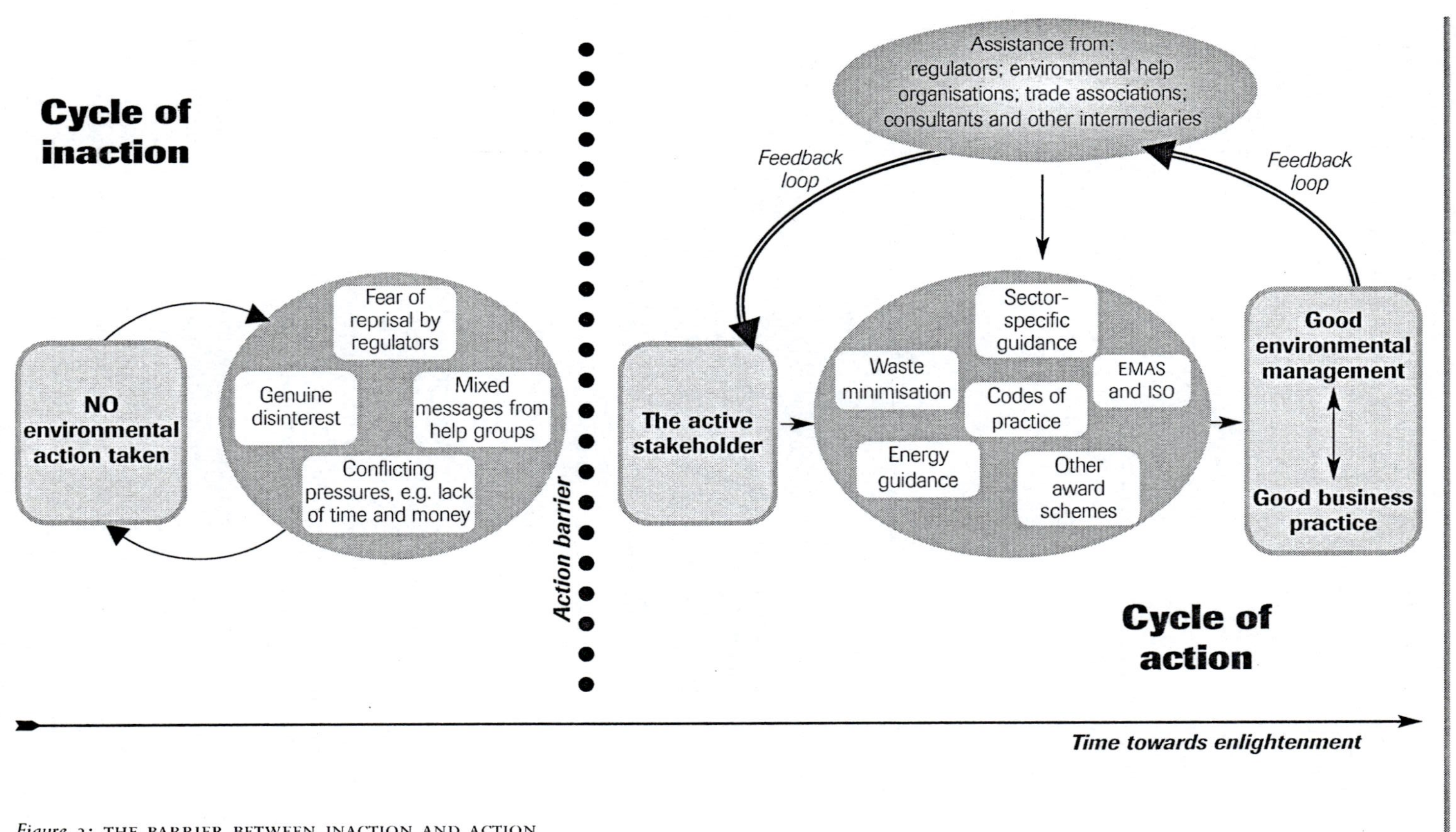

Figure 3: THE BARRIER BETWEEN INACTION AND ACTION

Figure 4 expands the model to include the awareness-raising information that is produced by regulators and environmental help organisations (depicted by four arrows pointing down on the cycle of inaction). The thick arrow, pointing from the far left of the figure to the active stakeholder in the middle, indicates the path that we would like SMEs to take to break out of the 'cycle of inaction' and move across the 'action barrier'. Making this happen will be dependent on a subtle mix of awareness raising about the environment, sending out consistent messages from all environmental intermediaries, proactive public relations to reduce the fear factor of regulators, and incentives to reduce the problem of conflicting pressures faced by SMEs. However, since most SMEs remain ignorant of their environmental responsibilities, perhaps the most important issue at the heart of good marketing must be to understand their needs.

It is worth noting that the requirement on smaller businesses to comply with environmental legislation is no longer limited to 'avoiding being caught by the regulator'. Many of the larger companies now require their suppliers to comply with environmental legislation and this is incorporated into any contract.[5] In addition, in 1998 the insurance sector developed a document, with Environment Agency support, giving recommendations for the underwriting of pollution risk.[6] These recommendations are likely to impact on business insurance with premiums reflecting environmental risk. It is evident, therefore, that SMEs need to understand the requirements placed on them and to take action to satisfy environmental needs as an integral part of business.

Figure 5 shows how such supply chain and insurance pressure can affect any SME, whether they are caught in the 'cycle of inaction' or actively pursuing good business practice. The difference between the two is that those caught in the 'cycle of inaction' are more likely to consider such things as additional and conflicting pressures. They may also be seen to be contributing to the mixed messages often received by SMEs. Either way, it is more likely that SMEs who are caught in the 'cycle of inaction' will view supply chain and insurance pressures in a more negative way.

The model described in the preceding paragraphs shows a very simplified view of the main interactions between the various players involved in helping SMEs comply with environmental legislation. It is worth spending a few moments looking in more detail at the specific role of the Environment Agency. As one of the largest regulatory bodies in Europe the Environment Agency is charged with applying environmental legislation in England and Wales and is committed to being a firm but fair regulator, employing a consistent approach to regulations that is appropriate to the level of risk to the environment.

All companies, however large or small, are expected to comply with the law. This applies to environmental legislation the same as it does to any other law. Over the last two decades the amount of legislation applicable to environmental protection has been steadily increasing, both in terms of the number of laws and their

5 See Powell on supply chain management (Chapter 18).

6 Joint Pollution Working Group, 'Recommendations for the Underwriting of Pollution Risks', Association of British Insurers at *http://www.abi.org.uk*

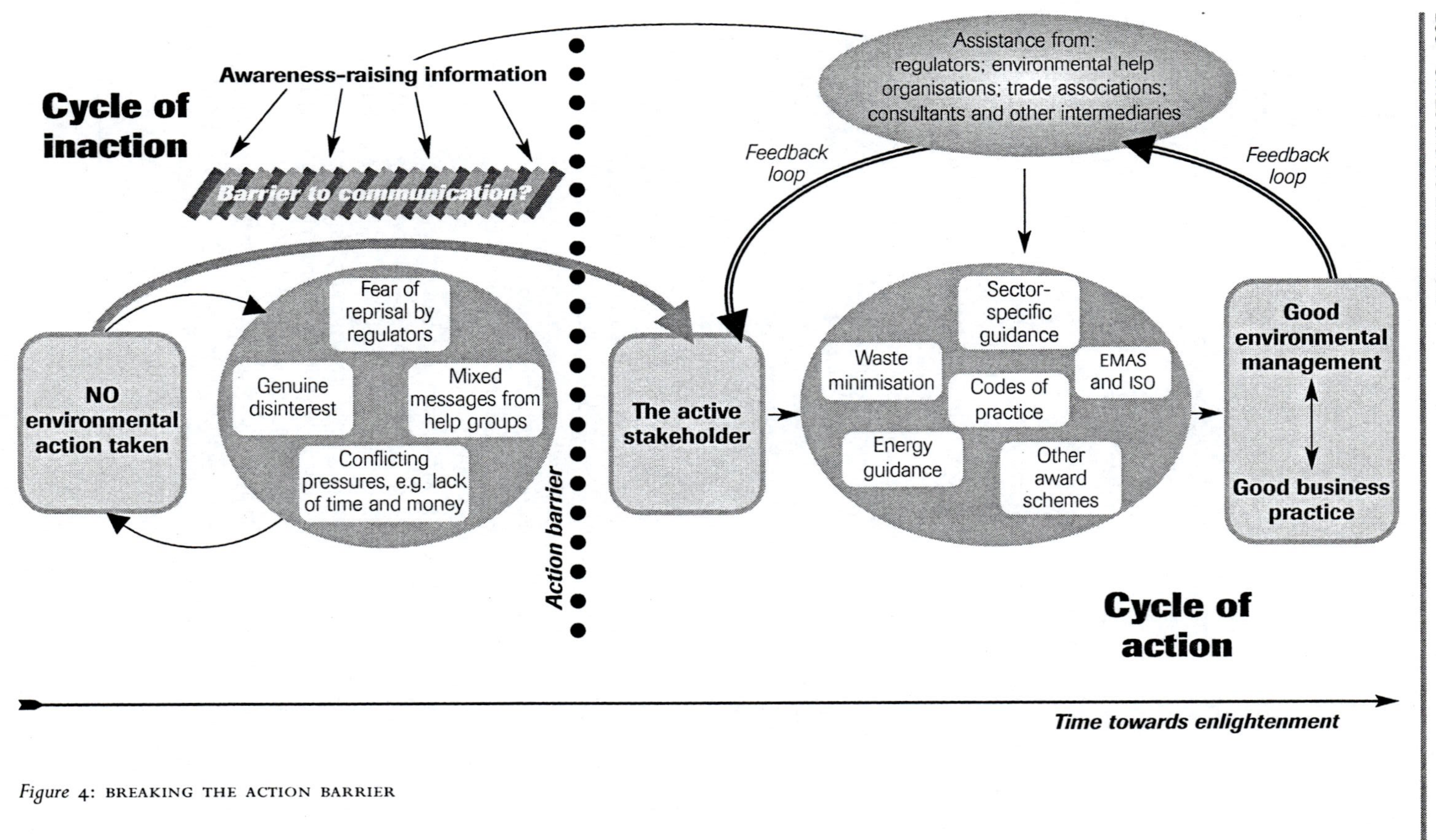

Figure 4: BREAKING THE ACTION BARRIER

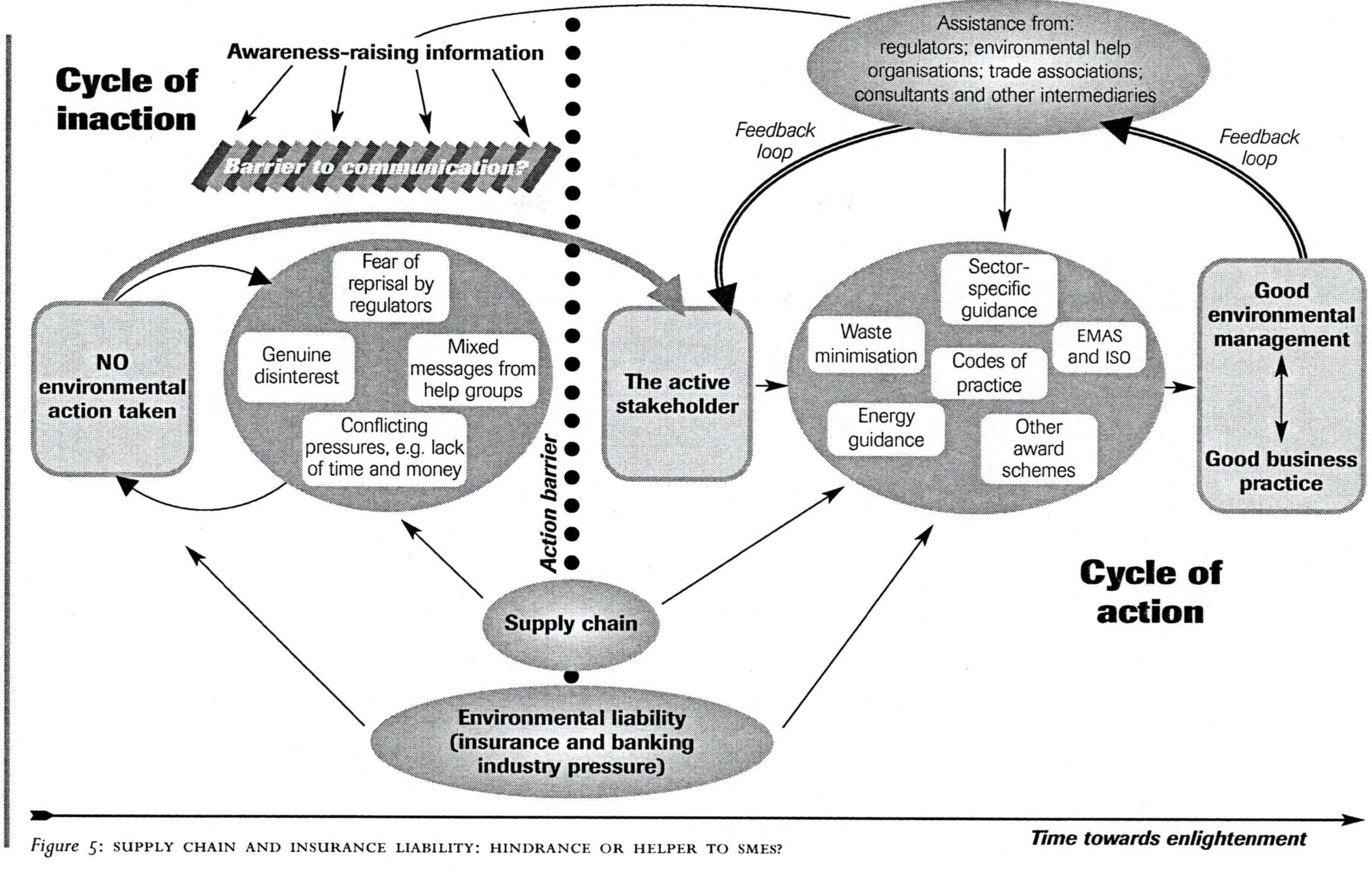

Figure 5: SUPPLY CHAIN AND INSURANCE LIABILITY: HINDRANCE OR HELPER TO SMES?

complexity. This trend is not expected to change. Although some of this legislation is very specific to particular industrial sectors, often there is no cut-off with respect to the size of the companies to which it applies.

A primary means of applying legislation in the UK is through the granting of a permit, e.g. consents to discharge to water, waste management licences and Integrated Pollution Control (IPC) authorisations. The Environment Agency issues such permits and has a routine inspection programme to ensure the conditions of the permit are met (including the holding of a permit where needed). The permitting and inspection regime is targeted at those who pose the greatest risk to the environment (within legislative constraints). Hence, larger companies may be contacted routinely by Environment Agency staff and visited on a routine basis, whereas many SMEs are visited less frequently, if at all.

Some parts of legislation are, however, all-encompassing. For example, it is illegal to allow polluting materials to enter a watercourse, whether accidentally, e.g. oil seepage from an old storage tank, or through an intentional discharge without consent. *All* businesses, regardless of size, are required and expected to abide by such environmental legislation. Where any requirement of these all-encompassing laws is not met, then the Agency will be obliged to act, but it will take into account all the circumstances surrounding the event. However, in responding to a pollution incident, the Agency can find itself in a position whereby it is required to take legal action on the first contact with a smaller company.[7] Where companies do not comply with the law, the Environment Agency will use its enforcement powers in a proportional manner to maximum effect.[8]

The Environment Agency aims to be a firm but fair regulator. However, the role of the Environment Agency as 'regulator' has evolved from that of purely an enforcer to place more emphasis on education. Evidence of the Environment Agency's commitment to such an approach can be seen in the recently published Agency policy on waste minimisation.[9] This states that 'the Agency will work directly, with government, and in partnership with others, to inform, educate and facilitate measurable reductions in the generation of all types of waste'. Although primarily focused at directly regulated industry, the benefits of promoting the sustainable use of natural resources will also impact on smaller businesses.

The Environment Agency is committed to addressing the needs of SMEs and is currently working towards developing a national strategy to bring more consistency to this area of its work. What the model in this chapter does not show is the lack of co-ordination and communication between all the intermediaries. This includes both government and regulators, environmental help organisations and trade associations, as well as the insurance and banking industries. Although some inroads have been made recently, the feedback from questionnaire surveys is that SMEs feel there is insufficient information available to help them, compounded by inconsistent and confusing messages.

---

7 See Petts's discussion on attitudes to compliance in Chapter 3.

8 See the Environment Agency's enforcement and prosecution policy at *http://www.environment-agency.gov.uk*.

9 For more information and a copy of the policy, contact Jon Foreman, tel: +44 (0)117 914 2779.

The reality, of course, is that there is an almost overwhelming amount of information available about almost every environmental subject. There are many examples of guides to legislation, codes of practice, good practice guides, computer software, leaflets, booklets, videos and Internet websites. Since these products only seem to reach a small percentage of their intended audience, the questions that need to be asked are: 'Does each product answer the specific needs of SMEs?' and 'Has the product been marketed effectively?'

To answer such questions requires a change in behaviour by all those involved in helping SMEs. We need better co-ordination and co-operation between organisations and less duplication of help products.[10] By limiting the number of sources of information, through collaborative work, there should be greater consistency of message. This should in turn enhance uptake of the message. However, we must make sure the message is correct. Those who help SMEs must also place much greater emphasis on 'targeting' audiences more effectively. The first step is to understand the overall problem: namely, that the environment often takes a low priority in the minds of SME managers and that there is little understanding of their legal obligations.[11]

The second step often appears to be missed or diluted. This involves the recognition that the whole audience (i.e. SMEs) may fall into discrete groups with specific needs and it might be relevant to ask certain questions before creating a help product. For example, what proportion of a business sector are members of a relevant trade association? Is the trade association proactive in supporting its members on environmental issues? What geographical coverage does the sector have? Is the sector predominantly 'techy' (i.e. computer-literate) or 'non-techy'? Is there a relevant trade magazine? These are just a few of the possible questions that could be asked. What should be clear is that, while technical language presented on a CD-ROM distributed nationally will suit one business sector, a more plain-language approach, e.g. using locally distributed posters, might be more relevant for another sector.

Once the second step is complete, writing the message and choosing the required marketing mechanism(s) should be straightforward. The fourth and final step is, of course, to launch whatever product has been created. However, there should also be the built-in expectation to both measure the product's impact and feedback to step one, in order to check current understanding of the problem and audience response. This should allow a refinement of the message and give the opportunity to select an alternative method of marketing—a continuous cycle of improvement in our understanding of business needs and how to address them.

Conclusion

In this chapter I have tried to demonstrate a model that reflects some of the interrelationships between SMEs and those organisations that try to help them understand their environmental responsibilities. The aim is not to shock readers with

10 See Chapter 14 for a model of information provision to SMEs.

11 See Chapter 4.

something new. Rather, the hope is to suggest that perhaps we, as environmental help organisations, have been too close to the 'problem' of non-compliance with legislation. Perhaps we have missed the real 'problem' of not understanding the audience or marketing effectively. Many of the 'good ideas' to help SMEs that have been developed to date do not fully address the specific needs of SMEs.

Similarly, although there are a growing number of partnership projects and instances of dialogue between the various intermediaries, perhaps it is naïve to believe that this automatically results in 'good communication'. Talking together is a positive step. However, there is a gulf of difference between 'talk' and the reality of achieving consistently good communication between organisations. Over the last two or three years, communication and marketing has become a regular and high-profile topic at seminars and conferences about SMEs, often dominating discussions. Perhaps now we need to move beyond discussion and work harder together to put into practice what we have been preaching.

The chief executive of a London Business Club summed up my thoughts better than I ever could at a recent seminar. He made it clear that the job of conveying messages to SMEs was not something that was achieved by one visit, telephone call or leaflet. It was the result of repeated visits, providing information, offering incentives and help, then doing it all again and, if necessary, keeping on doing it until you were successful. I think it is also worth adding that we must not make assumptions that people are interested in and care about the same issues, or that telling someone something once will change the way they think and act. Yet that is what we are trying to achieve. Helping SMEs make the 'journey' to reach the 'end-point' of full compliance and pollution prevention is all about changing what people believe is important and how they carry out their lives. We can only achieve this by truly working together.

20
Time to reassess the small business environmental advisory system in the UK

Erik Bichard

The National Centre for Business and Ecology (NCBE) recently completed a large environmental assistance programme for small companies in the north-west of England. The programme was designed to enable small companies to benefit from the resources and knowledge in the four universities in the Greater Manchester area,[1] and to use these skills to improve the environmental performance of the participating businesses. While the programme met the expectations of the funding body (the regional government on behalf of the European Commission), the NCBE identified a number of structural obstacles during the course of the programme.

The Centre had encountered these obstacles during earlier small company projects. This chapter is an abstract from the main report of the Business and Ecology Demonstration Project (BEDP). It details the problems surrounding the way in which environmental assistance schemes for small companies are administered in the UK. Solutions are proposed to enable these schemes to be more effective and efficient.

Environmental assistance to small and medium-sized firms in north-west England

Large-scale 'demonstration projects' for the environmental improvement of small companies have become an established point of delivery for regional strategists in the

1 The University of Salford, The University of Manchester, The Manchester Metropolitan University and The University of Manchester Institute of Science and Technology (UMIST).

north-west of England. Inspired by the success of the Aire and Calder Project (CEST 1995) in Yorkshire less than a decade ago, the region has hosted Project Catalyst (W.S. Atkins *et al.* 1994), the Merseyside Waste Minimisation Project (Merseyside Innovation Centre 1998), Environet 2000 (Nimtech 1999) and the Business and Ecology Demonstration Project (NCBE 1999a), all within the space of six years. Excluding many other smaller schemes, this represents over £15 million of investment solely devoted to the environmental improvement of small companies in the region.

Each of the projects followed a successful formula. A number of small companies (less than 250 employees), aggregated under the banner of the small to medium-sized enterprise (SME), were recruited to participate in a subsidised environmental improvement programme. The projects generally offered waste minimisation and/or the means to build an environmental management system into businesses that did not previously have the means or opportunity to take on such initiatives. Most of the projects were directed at particular industrial sectors and were limited to specific geographical areas within the north-west region of England.

The outputs of the projects varied depending on the source of funding, but all of the end-of-project reports highlighted the monetary benefits accruing to the participating companies. Over 2,500 companies have benefited from these projects, which must therefore be considered worthwhile based on the sheer number of individuals they have touched. However, the mark of success of a demonstration project should also be measured against its effects on a wider audience of businesses long after the project has concluded. The aim of a demonstration project is to illustrate that the methods and techniques that improve environmental performance will benefit *all* business.

It is notoriously difficult to gauge the post-project effects of demonstration projects, partly because the funding rarely allows research outside the immediate beneficiaries and also because the decision by businesses to implement environmental programmes can seldom be attributed to information from a single programme. Accepting that there is poor data on this subject, it is possible to use a number of surveys on the environmental needs of SMEs to conclude that there is rarely life in a demonstration project after the funds run out.

Polls and surveys carried out in the north-west of England by (among others) the NCBE (1999b), Manchester Chamber of Commerce and Industry (1998), Groundwork (1998) and Manchester Metropolitan University (Hooper *et al.* 1995) show that SMEs still rank environmental issues relatively low on their list of business priorities.[2] Many are still unsure about where to look for advice and most have never heard of the demonstration projects that have taken place in the region.

The report of the NCBE's BEDP programme describes a typical demonstration programme. The project qualified as an initiative within the 'Objective 2' area in England known as GMLC (Greater Manchester, Lancashire and Cheshire). The project received funding from the European Regional Development Fund (ERDF) and the European Social Fund. This was matched by a number of private-sector donors. More details about the project are set out in Box 1.

---

2 See Chapters 1, 2 and 4.

The Business and Ecology Demonstration Project (BEDP) ran from August 1996 until October 1998. Using nine postgraduate researchers, the project aimed to assist at least 250 SMEs to optimise competitiveness through improvements to environmental performance.

The project was funded by The Co-operative Bank plc, the ERDF, the BOC Foundation and the Environment Agency. It was managed by the NCBE. The catchment area for participating companies included Bolton, Bury, Manchester, Oldham, Rochdale, Salford, Tameside, Trafford Park, Brinnington and South Reddish (in Stockport) and Wigan.

The primary output of the project was to carry out detailed case studies with 50 companies and provide more limited assistance to a further 200. Companies were recruited from both the manufacturing and services sectors. The researchers spent a portion of the two-year project studying for an MSc, PhD or MPhil (part-time) and the remainder of the time working on case studies with participating companies.

Companies that participated in the case studies received up to 30 days of contact time from the postgraduate researchers. The projects included a formal report with recommendations and conclusions. In return for this help, each company was required to nominate a member of staff to co-ordinate with the rest of the employees involved in the project and to keep a log of staff time invested in the project. The co-ordinator would also give permission for a case study to be disseminated where it was considered that innovative solutions would benefit other companies.

Box 1: THE NCBE'S BUSINESS AND ECOLOGY DEMONSTRATION PROJECT

BEDP case studies

BEDP identified potential financial savings of approximately £172,200 per annum. Of the total value of opportunities identified, 68% of the monies saved would come from the use of modified technology or payback after acquiring new equipment. Waste reduction/recycling projects accounted for 30% of the savings. 'Green grants' were offered to some of the participants to allow them to buy clean technology equipment required to achieve the improvements set out in the project recommendations.

Many of the people who championed the projects within the companies were prepared to contribute valuable time and resources to prove to their colleagues, and others, that the effort was worthwhile.[3] However, these admirable efforts are in danger of being wasted if the means by which environmental projects for SMEs are administered is not significantly overhauled.

The following two examples illustrate the way small companies were assisted through the project.

Case Study 1: *Sustainable communications*

D.A. Millington is a small accountancy firm providing services to over 300 SMEs in the Salford and Trafford area of Greater Manchester. The study determined that the

3 See Walley's discussion on the environmental champion in Chapter 27.

company was dependent on knowledge-based resources on a daily basis that helped to keep its clients abreast of the latest legal requirements. Much of the information coming into the company was paper-based and some parts of the documents it bought were irrelevant to the business. The BEDP recommendations lead to an Internet-based information sourcing policy that has saved the company over 1% of its annual turnover. In addition, the company is planning to convert some of its commodity services, such as tax returns, to an e-mail-based service and is planning to join e-commerce discussion forums related to the accounting profession.

Case Study 2: *Water conservation and effluent control*

Simpson Ready Foods Ltd was founded in 1910 and is a small family-owned company producing canned and bottled food products. A significant amount of water is used for the preparation of the company's products. The BEDP team sought to determine whether the quantity of water used in the processes could be reduced and if the quality of the effluent produced could be improved. Initial findings included the need to install water meters to track the variable quantities of water used over a year. This enabled the company to identify inefficient high consumption activities. The meters were purchased with the aid of a BEDP green grant. Tests on the effluent quality showed that separation of certain organic solids from the liquid waste-stream would result in significant savings and produce low payback periods for the equipment purchased to implement the improvements. The company has also made a commitment to gain certification to ISO 14001 as part of a continuing environmental strategy.

Funding of assistance

There are over 180 organisations in the north-west of England purporting to offer small companies advice on environmental matters. These organisations make use of many different sources of funding offered by a wide range of bodies. The following bodies have funded just a few of the recent initiatives in the north-west:

- European Union (Regional Structural Funds, ADAPT, LIFE and others)
- UK national sources (Department of Trade and Industry [DTI], Department of the Environment, Transport and the Regions [DETR] and Single Regeneration Budget preceded by various 'challenge' initiatives)
- UK regional sources (Welsh Development Agency, The Welsh Office, English Partnerships and county councils)
- Local sources in the north-west of England (Landfill Tax Credit Schemes, ad hoc support from Business Links and Training and Enterprise Councils [TECs], and match funding from the private sector and local government)

The rules that govern the way bidding organisations can obtain these funds vary widely and combine a myriad of public/private match-funding formulae, together with differing outcomes or output criteria. In one sense, those that are committed to working with SMEs on environmental projects should be grateful for such a cornucopia of opportunities. Regretfully, the reality is that many of the resources on offer are inefficiently spent and some of the funds fail to be claimed.

The UK national government, in the form of the DTI, is aware of the current inefficiencies, but plans to rectify matters go no further than a commitment to review the way in which small businesses receive advice from government-sponsored organisations. Recent changes in the administration and management of EU Structural Funds make it even more difficult to gauge how much value is being added to the region through environmental programmes. While most funding bodies have some form of ex-post audit system, these are mainly designed to identify misappropriation of funds and rarely raise issues of substance on a more holistic scale.

There are three issues that need to be addressed before the radical improvements to UK environmental assistance programmes can begin to make significant inroads into current levels of performance. These are:

- The need for consistency of advice
- The need for co-ordination of effort
- The need for a unifying vision of achievement

Consistency of advice

With over 180 bodies offering environmental advice to SMEs in the north-west alone, it is hardly surprising that conflicting advice or variable quality of support is encountered from time to time.[4] Most service providers use experienced professionals or trained graduates to advise participating companies. For the majority of the time, participants benefit from a good standard of support.

However, there is some evidence that small company schemes and environmental consultants in general are being discredited by the poor work of a handful of advisers. A paper given to a Business in the Environment conference (Foster 1998) in London by the managing director of a small company highlighted some of his grievances:

- **Poorly quantified business advice.** Advisers often suggest that businesses invest in solutions with open-ended payback periods, or seek to extend the investigation without stating the likely outcome.
- **Poor environmental advice.** Advisers can fail to bring sufficient expertise to the client, often because they have an alternative speciality (such as health and safety). The lack of experience shows in suggestions that companies should do their own checks to determine whether they are breaking the law without providing the necessary skills to carry this out.

4 See Chapter 10.

- **Unfocused or formulaic advice.** Recommendations to produce an environmental policy or implement an environmental management system may be less appropriate than more direct and immediate measures.

It seems reasonable that a minimum standard of qualification and experience should be set for any organisation offering advice under a small-company scheme. Although there is no such requirement for those offering advice on the open market, individuals operating under a subsidised scheme are rarely under the same scrutiny.

Co-ordinating effort

One outcome of the many small company programmes in the north-west is a wealth of innovative and inspiring approaches to solving environmental problems. This produces an impressive number of satisfied participants, particularly where considerable sums of money have been saved through waste minimisation or energy conservation measures.

The disappointing aspect of these success stories is that so few companies benefit from this enterprising work. The ripples of innovation that should travel through the region rarely make it further than the participant's site boundary.[5] This is sometimes due to a unique solution that is sector-specific or can only be applied to certain process technologies. Most of the time, however, it is simply due to a lack of dissemination funds at the end of the programme or unimaginative communication methods. Linked to this is the knowledge that many companies would like to participate in certain programmes but are excluded because of their location or sector.

There needs to be some mechanism to maximise efficiency and to manage and maintain the momentum of small-company schemes. The best way to achieve this is to appoint an overseeing or co-ordinating body responsible for identifying gaps in funding and the dissemination of best practice. Ideally, this body should not just concentrate on environmental issues but should have the remit to consider the larger task of building industrial competitiveness. It should have sufficient financial and political weight and require service providers to take note of its views.

Funding for a co-ordinating body could come from a percentage of the funds directed to each small-company programme. Whatever the administrative arrangements, small-company environmental schemes will continue to scatter both light and shadow across potential beneficiaries until a single body illuminates the whole of the region.

The obvious candidates to perform this function are the newly created Regional Development Agencies (RDAs). The RDAs have a number of responsibilities that would sit well with the role of co-ordinating body including:

- Promotion of inward investment
- Promotion of commercial competitiveness

---

5 See Meredith's discussion on environmental innovation in Chapter 13.

- A mandate to adopt an integrated policy approach linking economic, social and environmental objectives
- An advisory role for TECs and Business Links

The regional focus of the RDA fits with the geographical delineation of EU Structural Funds' regional assistance from central government. Whitehall could direct all regional fund managers (not just those in government offices allocating European regional funds) to work with the RDA in an attempt to co-ordinate the diversity of programmes.

A unifying vision

Many schemes that are designed to improve the environmental performance of small companies cite job protection as a main output of, and justification for, the project. Although important, this is hardly a progressive aspiration. Similarly, while accepting that it should be a societal goal to increase the level of social and environmental responsibility in the business world, this is unlikely to be the prime motivation that encourages small business to invest time and resources in sustainability initiatives.

Without diminishing the diversity of small-business schemes, there should be a better vision that links all initiatives in the future. This vision should seek to reduce environmental impacts and promote social responsibility within the context of increased regional prosperity. These three elements, taken together, form the basis for a vision of a sustainable future.

The vision should be embedded in guidance for fund managers and providers. Even the prescriptive conditions attached to Structural Funds allow additional local criteria to be added to the programme. A common objection raised when the Environment Agency was proposing environmental screening criteria for EU Structural Fund Programmes in the north-west was that it hindered the ability of fund managers to allocate the money to needy beneficiaries. This is a short-sighted view if the waste of resources caused by poor co-ordination and an unfocused vision is included in the calculation of value for money.

Better and more imaginative indicators of sustainable progress should be developed and built into the early stages of new projects. This strategic approach will allow service providers to promote bottom-line benefits to potential beneficiaries, while being aware of their more far-reaching duties to the prosperity of the region.

Conclusion

A wide range of approaches designed to help SMEs improve their environmental performance has been tried in the north-west of England over the past decade. These vary from the ad hoc assistance offered by government-sponsored assistance bodies to multi-million-pound demonstration projects making use of European funding initiatives.

The main lessons that have been learned during this period are that the majority of the SMEs that are the focus of this attention have yet to fully appreciate the reasoning behind the initiatives. This would explain why recruitment for demonstration programmes is so problematic and why assistance bodies and representative organisations, such as the chambers of commerce, report such low proactive interest in environmental matters. This is partly due to the limited relevance of environmental protection legislation for small companies but is also due to the supply-led nature of funding for SME assistance. Much of the funds are awarded to bodies that understand how to obtain these funds. A demand-led system should ensure a much higher take-up rate from companies that have taken the trouble to explain their needs to fund managers.

A combination of consistency, co-ordination and vision will radically transform small-business environmental improvement programmes in the UK. A regional body that has joint responsibility for sustainable development and economic growth, including inward investment, would be able co-ordinate such a programme, providing that it adopted a progressive style of management. While this would require widespread co-operation among the diverse range of fund managers, all parties would ultimately gain from co-operation and the pooling of experience and knowledge. This would allow projects such as the NCBE's BEDP programme to cause ripples far beyond the businesses that directly benefit from the advice.

21
Driving small and medium-sized enterprises towards environmental management

Policy implementation in networks

Theo de Bruijn and Kris Lulofs

While much of the attention in the past has been focused on large firms, nowadays there is a growing recognition of the need to transform small and medium-sized enterprises (SMEs). After all, SMEs are collectively responsible for a significant portion of the total environmental burden. An important question, therefore, is how to engage with SMEs and convert them to environmental management.

In this chapter, the basic policy approach that has been chosen for SMEs in the Netherlands is analysed and evaluated.[1] Adopting environmental values and environmental management systems are thought to be important steps in the transformation process. Instead of dealing with SMEs directly, the Dutch government tries to facilitate and manage the formation of networks. More emphasis is laid on the organisational structure of policy implementation than on the policy instruments themselves. Within policy networks intermediary organisations, such as trade associations, are partners in policy-making and policy implementation. These intermediary organisations are supposed to convince and help SMEs by adapting and implementing concepts of environmental management which in turn should raise responsiveness, capacity and capability among SMEs. The basic question we ask in this chapter is: 'Is the approach chosen valuable in achieving sustainability within industry?'

1 Our definition of SMEs is related to our research population. Although other definitions are possible, here an SME has fewer than 200 workers.

We have studied the effectiveness of the Dutch approach between 1990 and 1996. In this chapter we highlight some of our findings. We start by giving a short introduction on the way in which environmental policy in the Netherlands is implemented. Special attention is given to the target-group policy. We then focus on the stimulation of environmental management in SMEs. This can be seen as a good example of the consensual, indirect steering model in which policy networks play an important role. We end the chapter by evaluating the Dutch policy approach in the light of its effectiveness for transforming SMEs.

Environmental policy in the Netherlands

The Netherlands is a very densely populated country. It has a comparatively large amount of industry, intensive farming and a fast-growing infrastructure. For these reasons there is a considerable degree of environmental degradation. Furthermore, the Netherlands may be described as a highly consensus-based community with a planning tradition covering a wide range of social aspects. A Dutch characteristic is the long tradition of governmental consultation with various groups in society (VROM 1997). This is reflected in the way our economic system and national economic policy works. The so-called 'polder-model' relates to negotiations and agreements between government, industry and trade unions on the sensitive balance of productivity, returns, competitiveness, wages, tax systems and currency inflation. Current practice in the environmental field continues and reinforces this custom of co-operation and shared responsibility. Co-operation between government and industry, negotiations within industry sectors, reasonableness and covenants (instead of 'command-and-control') are key factors in the greening of industry in the Netherlands.

Dutch environmental policy is created through a process of close co-operation between government, the business community, non-governmental organisations (NGOs) and other stakeholders (Bressers and Plettenburg 1996). Of course, direct regulation also plays an important role in Dutch environmental policy. But, even here, consultation with all the relevant parties is important. Often, however, voluntary agreements are also concluded with the business community. Instead of simply imposing legislation, the Dutch government may negotiate agreements with environmentally important sectors of industry with the aim of implementing environmental objectives.

The 1970s and 1980s

The basis for environmental policy in the Netherlands was laid in the early 1970s. In 1972 the Ministry of the Environment (VROM) was established. With some urgency environmental laws were formulated. The policy during this time had two important characteristics:

- **Sectoral approach**. Environmental problems were seen as mainly 'hygiene' problems within separate 'compartments'. Each environmental compartment (water, air, soil, waste, etc.) was seen as a different and isolated problem area. Different laws and policies were formulated for each compartment.

- **Use of permits.** The central instrument in Dutch environmental policy was the ban on performing any environmentally harmful activities without a permit. Permits were therefore the most used policy instrument. Lower authorities (mainly municipalities) were responsible for issuing permits.

The complex, fast-changing rules were not sufficiently clear or specific enough for SMEs. Furthermore, the monitoring of the rules by the public authorities, and their enforcement efforts in cases of non-compliance, were not very effective.

During the 1980s an additional strategy was used that was grounded on incentives. Soon it became clear, however, that, within the framework of free competition and free trade within the EU, the ability to raise or lower costs is limited. And, of course, industry has been willing and quite successful in opposing additional financial burdens. Moreover, evaluation research on the effectiveness of economic instruments proved that this strategy often resulted in the installation of end-of-pipe technology.

Both regulation and incentives were implemented in a hierarchical manner with little attention to the capacity or capability of SMEs or the regulators. These strategies were, therefore, not successful in driving SMEs towards improved environmental management.

Another strategy was called for, aimed more specifically at eliciting private initiatives and thus shared responsibility. This approach is not only designed to achieve more broadly based support for government policy, but also recognises that the know-how necessary to reduce environmental pollution can be largely found among the polluters themselves. Below we discuss some characteristics of this new strategy.

Towards an integrated policy approach

The cornerstone of current Dutch environmental policy is the National Environmental Policy Plan (NEPP) which was first published in 1989 and subsequently at four-year intervals. The plans aim for radical changes in order to make environmental problems manageable within the next 25 years. This means, for instance, that emissions of the most heavily polluting substances must fall by 80%–90%. This cannot be achieved by conventional policy instruments alone.

The sectoral approach was left behind and replaced in the NEPP by a thematic one (see Table 1). For each theme, the objectives for the next 25 years are set and relevant 'target groups' are identified that contribute to the problems described within the themes. Industry, agriculture, transport and consumers are some of the key target groups. Wherever possible, negotiated agreements are concluded with these groups which specify their contribution to dealing with the problem. The voluntary approach is, however, part of a policy mix in which there is also room for other strategies, such as direct regulation.

To transform SMEs, two main paths are followed: target-group policy; and stimulation policy on environmental management. Below we describe some basic

Policy themes	Targets (simplified)
Climate change	Reduction of CO_2 emissions by 20%–30%
Acidification	Reduction of SO_2 emissions by 80%–90%
Eutrophication	Reduction by 90%
Diffusion of toxic and hazardous waste	Reduction of emissions by 50%–90%
Waste disposal	Reduction of amount of waste dumped by 60%
Disturbance	Reduction of the number of people significantly affected by noise or odour by 70%–90%
Groundwater depletion	Reduction of affected areas by 25%
Dissipation	Increased material intensity

Table 1: DUTCH POLICY THEMES AND TARGETS

characteristics of the target-group policy. The subsequent section is devoted to the stimulation of environmental management.

Target-group policy: consultation and negotiations

A fundamental principle underlying the NEPP is that responsibility for reaching the environmental targets lies primarily with the target group itself (Suurland 1994). The setting of targets at the state level remains the exclusive responsibility of the government (as shown in the NEPP). After the targets have been set, the target groups have a strong say in all further stages of the policy process. Under the target group policy several sectors of industry participate in a consultation process with the authorities. During this process all of the relevant goals from the NEPP are translated into sector-specific goals. These reflect the contribution of the sector to the solution of the problems. After this, a 'declaration of intent' is formulated. This is a covenant between the parties involved and it contains agreements on specific targets for the relevant sector.

In homogeneous sectors of industry the sector goals are then directly translated into an implementation plan. The idea is that companies resemble each other so much that the measures needed can be generalised at sector level. An example of a homogenous sector in the Netherlands is the printing industry. In heterogeneous sectors, however, the targets need translation and specification at company level. Companies in these sectors are required to draw up their own environmental plan.

The trade association is an important player both during and after the negotiations, often acting as the sector's representative. The promotion of the terms of the agreements among companies within the sector is also the responsibility of the trade association.

Basically, the new strategy adopts a multi-level approach. Besides the macro level (national government) and the micro level (SMEs), the meso level (intermediary organisations) is involved in order to obtain access to players at the micro level.[2] The fact that the system is multi-level is not that new. What is special is the way in which the levels communicate, adjust to each other, produce agreements and implement those agreements. The players at the meso level are responsible and have to direct their influence, resources and power towards those at micro level.

The success of this strategy depends heavily on the capability of the sector to organise itself. If the sector fails to live up to the agreements by implementing the measures needed, there is always the threat of direct regulation. For companies this means loss of flexibility. This puts pressure on intermediary organisations (the trade associations) to influence, assist and, if necessary, force individual firms towards the negotiated adjustments and outcomes.

The target-group policy in practice

The target-group policy, with its emphasis on consultation and negotiated agreements, specifies what companies need to do. Thirteen sectors of industry were selected based on their 'share' of the total environmental burden. By the end of 1997 only nine of the sectors had concluded a negotiated agreement. The individual translation into an environmental plan was also delayed. Some 63% of the companies in the chemical sector, for instance, had concluded this plan on time (Overleggroep Chemische Industrie 1995). Taken together, the individual environmental plans do not always lead to the desired reduction for the sector as a whole. In conclusion, the consultation, negotiation and implementation took more time than expected and, in some cases, led to a watering-down of environmental targets. During the process that led to the various agreements and the activities of the companies that followed the agreement, there were some complaints about a lack of transparency (e.g. Biekart 1994). The agreements, however, all have the option of imposing sanctions on 'free-riders'. If a company fails to live up to the agreement, the enforcing authorities are obliged to apply supplementary conditions regarding the company's permit. Local authorities have to take the declaration of intent (one of the outcomes of the target-group policy) into account when issuing a new permit. In this way free-riding can be forced back. Companies that have neglected the chance to translate the demands as laid down in the declaration in their own way (by taking into account company-specific circumstances) will eventually be forced to do so by means of the permit system. Laggards can also expect more stringent inspections. Regulation has been announced that opens up the possibility of imposing some special requirements concerning mandatory analysis of the environmental performance of a company by third parties.

◻ ***Stimulation of environmental management***

The central concept underlying environmental management is to stimulate the companies' own responsibility and actions. This should improve the responsiveness,

2 See Chapter 14.

capacity and capability of SMEs. Where the target-group policy specifies what needs to be done, companies are expected to learn how to implement these requirements via environmental management.

In 1989, the Ministry of the Environment in the Netherlands issued the Memorandum on Environmental Management (Tweede Kamer 1989). The objective of this was to make Dutch companies introduce an environmental management system by 1995.

To achieve this aim, the Memorandum presented a programme of activities based mainly on the acquisition—through research, stimulation and demonstration projects—of specific knowledge which would then be disseminated among the business community through guidance and education. Intermediary organisations were asked to play a special role in this. They were responsible for keeping in touch with the individual companies. By supplying information, attempts were made to stimulate the introduction of environmental management into the companies. The underlying idea was that, by offering support, the uncertainty and, therefore, the cost to the individual company could be reduced to such an extent that they will actually proceed to introduce the system.

Instead of dealing with SMEs directly, the government tries to facilitate and manage the formation of networks and to enable those networks to promote and contribute to the implementation of the negotiated goals and rules. These networks, in which trade associations play an important role, are believed to be more capable of reaching and helping SMEs. A multi-level approach is therefore seen here too.

Different functions within a policy network

Intermediary organisations within a policy network can perform three functions. We talk about a **will-influencing function** when a network organisation tries to influence the motives of the dominant coalition within companies with regard to environmental management. Network organisations have to convince this dominant coalition of the relevance of environmental management and improvements in their environmental performance. We talk of a **supporting function** when a network organisation tries to help companies implement environmental management systems and introduce measures to improve their environmental performance. This involves supplying resources such as model approaches, guidelines and manuals, and offering courses and training. We talk of a **repressive function** if the network organisation intends to steer the unwilling members of the target group in the desired direction through gentle or hard pressure. Performing a repressive function means acting out a position of power. This is summarised in Figure 1.

The basic assumption of consensual steering in a network approach is that as the intended network relations are exercised to a greater extent, companies will show a more positive attitude and carry out more of the activities asked for.

Network organisations surrounding SMEs

Implementation of the policy programme on environmental management was aimed at the formation of a policy network as a first step. The network surrounding companies with regard to environmental management consisted of the trade

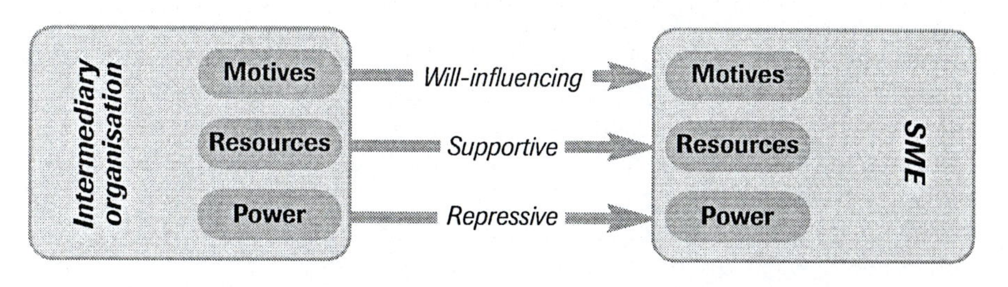

Figure 1: THE THREE NETWORK FUNCTIONS BETWEEN SMES AND INTERMEDIARY ORGANISATIONS

association, industrial environmental agencies, central government, the municipalities, employee organisations and consulting firms. The main role was for trade associations. They had to convince their members (and possibly the remaining companies within their sector) to help them actively: for instance, by providing handbooks and courses, and to use their position of authority and power to force members to take environmental management seriously. Municipalities were expected to support these activities for the companies within their borders.[3]

A second important category was formed by the industrial environmental agencies. These were set up specifically for the introduction of environmental management systems in SMEs. They had a regional perspective and were mainly intended to motivate and support companies. Trade unions were expected to inform their members, thus creating support on the shop floor. Consulting agencies could be valuable in assisting companies with the implementation process of environmental management. Finally, central government only saw a motivating task for itself. And, of course, it co-financed some of the activities of the other network organisations. Table 2 summarises the network organisations and their expected functions.

Function / Actor	Will-influencing	Supportive	Repressive
Trade association	X	X	X
Municipality	X	X	X
Industrial environmental agency	X	X	
Trade unions	X		
Consulting agencies		X	
Central government	X		

Table 2: NETWORK ORGANISATIONS AND EXPECTED FUNCTIONS

3 See Chapter 16 on the role of local authorities.

Environmental management in companies

To evaluate the success of the stimulation policy we looked at the following industries: chemical, printing, synthetics processing and concrete products. Random samples were taken from these four sectors. Out of the 343 firms we contacted, 141 (41%) co-operated in our research. Our main research question was whether the new policy approach was successful.

First, we looked at the degree of penetration of the different activities that were carried out through the network. Table 3 summarises the main findings. These outcomes show that the respondents were certainly aware of the network's activities.

Network indicators	Percentage of companies
Familiar with terminology	92%
In possession of the supporting material	62%
Attended informative meetings	54%
Implementation supported by the network	43%

Table 3: THE DEGREE OF PENETRATION

A second aspect we looked at was the level of progress of the companies. Table 4 shows the distribution of the companies on this variable.[4] The level of progress found is reasonable compared with the descriptive data found by the official evaluation studies performed (commissioned by the Ministry of the Environment) in 1991, 1992 and 1996 (Calkoen and ten Have 1991; Van Someren *et al.* 1993; Heida *et al.* 1996).

Level of progress	Percentage of companies
Inactive	6%
Orienting	39%
Initiating	51%
Advanced	4%

Table 4: THE PROGRESS OF ENVIRONMENTAL MANAGEMENT IN COMPANIES

4 The variable 'progress' comprises two aspects. First, we determined to what extent the companies had implemented an environmental management system. Driven by the research design, we

Policy effectiveness

At first glance, it could be said that the stimulation policy has been quite successful, given the fact that these results were achieved in only a few years. The main question, however, is whether the stimulation policy, implemented in a network configuration, is accountable for this. We studied the effectiveness of this configuration in the Netherlands between 1990 and 1996. We used an inter-organisational explanatory model to conceptualise and make empirical explanatory research possible. In order to be able to measure the level of transformation of SMEs between 1990 and 1996 we used the attitude of the management towards environmental management and the level of progress in building environmental management systems and improving environmental performance as variables. The research model is shown in Figure 2.

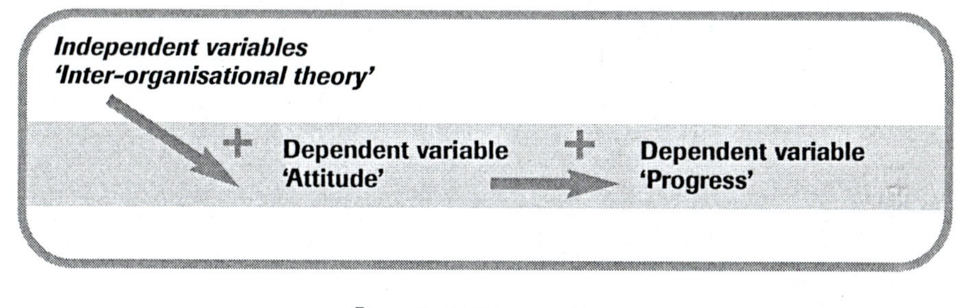

Figure 2: RESEARCH MODEL

The implementation of environmental management requires changes within the organisation, or in other words changes in the patterns of behaviour of members of the organisation. These changes are, to a certain extent, determined by attitudes towards environmental management within the company. The dependent variable 'attitude' regarding changes in the rules of behaviour can be placed on a continuum ranging from acceptance to rejection of change. The dependent variable 'progress' measures stable patterns in the behaviour of members of the organisation.[5]

Research expectations were formulated about the positive relations between the attitude of the dominant coalition and stable patterns in the behaviour of members of the organisation. These research expectations were proved to be true (Kendall τ-c = 0.42, T-value 5.84).

used as indicators for an environmental management system the requirements posed by the Dutch Ministry of the Environment. These requirements are, however, very similar to those of ISO 14001. The existence of a management system is seen as an indicator for progress on an organisational level. Second, we determined the amount of measures taken on environmentally relevant issues within a company. After all, *sustainable* environmental management is more than just an organisational matter. We combined these two aspects into an index for the level of progress on environmental management within the company. The 'inactive' companies have developed few or none of the elements of an environmental management system and have taken hardly any concrete measures to minimise their environmental impacts. The 'advanced' companies are the opposite of this: they have developed all, or nearly all, the elements and have also taken a lot of measures.

5 See footnote 4.

The next question, of course, is whether these results can be explained out of the efforts of the intermediary organisations (i.e. the organisational setting in a policy network). Our inter-organisational theory proposed that it was the organisational framework (at the meso level of intermediary organisations) that can explain the results. The basic assumption was that as the intended network relations are exercised to a greater extent, companies will show a more positive attitude (see Fig. 2).

To test this assumption, we first measured the frequency of contacts between network organisations and the company. Furthermore, we looked at the relations between the different network functions and attitude. Each relation proved to be positive and significant (Table 5). The chance that all four relations point in the same direction (as in our case) by accident is extremely small. This means that, as network organisations are more active, companies show a more positive attitude towards environmental management and are more advanced with their implementation of it.

Indicator	Relation
Frequency of contacts	Kendall τ-c = 0.18; T-value 1.99; significance: α approximately 0.02
Will-influencing function	Kendall τ-c = 0.27; T-value 3.08; significance: α approximately 0.00
Supportive function	Kendall τ-c = 0.24; T-value 2.29; significance: α approximately 0.01
Repressive function	Kendall τ-c = 0.16; T-value 1.71; significance: α approximately 0.04

Table 5: NETWORK RELATIONS

The second step consisted of an analysis of the explanatory power of the inter-organisational theory in total (and thus the activities arising from the network), via regression analysis. When we take a look at the explanatory power of the independent variables, we can explain some 22% of the variance of the dependent variables. Best explanations are found in the activities of the trade associations and, surprisingly, those of central government. As described earlier, the government did not deal with SMEs directly. Its role was to facilitate and manage the formation of networks and to assist them in their attempts to encourage more environmental management within SMEs. The empirical outcome, that central government is an influential actor in this configuration also, is therefore surprising. It might be explained by normative effects of the policy position taken. Nevertheless, trade associations are the most important actors in the network.

Although our analysis leads us to the conclusion that more than 20% of the variance in attitudes can be explained out of the activities of the policy network (which we consider to be quite successful),[6] it is obvious that other company-specific factors can specify the exact influence of network relations. Nevertheless our general conclusion is that the approach used was rather successful (de Bruijn and Lulofs 1996).

Conclusions

In this chapter we have taken a look at the way Dutch environmental policy aiming at SMEs is implemented nowadays. To engage with SMEs, an indirect, consensual steering model is used in which policy networks play an important role. Within this policy network intermediary organisations (for instance trade associations) are partners in policy shaping and refining, and in policy implementation. A first argument in favour of such a steering arrangement is that organisations that are more closely involved with the eventual target group, are better able to shape and implement the central government's policy in an effective and practically applicable way. In addition, intermediary organisations would be able to get the message across to the policy's target group more convincingly. Finally, the target group would tend to justify its own behaviour towards some intermediary organisations. For instance, in the Netherlands, the trade associations are believed to be the suitable intermediary organisations to shape and implement policies on pollution prevention and environmental management for SMEs.

Although it is still a little early for a final judgement of the new policy approach, some conclusions can be drawn. For the time being the score is positive. A lot of specific knowledge that is of use to companies has been developed, both as a result of the target-group policy and the activities of the policy network. The formation of this network was also a success. Our research has shown that a real network has developed over the years. The policy mix also offers options for governments to co-operate with proactive companies while at the same time employing a strict regime towards laggards. A further positive development is that the attitude of companies towards environmental affairs has improved. There are also signs that the relationship between the legal authorities and the companies seems to have improved (e.g. Stuurgroep Grafische Industrie en Verpakkingsdrukkerijen 1997).

Therefore, we conclude that a consensual steering approach, based on co-operation between government and industry, sectoral negotiations, reasonableness, and covenants (instead of command-and-control), can really reach SMEs and influence their behaviour positively as far as environmental management and environmental performance is concerned. Nevertheless, the explained variance is not high enough to lead us to believe that consensual steering with the use of networks can cope with the increasingly intractable nature of environmental problems that need to be tackled. We should realise ourselves that the easy reduction percentages are already realised or will be realised in the near future (i.e. the 'low-hanging fruit').

---

6 In other situations we judged the theory could explain up to 56% of the variance in the dependent variables.

A second conclusion is in line with that of Angel and Huber (1996). They conclude that private organisations are influenced by external forces as far as entering new items on the agenda is concerned. Characteristics of the organisation itself tend to be more influential than external pressures. This conclusion means that, within a configuration chosen by the Dutch government, effectiveness could increase considerably when the characteristics of the organisation itself are taken into account when trying to increase pressure. Help offered from outside should therefore be tailor-made to the sector of industry and deal with organisational characteristics.

As a third and last conclusion, we can say that the explanatory power of inter-organisational activities is stronger when a long-term relationship exists between the intermediary organisation in the network and the SME and where the SME sees the intermediary organisation as reliable and credible. New contacts or organisations do not work. As mentioned earlier, best explanations were found in the efforts of trade associations that were studied. New contacts or organisations such as the new industrial environmental agencies do not work effectively. In our research towards inter-organisational relations there is no such thing as 'love at first sight'.

The new policy approach is, therefore, certainly not 100% successful. The main question, however, is whether there is an alternative. The stimulation policy of environmental management is aiming at target groups in a sense that they learn how to implement environmental management. The question on what to achieve in terms of environmental performance is dealt with in the target-group policy. The results of the target-group policy are the minimal requirements. The local authorities are able and obliged to use the results of the target-group policy as a basis for issuing the permit. Environmental management is the tool that enables the target groups to reach the minimal requirements set by the target-group policy or to reach even higher goals. The greatest benefit in this approach lies in the regained flexibility. The current policy mix gives some room to proactive companies and supports them actively. At the same time, the permit system can tackle laggards and secure developments in other companies. Therefore, we feel that the consensual steering approach is a promising new supplement to engage with SMEs, but it needs to be used in combination with various other instruments.

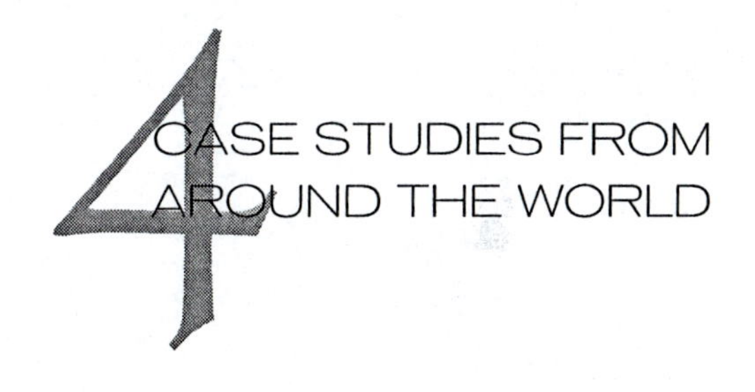

4 CASE STUDIES FROM AROUND THE WORLD

22
Small-scale enterprises and the environment in developing countries

Andrew Scott

In developing countries, the economic and social significance of small-scale enterprises is well recognised. They provide employment to very large numbers of men and women and, in many places, offer the only opportunity for the poor to secure some form of livelihood. By one estimate, between 17% and 27% of the labour force is employed in micro and small-scale enterprises (Mead and Liedholm 1998). In Africa and Asia the majority of the population lives in rural areas where small-scale (non-farm) enterprises provide 20%–45% of full-time employment and 30%–50% of rural household income (Haggblade and Liedholm 1991). Latin America, which is more urbanised, has an estimated 50 million micro and small-scale enterprises, employing 120 million people (Berger and Guillamon 1996).

Enterprises employing up to 50 workers are usually categorised in developing countries as small and medium-sized enterprises (SMEs).[1] The vast majority of these are at the lower (micro) end of the size spectrum, and employ fewer than five workers. They are characterised by low returns and generally use simple, low-cost technologies. They serve local, low-income markets, and are often part-time or seasonal concerns. Many are in the informal sector (i.e. they operate outside any regulations).

Small-scale enterprises in developing countries tend to be concentrated in a small number of industrial sectors. The majority are engaged in trading activities of one kind or another (retailing, hawking, vending), but they also account for a large proportion of those employed in the manufacturing sector. In India, for example, 27%

1 The terms 'SME' and 'small-scale enterprise' are used interchangeably here.

of the 12.6 million micro and small-scale enterprises are in manufacturing (Fisher *et al.* 1997).

Traditional agricultural activities and formal-sector employment in medium and large-scale enterprises cannot provide sufficient employment opportunities for the large numbers in developing countries entering the labour force each year. Small-scale enterprises are increasingly seen as an important means of addressing the problem of unemployment. SME promotion is now entrenched in national development plans and strategies and, over the last decade, small-scale enterprise development has received substantial support from the donor community, notably through credit programmes (OECD 1997).

Small-enterprise development programmes have been established in most countries, but they have scarcely addressed SME environmental sustainability. The environmental impacts of small-scale industries have tended to be ignored by small-enterprise promotion organisations, including the donor agencies of industrialised countries. This is partly because they have assumed that such enterprises have little impact due to their small scale. Many small-scale industries do adversely affect the local environment, however, and one study claims that 'the bulk of pollution in urban areas is the result of dispersed medium/small-size industries' (Hamza 1991).

Given the very large numbers of small-scale industries in developing countries, there is clearly a need to consider their aggregate contribution to environmental degradation as well as the local impacts of individual units. Increased emphasis on the development of such enterprises suggests that there is a growing need to address questions arising from pollution by small-scale enterprises, efficiency of raw materials use, energy efficiency and health and safety hazards.

In order to understand how SMEs can contribute to environmentally sustainable employment creation strategies, it is important to understand how the factors affecting environmental impact differ between large and small-scale enterprises, and how these differences will determine the effectiveness of options for improving the environmental performance of small-scale producers.

With this in mind, the following sections review the literature about small-scale enterprises and their environmental effects. A summary of the findings of recent empirical research by the Intermediate Technology Development Group (ITDG)[2] is then presented and, finally, some conclusions are drawn about the significance of the problem and how to tackle it.

Literature review

In the first major survey of the literature on small-scale enterprises and the environment, Lawrence Kent noted that: 'Most literature on small-scale enterprises emphasises issues of efficiency and employment, ignoring environmental concerns'

2 ITDG is an international non-governmental organisation (NGO) whose mission is to build the technical skills of poor people in developing countries, enabling them to improve the quality of their lives and that of future generations.

(Kent 1991). The number of studies that consider the environmental aspects of SMEs remains small. The few there are present some data on the pollution caused by small-scale industries and highlight policy and regulatory issues, particularly in relation to SMEs in urban areas.

One reason for this neglect of environmental impact, despite heightened awareness of the environment and development, might be that the great majority of SMEs are engaged in trading and commerce that do not have significant direct environmental effects, although Pallen (1997) notes that: 'Non-manufacturing sectors, such as roadstand shops and restaurants, can contribute to environmental problems through inefficient use of natural resources such as firewood.' Large numbers of SMEs, however, are engaged in manufacturing which inherently has some effect on the environment through the transformation of raw materials and consumption of energy.

Some small-scale industry sub-sectors are generally recognised as having a greater environmental impact than others, because of the nature of their processing or because of their total contribution to production in their sub-sector or location. Textiles, tanneries and brick-making are all cited as serious polluters by most authors. Metal-working, foundries, electroplating and the associated activities of vehicle repair are also frequently mentioned. Other significant polluters are food processing, mining, paint and print shops, battery production and recycling, soap making and wood processing (Kent 1991; Biller and Quintero 1995; Pallen 1997; Bartone and Benavides 1997; van Diermen 1997).

Though there has been some consideration of the practical problems in studying the environmental effects of small-scale producers (Frijns *et al.* 1997), the data that is presented is not readily comparable between studies. Comparisons have been made on the basis of 'pollution units' (van Diermen 1997) and pollution per employee (Dasgupta *et al.* 1998b), but none provide data on impact or pollution per unit of output as advocated by Kent (1991).

The scope of the limited research to date has been confined to the effects of the production processes that take place within small-scale industrial units. The impact of the associated transport of raw materials and finished products, and the advantages and disadvantages of the dispersion (e.g. in rural areas) or concentration (e.g. in urban informal sector areas) of small-scale producers have not been considered, except in relation to centralised effluent treatment plants. Nor has there been much attention paid to the environmental health and safety of workers in small-scale industries, though it is clear from the limited evidence available that the health of many thousands of SME workers is affected by their working conditions (Scott 1998).

The literature tends to focus on the pollution and waste generated by small-scale industries. Common issues highlighted include location, mitigation and waste reduction, and regulation.

Location

The dispersed nature of small-scale industries, with the majority in rural areas, lends weight to the general assumption that they have little environmental impact. Because the output from each enterprise is small, the receiving environment is assumed to be able to absorb the waste generated. Apart from the fact that scale is generally measured by number of workers, and this assumption rests on the further assumption of a direct relationship between number employed and output, it clearly does not take account of industries in environmentally sensitive locations. Even if small-scale industries are not a major source of pollution at national level, there is a need to address them because they contribute to local problems (Biller and Quintero 1995).

The assumption does not apply where small-scale industries are concentrated together, as is found in urban areas. The environmental effects of small-scale industries are more noticeable in urban areas because their geographical concentration results in a greater intensity of pollutant load and because they are frequently located close to residential areas so affected populations are greater. There can also be effects on the public infrastructure (sewers, water supplies, etc.), although the physical clustering of production units in urban areas permits options for the collective treatment of wastes and mitigation options which would not be feasible for dispersed, rural industries. It is not surprising, therefore, that much of the attention that has been given to small-scale industries and the environment has focused on problems associated with urban areas (Biller and Quintero 1995; Bartone and Benavides 1997; Frijns *et al.* 1997; van Diermen 1997).

Scale and technology

When it is not assumed that small-scale industries have insignificant environmental effects, the almost opposite assumption is made instead—that small-scale enterprises are in fact worse for the environment than their large-scale counterparts. Although the total pollution load from small-scale units might be low because they account for a small proportion of total output, the pollution per unit of output is assumed to be higher.

This assumption is based on several further assumptions: that small-scale enterprises tend to use 'outdated' technology which is inefficient; that the skills and knowledge available to them result in less efficient use of raw materials and energy; that they tend not to undertake any kind of effluent or waste treatment; and that they tend to avoid inspection and the enforcement of regulations by the authorities.

There is a distinct lack of empirical information about small-scale industries and their environmental effects, and little in the literature that actually attempts to compare across scales of production. Thus, there is very little on which to substantiate this assumption and, it should be noted, it overlooks other factors, such as the poverty of most people employed in SMEs which means they cannot afford to waste anything that might have some value.

Approaches to abatement

Much of the literature that does exist on SMEs and the environment is concerned with ways of reducing pollution. Approaches have usually been described hypothetically because there is little documented experience.

The 'command-and-control' approach, which is the basis of environmental regulation in many developing countries, includes direct measures such as land use restrictions and penalties for contravening standards. Where regulations do exist they can be inadequate or exclude small-scale enterprises, and there is often lack of capacity to enforce them (van Diermen 1997). When they are enforced, the intended gains are not always achieved. In Delhi, for instance, enforcement of zoning regulations in 1996 closed 1,328 small-scale enterprises with a view to relocation, putting out of work an estimated 125,000 workers. This resulted in dispersal of pollution rather than a reduction (Dasgupta 1998).

Voluntary compliance or 'informal regulation' has been advocated for pollution control in developing countries (e.g. Pargal and Wheeler 1996b), but has rarely been tested with the SME sector. In one instance where it has been tried successfully, the approach depended heavily on facilitation by a development agency (Blackman and Bannister 1998).

Similarly, the promotion of 'cleaner production' approaches has not been widely attempted among SMEs. A United Nations Industry and Development Organisation (UNIDO) study (1995) demonstrates clearly the potential for pollution abatement in small industry through 'cleaner production' approaches.[3] The financial returns to medium-scale enterprise owners were also demonstrated in the Environmental Pollution Prevention Project (EP3) supported by the United States Agency for International Development (USAID) (USAID 1997).

Empirical evidence

Although there is increasing attention being given to environmental issues by micro- and small-enterprise development organisations, there continues to be little empirical information available to inform decision-making. One of the objectives of recent research by ITDG was to address the problem of lack of detailed and reliable information. For this research, data for three sub-sector case studies—brick-making, mining and textile dyeing—was obtained by a variety of methods, including fieldwork on enterprises in Bangladesh and Zimbabwe.[4]

Brick-making in Zimbabwe

Current trends in housing demand in Zimbabwe show a general shift in preference of housing structure, with more people opting to build brick houses instead of traditional pole and dagga houses. The demand for burnt bricks started to outstrip supply in 1991–92 when demand stood at about 800 million bricks and production stood at only 300 million. The demand for bricks is now estimated at over 1,200 million a year, with annual production still well below this.

The burnt brick industry in Zimbabwe has been dominated by a small number of large-scale plants, located within or close to urban areas. As a result of the increased

3 See Chapter 11.

4 The research was supported by grants from the Global Environmental Change Programme of the Economic and Social Research Council and the Environment Research Programme of the Department for International Development.

demand for bricks, many small- and medium-scale production units have been established. There are now six large-scale brick manufacturers, at least eight medium-scale and several hundred small-scale plants operating on a commercial footing. The sample included in the study by ITDG comprised two large-scale producers, two medium-scale and seven small-scale brick-makers.

There are wide differences in production techniques between small-scale, medium-scale and large-scale brick manufacturers, though they all follow the same basic process. The characteristics of different scales of brick-maker are summarised in Table 1. In small-scale units, brick-making is manual with rudimentary tools in use, while in the large-scale manufacturers the process is largely mechanised. The quality and durability of the product also differs with, generally, a better-quality and more durable brick produced by the large-scale brick-makers.

Characteristic	Small-scale	Medium-scale	Large-scale
Capacity	<10,000 bricks per 8-hour day	>10,000 <30,000 bricks per 8-hour day	>30,000 bricks per 8-hour day
Degree of mechanisation	Low to non-existent (labour-intensive)	Medium to high (fairly labour-intensive)	Highly mechanised
Investment cost	Low to several hundred thousand dollars	Medium to several millions of dollars	High: beyond five million dollars
Type of process	Batch	Partially continuous	Largely continuous
Type of firing technology	Wood/coal-fired scove kilns/clamp	Coal-fired scove kilns	Coal-fired scove kilns, Hoffman, beehive, bull's trench kilns

Table 1: CHARACTERISTICS OF BRICK-MAKING UNITS BY SCALE

A typical small-scale brick-maker is likely to use 157.5 m^3 of anthill clay a year, having a ground disturbance effect over an area of about 80 m^2. The water required (66,150 litres a year) would be sourced from rivers and transported by animal-drawn cart. Wood consumption for fuel was found to be between 30 and 60 trees (depending on size) to burn 90,000 bricks, or 330 kg per 1,000 bricks. The kilns employed by small-scale brick-makers were generally, however, found to have good energy efficiency.

Large-scale brick manufacturers, on the other hand, would typically use 15,000 m^3 of clay a year to produce 6.3 million bricks, for which they would use almost 1,000 tonnes of coal in firing. In addition to emissions of CO_2 and SO_2, large-scale manufacturers generate significant amounts of dust, often above permitted levels.

The consumption of wood-fuel by small-scale brick-makers contributes to deforestation and, because of the preference for particular tree species, loss of biological diversity. Total firewood consumption by small-scale brick producers in Zimbabwe was estimated at around 30,000 tonnes a year, equivalent to less than 0.5% of total annual firewood consumption. This would account for approximately 750 hectares of (clear-felled) woodland. In other words, brick-making does not amount to a significant cause of deforestation in Zimbabwe.

Land degradation results from clay extraction, with a total area of about 12 hectares affected annually in Zimbabwe. The land degradation caused by large- and medium-scale brick-making differs from small-scale operations because they excavate more deeply, rather than widening the excavation area.

The air pollution from brick-making, which includes particulate matter (dust, smoke, fumes), oxides of sulphur, nitric oxide, hydrocarbons and carbon monoxide, can have an impact on human health and on vegetation. In urban areas the air pollution problem is significant since the brick-makers use coal and the large-scale nature of the units intensifies the problem. Gas emissions contribute marginally to global climate change.

Reducing the environmental impacts of brick-making might be tackled on the energy side through improved efficiency (for example, through kiln design changes, improved maintenance of kilns during firing, and reduction of losses), or fuel switching to alternative forms of energy, such as agricultural wastes, boiler waste and coal. Rehabilitation of land could be achieved by filling holes dug in the ground with debris, river sand and broken bricks, and replanting trees would help reduce soil erosion.

Improvements in the environmental performance of small-scale brick-makers will depend on the extent to which they can afford to make improvements and the incentives to do so. Training of brick-makers in the most efficient methods of brick production will be required.

Mining

Mining, almost by definition, causes environmental damage and small-scale mining has attracted particular attention because of its environmental effects. Although small-scale mining in developing countries provides employment and incomes to an estimated six million poor men and women (Ghose 1997), it is often perceived only as an illicit and environmentally damaging activity.

In Zimbabwe, the incidence of small-scale mining increased rapidly after independence in 1980. There are now about 2,000 active small mines, 90% of which are gold mines, employing approximately 10,000 people. There are also between 100,000 and 200,000 people engaged in alluvial gold panning along some 3,000 km of river-bank. For the mining sector case study, the research by ITDG was based on a sample of eight large mines, eleven small-scale mines and eight gold panning sites.

The principal environmental effects of small-scale mining are related to the mining operation (land degradation) and ore processing (chemical pollution). As far as land degradation is concerned, mining operations in Zimbabwe move an estimated total of 55 million tonnes of rock each year, most of it from underground. Small-scale mines, although numerically greater, account for about 10% of this. On average, a small-scale miner produces 20 tonnes per month. The local effects of small-scale mining can be significant, with erosion, riverbank destruction and dam siltation all being effects of alluvial gold panning. But the contribution of mining to overall land degradation in Zimbabwe needs to be considered in relation to the 750 million tonnes of soil eroded each year by agriculture.

The processing of ore by small-scale gold miners entails grinding it into a fine powder which is mixed with water and mercury. The resulting amalgam is then heated or retorted and the gold separated. Up to two tonnes of mercury are used per tonne of gold produced and an estimated 25% is lost during amalgamation, equivalent to a total loss of about 4.5 tonnes of mercury a year nationally.

The mercury tends to be handled with little care to its toxicity and usually without protective clothing, such as gloves. During retorting, mercury vapour escapes into the atmosphere and is inhaled by the workers. Mercury poisoning among small-scale miners has been confirmed by medical evidence (Sunga and Marima 1998).

The use of cyanide for gold separation by small-scale miners is not widespread because they are normally involved with the recovery of 'free' gold from their primary ore. Vat-leaching or cyanide treatment ponds to treat tailings also require considerable capital investment which small-scale miners cannot individually warrant. As a whole, however, the mining sector in Zimbabwe uses between 250 tonnes and 350 tonnes of cyanide per tonne of gold produced (Hollaway 1993).

Other notable environmental effects of mining are water pollution (mercury and cyanide from small-scale mines, heavy metals and acids from large-scale mines), and dust and noise pollution, especially from the machinery used in large-scale mines. The health and safety of workers is another major concern, as is the social disruption resulting from the establishment of temporary communities at gold panning sites.

The lack of awareness about the environmental effects of their activities found among small-scale miners is a major constraint on improving environmental management in the sector. Training programmes for small-scale miners to increase their awareness and knowledge of the environment and the effects of mining are needed, with support from the public sector (government and development agencies). The capacity of existing associations of small-scale miners, such as the National Miners' Association, the Zimbabwe Women Miners' Association and the Mine Developers' and Prospectors' Association, needs to be strengthened to enable them to support their members in improving their environmental performance, advising on environmental impact assessment and monitoring practice.

Arrangements for the processing and marketing of gold could also be reviewed, as these appear to encourage the use of mercury for amalgamation. Collective or centralised processing, using cyanide, would reduce mercury pollution and contribute to health improvements among those engaged in small-scale mining.

Textile dyeing

The textile industry is a significant contributor to the economy of Bangladesh. There are 4,886 textile-processing units, of which 1,254 are weaving and spinning units and 314 dyeing, printing and finishing units (BCAS 1997). The case study research sampled a total of 42 units, including ten cottage-level textile units employing fewer than ten workers.

Data from the Department of the Environment indicates that the total waste-water production and total BOD (biological oxygen demand) load from the textile industry in Bangladesh amounts to 40,000 m^3/day and 26,000 kg BOD/day respectively. For a typical textile mill, the waste-water volume is 70 m^3/tonne and organic load is 115 kg COD/tonne (chemical oxygen demand per tonne) of cloth. The waste-water volume from dyeing and printing and final treatment amounts to approximately 55 m^3/tonne of cloth. The organic waste load is about 49 kg/tonne BOD (BCAS 1997). Survey data on the waste-water produced by average units at different scales of production is shown in Table 2.

Scale of production	Quantity of dyes (kg/day)	Quantity of chemicals (kg/day)	Quantity of waste-water (m^3/day)
Cottage-level	<1-35	10–50	<1-8
Small-scale	5–35	22–76	6–14
Medium	15–82	220–262	12–25
Large	20–118	270–1,564	70–400

Table 2: QUANTITY OF DYES AND CHEMICALS USED AND WASTE-WATER BY SCALE

Analysis of the waste-water showed that it contains a high concentration of organics and chemicals (see Table 3), and that waste-water from small-scale and cottage units has a higher pollutant content. Large-scale dyers, on the other hand, generate large volumes of effluent, with lower pollutant content per m^3.

	Survey averages			
	Cottage	Small-scale	Medium	Large
pH	11.7	7.52	7.32	7.25
BOD[1]	456	360	332	315
COD[2]	690	576	517	469
DO[3]	0.15	2.68	2.90	3.10
TOC[4]	715	864	774.8	314
Alkalinity	41	149	132	133
Total dissolved solids	3,900	2,410	2,120	1,290

(All units mg/litre, except pH)
1 Biological oxygen demand; 2 Chemical oxygen demand; 3 Dissolved oxygen; 4 Total organic compound

Table 3: TEXTILE WASTE-WATER SURVEY RESULTS

Wastage of dyes and chemicals is a major problem for cottage-level producers. Cottage units do not, however, have a significant adverse impact on soil characteristics, crop production and wetlands. Wastage of dyes and chemicals was found to be at a minimum in large-scale units because of their dyeing techniques and good management practices.

Almost all textile processing enterprises discharge their waste-water into the nearest body of water or low-lying areas of land. Few textile units in Bangladesh have effective waste-water treatment facilities, and their discharges may affect the aquatic environment, soil fertility, and the quality of surface and groundwater. The significance of the impacts of small or large producers depends on the local intensity of discharges. Clustered small-scale and cottage units were found to be as significant as large units in terms of pollutants and environmental degradation.

The technological options open to small and cottage units for mitigating the impacts of their effluent are limited to biological treatments—such as aerated ponds or activated sludge—on grounds of cost and availability of technical skills. Central effluent treatment plants for clusters of small and cottage units might be considered, but there is a need for research to determine the feasibility of this option. As well as considering waste-water treatment systems to reduce pollution from small-scale textile units, cleaner production approaches, including waste minimisation, good housekeeping, alternative chemicals, process modification, and conservation of water and energy, might also be applied to reduce environmental pollution.

Common issues

Caution should be exercised in drawing general conclusions from this research by ITDG about the environmental impact of small-scale industries in developing countries. The detailed case studies covered only three of a multiplicity of industrial sub-sectors, in only two countries. Production in these sub-sectors has different environmental effects and the two countries have quite different physical and climatic characteristics and population densities. Nevertheless, the findings do lend empirical support to the argument that the environmental impacts of SMEs can be locally significant. The findings also suggest an approach to the mitigation of the impact of small-scale industries in developing countries where there is weak enforcement capability and a need for low-cost options.

While assessment of environmental effects has been shown by the research to be possible for small-scale industries, the *impacts* on the environment were less readily measured. Crucially, this would require understanding and definition of the receiving body for pollutants, discharges and other waste. A full environmental impact assessment, as conventionally carried out, would have required more extended fieldwork to collect the additional information needed about the receiving environments for each production unit. When the receiving environment is defined at the national (and more so the global) level, however, it seems clear that the environmental impact of small-scale industries in developing countries, even allowing for the difficulties of aggregation, is not as significant as that of large-scale producers and other human activities.[5]

5 A similar conclusion was drawn by Dasgupta *et al.* 1998.

Pollution per unit of output is a convenient indicator that allows ready comparison between producers in the same sector, and the research demonstrates that measurement in these terms can be made at reasonable cost and without highly sophisticated techniques. The pollution per unit of output measurement should focus on the main emissions, which can be identified by means of a materials balance or inventory. In the case of the brick-making producers, for example, clay extraction (in volume and area terms) and CO_2 emissions were all estimated per thousand bricks produced (see Table 4).

Effect	Scale of production	Pollution/unit of output
CO_2 emissions	Small	763 kg/1,000 bricks
	Medium	454 kg/1,000 bricks
	Large	929 kg/1,000 bricks
Land degradation	Small	1.06 m^2/1,000 bricks
	Medium	1.07 m^2/1,000 bricks
	Large	0.26 m^2/1,000 bricks

Table 4: POLLUTION FROM BRICK-MAKING PER UNIT OF OUTPUT

Though a useful measure for assessing or comparing impacts, the standards set by the regulatory authorities are not usually defined in terms of pollutant per unit of output. A system of benchmarking based on this indicator would provide a useful guide to SMEs, to financial institutions and to enterprise promotion agencies. The usual indicators of scale (number of workers, turnover, or capital invested) used by small enterprise development organisations are not helpful for assessing environmental effects and would require greater attention to volume of production.

Scale

On the question of scale, the evidence from these three case studies suggests that the assumption that small-scale industries pollute more per unit of output than their large-scale counterparts does not necessarily hold. In brick-making, for example, small-scale producers, who mainly use wood to fire their bricks, contribute to deforestation, while large-scale producers who use coal, do not. In terms of greenhouse gas emissions, small-scale producers, which account for about half of total brick production in Zimbabwe, emit about half the industry's CO_2; but per thousand bricks produced their CO_2 emissions are on average lower than large-scale producers. Large-scale producers generate most of the industry's SO_2 emissions from their coal combustion, but SO_2 emissions are negligible for small-scale producers. The large-scale brick-makers have more dust in the workplace (the latter often exceeding standards), because of mechanisation and volume of material processed, and they show greater local land degradation.

In the case of textile dyeing, on the other hand, the small-scale dyers generate more pollution per unit of output, although the volumes of waste-water discharged into individual water bodies are much less. While in the mining sector, small-scale gold miners and panners discharge significant quantities of mercury into the environment, but large-scale mines do not because they process the ore in a different way.

The conclusion must be that comparisons of environmental effects between scales of production is more complicated than the simple question of scale. While the basic industrial processes are the same between scales of production, the scale difference implies differences in the technologies used, which has implications for efficiencies and wastes.

Regulations

Existing regulations and environmental legislation in Zimbabwe and Bangladesh were found to be inadequate for the control of pollution by small-scale industries. Zimbabwe, for instance, had 18 separate Acts of Parliament controlling the use of natural resources. The legislation that does exist tends to be reactive rather than proactive, while enforcement capacity is extremely limited and is focused on the large polluters. Although the environmental legislation is undergoing revision in both countries, reliance on the enforcement of a regulatory framework to bring about improved environmental performance by small-scale producers is likely to be ineffective.

If regulations are ineffective because of the limited resources available in developing countries for enforcement, alternatives need to be considered. Measures to improve environmental performance, in terms of reduced emissions and waste and better working conditions, are more likely to be introduced by small-scale industry if there is a direct incentive or financial return to the entrepreneur. Such incentives might come through the 'cleaner production' approach (including waste minimisation, energy efficiency, cleaner technology, etc.), which has the potential to provide financial returns to the entrepreneur.

Greater use could be made of 'voluntary' compliance methods—'informal regulation'—as a means of encouraging improved environmental performance. Stakeholder involvement, through greater environmental awareness among affected parties and information sharing about pollution impacts, is a potentially effective tool. For the stakeholder approach to enforcement and for the adoption of cleaner production methods to work, however, there is a need for training and awareness raising programmes. Non-governmental organisations (NGOs) could play a critical role in mobilising stakeholder involvement in the enforcement of standards, through public or market pressure.

Conclusions

Recognition that small-scale industries can have significant environmental impacts points to a potential conflict between the development objective of employment creation in small-scale, labour-intensive enterprises and the objective of sustainability and reduction of environmental degradation. A lack of in-depth knowledge of the actual environmental impact of small-scale producers and of the ways in which environmental regulations affect their operations, limits the understanding of policy-makers on how to balance these two policy objectives.

Small-scale industries currently provide, and will continue to be, a source of employment and income for many millions of poor people in developing countries. From a national perspective their activities are usually not the most significant sources of environmental degradation, and these costs might be regarded as being outweighed by their social and human benefits. At national and donor policy level, therefore, it is perhaps understandable that the question of their environmental impacts has been overlooked.

It is likely that small-scale producers will be increasingly called to account for their environmental performance. For entry into some (export) markets or in order to access credit, for example, small-scale industries will become expected to be able to demonstrate their environmental impact. Development agencies and financial institutions that support small industry or small enterprise are likely to demand environmental impact information in order to account to their constituents (as evidenced by the growing number of guidelines and manuals). The need for appropriate methods and indicators to collect and present this information will therefore increase.

23
Small plants, industrial pollution and poverty

Evidence from Brazil and Mexico*

Susmita Dasgupta, Robert E.B. Lucas and David Wheeler

Small enterprises, often defined as employing 20 people or fewer, are controversial in the literature on environment and development. Noting that pollution from large factories may overwhelm environmental absorptive capacity, Schumacher (1989) touts small plants as the agents of choice for sustainable development. In contrast, Beckerman (1995) argues that small factories are pollution-intensive, costly to regulate and, in the aggregate, far more environmentally harmful than large enterprises. Recent policy reports from the World Bank and other international institutions have tended to side with Beckerman, at least in noting the potential gravity of the small-enterprise pollution problem (World Bank 1997a, b).

Economic theory provides no clear guidelines as to whether plant size is likely to rise monotonically with development (Mills 1990). Smaller plants may remain competitive because of the ability distributions of principals or managerial agents and because they operate in niche markets and may offer greater flexibility under

* This chapter would not have been possible without the generous assistance of our colleagues in Instituto Nacional Ecologia (INE) in Mexico's Environment Ministry (SEMARNAP); IBGE, Brazil's National Census Bureau; and the World Bank's Environment Department. The chapter and supporting data on Mexican industrial air pollution intensities are available at the website, 'New Ideas in Pollution Regulation' (NIPR), *http://www.worldbank.org/nipr*. Financial support was provided by the World Bank's Research Committee, the Poverty, Growth and Environment Trust Fund, and by operational support funding from the World Bank's Brazil and Mexico Departments. For useful comments and suggestions, we are grateful to Joachim von Amsberg, Peter Lanjouw, Gordon Hughes, Richard Ackermann, Kseniya Lvovsky, Muthukumara Mani, Sergio Margulis and Paul Martin.

uncertainty or, simply, because scale economies are unimportant (Manne 1967; Lucas 1978; Caves and Pugel 1980; Oi 1983; Mills and Schumann 1985; Gilbert and Harris 1984). Nonetheless, the limited available empirical evidence does provide a fairly uniform picture of rising plant scale as development proceeds, even though concentration tends to decline with development (Banerji 1978; Little 1987).

If small plants are dirtier and plant size increases with incomes, does this imply that industrial pollution is more problematic in lower-income areas? This is the central question posed in this chapter, using newly available large data sets on enterprises in Mexico and Brazil. This data is described in the third section of this chapter, following a brief review of some of the issues surrounding industrial pollution in relation to plant size and local incomes. The analysis in this chapter focuses on air pollution because this poses the most serious threat to human health. We believe that this evidence provides the first systematic assessment of the relationships linking income levels, distribution of plant sizes and industrial pollution. In closing, the chapter accordingly draws together both a summary and some broader implications.

Industrial pollution, plant size and local incomes: a review of issues and evidence

Three issues are reviewed in this section: factors likely to influence the pollution intensity of smaller plants; factors likely to influence the location of smaller plants in relation to local incomes; and factors likely to influence the location of dirtier industries in relation to local incomes.

The pollution intensity of smaller enterprises

A clear distinction needs to be made between the pollution intensity of small versus large plants *within* the same industry,[1] as opposed to the relative pollution intensity of small and large plants *across* industries. Since some industrial activities are inherently more pollution-intensive than others, the principal question to be addressed here is the relative pollution intensity of small as compared to larger plants within the same industry.

At least five reasons may be cited as to why pollution intensity may vary with plant size within an industry:

- Abatement costs may be subject to (dis)economies of scale.
- The average costs of formal regulation and monitoring of smaller plants is presumably higher. However, there is mounting evidence that informal community pressure may provide an important alternative regulatory mechanism which may be better suited to monitoring small plants (Pargal and Wheeler 1996a; Hettige *et al.* 1996).

1 See Chapter 22.

- Small plants may be producing goods that are differentiated from those of their larger counterparts, using different technologies.
- Even where products are identical, the choice of production technique and age of technology may well differ with plant size.
- In particular, small plants tend to use shorter smokestacks, resulting in greater local impact on air quality.

Given this range of possibilities, it is not self-evident what the outcome is likely to be with respect to relative emission intensity of smaller enterprises within specific sectors. Recent empirical research suggests that plant size is inversely correlated with emissions intensity (emissions/output) in developing countries because of private-scale economies in pollution control and public-scale economies in regulatory monitoring and enforcement (Dasgupta *et al.* 1996; Pargal and Wheeler 1996a; Hartman *et al.* 1996; Dasgupta *et al.* 1998a; Hettige *et al.* 1996). However, problems of data availability have limited most empirical research to analyses of medium- and large-scale industrial water polluters.[2] Little work has been done on problems related to air pollution in general and on emissions by small plants in particular.[3]

Income levels and the role of small enterprises

The relative role of small enterprises may be linked to income levels across regions or through time by a number of factors:

- At lower wage levels smaller plants may be more competitive, to the extent that labour intensity and small scale are positively correlated.
- For normal goods, the size of market rises with income levels. Larger markets may then result in larger plants where scale economies are important. However, this link between local incomes and market size is much weaker for traded goods or goods that are close substitutes for traded goods.
- The evolution of financial markets and of managerial skills at higher income levels may alter the governance of firms from family-held operations to public corporations, a shift that may result in an increasing role for larger plants.
- Limitations on regulatory capacity in lower-income settings often leave small enterprises exempt (*de facto* if not *de jure*) from taxation, licensing requirements, minimum wage laws, collective bargaining and other regulations. Some observers have touted the small-scale sector as the dynamo of

---

2 Industrial water pollution has traditionally been the first focus of regulation in developing countries. Water pollution is also easier to measure than air pollution. For both reasons, the available data is much more plentiful for water polluters.

3 However, Blackman and Bannister (1996) find that small-scale brick-makers in Ciudad Juarez, Mexico, have responded significantly to informal community pressure for improved environmental performance.

growth precisely because of this lack of effective regulatory coverage (Desoto 1989). However, any successes of the small-scale sector may also reflect restrictions placed on the larger enterprises; such restrictions limit competition and frequently result in subcontracting work out to the small-scale sector to evade taxation and regulation.

Most of these arguments suggest a declining role for smaller enterprises as income levels rise, although this may be offset, as previously noted, by the continued importance of niche markets and, perhaps, the greater flexibility of smaller enterprises. Surprisingly, systematic empirical evidence on the role of small enterprises is rare. Little (1987: 229-30) notes:

> Cottage shop manufacturing still accounts for over half of manufacturing employment in poorer countries (India, Indonesia, the Philippines, and most of Africa). There has also been a relative fall in employment in small to medium size factories in the more rapidly industrialising economies (Colombia, Korea, Malaysia, Singapore, and Taiwan).

This pattern is not inevitable, however, for Little goes on to observe:

> The relative decline of small-scale enterprises in most developing countries . . . has been accelerated by the industrialisation policies adopted in these countries . . . In most countries there have been countervailing measures in support of [small enterprises]. But these have only scratched the surface . . .

Where such countervailing policies are more substantial (notably in India), the role of small enterprises has actually increased.

Dirty industries and poor regions

The sectoral composition of industry is a key determinant of environmental quality because some industrial processes are much 'dirtier', or more emissions-intensive, than others. For example, the wood pulping and metals sectors generate far more emissions per unit of output than saw-milling and electronics. Emissions abatement requires factor inputs subject to diminishing returns, so cost-minimising firms in dirty sectors should have higher emissions intensities, *ceteris paribus*. But how is the dispersal of dirty industries likely to be associated with income levels?

- As income rises the demand for a cleaner environment presumably rises too. This may result in increased pressure to regulate dirty industries as well as relocation to avoid dirty industries. At higher income levels, communities and countries may also possess greater ability to regulate pollution effectively. Thus, a number of recent contributions indicate that pollution regulation is weaker in poor regions (Pargal and Wheeler 1996a; Wang and Wheeler 1996; Dasgupta *et al.* 1995). This is partly because the poor assign lower relative value to ambient quality, and partly because low-income communities may be poorly informed or unable to regulate pollution effectively. Weaker pollution regulation lowers the 'price of polluting' for emissions-intensive industries, providing an incentive to locate in poor regions.

- In addition, recent research suggests that pollution and labour are complements in production (Lucas 1996; Hettige *et al.* 1998). Dirty industries should therefore have a tendency to locate in low-wage regions.
- On the other hand, as income rises so does the value of time. Higher-income workers may then be concerned to reduce their commuting time. Whether this results in proximity to dirty industries then depends on whether the dirty industries are intensive in their use of more highly skilled workers.
- Many dirty sectors transform bulk raw materials into semi-finished products, producing large volumes of waste in the process, e.g. mines, metals smelters, pulp mills and sugar mills. Transport cost considerations frequently dictate that such 'weight-reducing' industries locate near extraction sites for heavy raw materials. Where such extraction processes happen to be isolated in more remote rural regions of developing countries, poverty and dirty production will reveal an inherent, if accidental, spatial correlation.

The combined outcome is ambiguous, although a common postulate is that the pollution intensity of industry may at first rise with incomes then decline at higher incomes (Hettige *et al.* 1998).

Summing up: empirical questions

From this brief review of issues, three questions may be formulated to bring to the data:

1. Are small plants more pollution-intensive than large facilities?
2. Is the share of small enterprises in industry lower in higher-income regions?
3. Is industry less pollution-intensive in higher-income regions?

Data

Two cross-sectional data sets are deployed in the following sections to explore these questions—one from Brazil the other from Mexico.

Mexico

The SNIFF (Sistema Nacional de Informacion de Fuentes Fijas) database, maintained by the Instituto Nacional de Ecologia (INE) in Mexico's Environment Ministry—Secrataria del Medio Ambiente, Recursos Naturales y Pesca (SEMARNAP)—records emissions of conventional air pollutants (particulates, SO_2, etc.), sector of production and number of employees for approximately 6,000 plants. This data is unusually rich in covering very small manufacturing plants as well as medium and large. In fact, three classifications of plant size, according to level of employment, are distinguished here:[4]

Range of employment	*Number of plants in sample*
Small 1–20	2,346
Medium 21–100	2,143
Large 101+	1,310

Research on industrial emissions must consider some subset of pollutants because factories discharge hundreds of waste products which are potentially harmful to human health or ecosystems. In this chapter, we focus on emissions of airborne suspended particulates from the Mexican database. Environmental scientists generally agree that air pollution from fine particles (those with diameters less than 10μ) has the greatest impact on morbidity and mortality. Although fine particulate emissions are not yet commonly measured in developing countries, they are known to be highly correlated with total particulate emissions on which the present study focuses.

Brazil

Although the Mexican data is very rich, no information on plant location is available. Thus, it is not possible with this data to explore hypotheses relating to local levels of development. The Mexican data is therefore supplemented by a database of 156,000 factories in 4,972 Brazilian *municipios*,[5] made available by the Instituto Brasileiro de Geografia e Estatistica (IBGE). The factories in this database are categorised by 266 four-digit Classificacao Nacional de Atividades Economicas (CNAE) codes, employment size, and location.

Information on total population for each *municipio* is drawn from the 1991 demographic census. To separate Brazilian *municipios* into ten income classes, median wage of the head-of-household from the same source is used for each *municipio*. Median wage data and population is available for 4,319 of the 4,972 *municipios*, within which positive industrial employment levels are reported for 3,137 *municipios*.

In the ensuing analysis, emissions for the Brazilian plants are imputed according to pollutants per employee within a specific sector, and for plants in a given size category from the Mexican database. To achieve this matching, both the Mexican and Brazilian data are first aggregated to the three-digit level of International Standard Industrial Classification (ISIC).

---

4 Small, medium and large enterprises are defined in various developing economies by value of fixed assets or value added, as well as by employment size. In the present context, only employment data is available, and the cut-offs adopted to define the three categories are chosen to be consistent with conventional practice in many developing countries.

5 A *municipio* is somewhat comparable to a county or district in other countries. Data on area is available for 4,487 of the *municipios* which have a mean area of 1,916 km^2.

Results

Pollution intensity, scale and dirty sectors

The results from the Mexican data are presented first, followed by the results of applying the Mexican emission intensity measures to the Brazilian plant data.

Scale and pollution intensity: evidence from Mexico

The first empirical exercise measures the relative air pollution intensity of small and large plants in the Mexican database. Intensities are computed as particulate emissions per unit of labour because comparable output data are not available.[6] The sample of 6,000 plants suggests that plant-level emissions per unit of labour decline by 65% for each 1% increase in employment.[7]

Table 1 shows the average emission intensity for each sector and plant size category. While the general pattern is consistent with a negative relationship between pollution per employee and plant size, it also reveals substantial variation. Indeed, in several sectors (such as basic foods, petroleum products and industrial chemicals), large plants are more pollution-intensive than small ones.

So far, emissions intensity has been measured by the physical volume of emissions per employee. A true measure of pollution intensity, however, must also consider the impact of emissions on air quality. Technical models of atmospheric dispersion treat large and small facilities separately because of differences in average stack heights. Large plants have higher stacks, producing substantially lower pollutant concentration per unit of emissions. The standard dispersion model used for the World Bank project analysis[8] implies that each unit of particulate emissions from a small plant increases air pollution approximately 14 times more than a unit of emissions from a large facility because of the difference in stack size. In combination with the broad pattern noted in the regressions, and even relative to almost all of the sector-specific patterns in Table 1, this implies much greater pollution per employee for small plants.

The role of small plants in lower-income areas: evidence from Brazil

Table 2 presents the distribution of industrial employment in Brazil, by plant size, within each wage decile of *municipios*, ranking *municipios* from low to high median wage of heads of household.[9] Since the principal concern here is whether the population in lower wage deciles is more exposed to pollution from small or larger

6 If small enterprises are more labour-intensive (per unit of output) than large facilities, then a negative relationship between plant size and the pollution/labour ratio implies that pollution/output declines more rapidly than labour/output.

7 In particular, in a simple regression of the logarithm of emissions per employee on logarithm of actual plant level employment, with dummy variables to control for sectoral differences, the estimated elasticity of emission intensity with respect to employment is –65, with an extremely small standard error such that the t-ratio for a zero null hypothesis is 34.19.

8 This model is embedded in the environmental 'decision-support system' (DSS), which is available at: *http://www-esd.worldbank.org/pollution/dss.*

9 Each wage decile contains approximately 10% of the population, although in constructing Table 2 *municipios* with no reported industrial employment are excluded.

	Large	Medium	Small
Basic foods	0.226	0.103	0.018
Other foods	0.011	0.213	0.248
Beverages	0.046	0.019	0.538
Tobacco products	0.007		
Textiles	0.014	0.017	0.010
Apparel	0.000	0.001	0.007
Leather products	0.008	0.010	0.027
Footwear	0.000	0.001	
Wood products	0.088	0.053	0.525
Furniture	0.001	0.001	0.009
Paper	0.053	0.049	0.027
Printing	0.000	0.002	0.001
Industrial chemicals	0.139	0.069	0.060
Other chemicals	0.015	0.023	0.016
Petroleum refining	0.025	0.040	0.087
Petroleum products	0.319	0.166	0.221
Rubber	0.010	0.010	0.084
China and pottery	0.002	0.007	0.011
Glass	0.034	0.031	0.005
Other non-metallic	0.004	0.139	0.091
Iron and steel	0.062	0.123	0.248
Non-ferrous	0.088	0.040	0.022
Metal products	0.011	0.011	0.042
Machinery	0.147	0.021	0.086
Electrical apparatus	0.023	0.009	0.010
Transport equipment	0.004	0.003	0.007
Professional equipment	0.000	0.002	0.021
Other manufacturing	0.001	0.003	0.010

Note: Gaps in the table indicate a lack of observations in a particular category. The value 0.000 appears where emissions are <0.0005.

Table 1: ANNUAL PARTICULATE EMISSION COEFFICIENTS (TONS/EMPLOYEE)

Source: SNIFF (INE/SEMARNAP)

plants, the ratios reported in Table 2 weight employment in each category by the population in that particular *municipio*.

Wage decile	Plant size			
	Small	Medium	Large	Total
1 Low wage	70.7	13.7	15.6	100.0
2	55.4	20.4	24.2	100.0
3	41.5	23.7	34.8	100.0
4	27.5	23.2	49.3	100.0
5	19.5	24.6	55.9	100.0
6	14.3	22.2	63.4	100.0
7	9.7	16.6	73.7	100.0
8	2.4	6.5	91.1	100.0
9	3.8	13.1	83.1	100.0
10 High wage	1.3	5.6	93.1	100.0

Table 2: INDUSTRIAL EMPLOYMENT BY PLANT SIZE: MUNICIPIOS OF BRAZIL BY DECILE OF MEDIAN WAGE OF HOUSEHOLD HEAD

Small industry clearly dominates the poorest regions of Brazil.[10] In the lowest wage decile areas, 71% of industrial employment is in small plants, dropping to 1.3% in the highest wage decile. Large plants obviously dominate in the top three wage deciles.

Dirty industry concentration and local incomes

Numerous studies have identified six industry sectors as exceptionally pollution-intensive: iron and steel, petroleum and coal products, metal products, pulp and paper, chemicals, and food products (Robison 1988; Tobey 1990; Mani 1996; Mani and Wheeler 1997). In this section, the industry share of these six sectors in each Brazilian *municipio* is used as a proxy for the concentration of dirty industry.

In fact, the Brazilian data suggests that the share of these six dirty industries—no matter whether based on share of employment or share in number of enterprises—declines across wage deciles. This may be seen in Table 3 which shows the share of

10 The database provided by IBGE is more comprehensive than any comparable information source we have seen. Nevertheless, it is entirely possible that small enterprises are undercounted in the database. For this exercise, what matters is the effect of income level on the propensity to undercount. Data-gathering is probably less efficient in poor regions, so undercounting of small plants should be more serious there. Thus, the results in the text probably *underestimate* the decline in small-enterprise share as income increases.

the six dirty industries in employment falling from just over 75% in the poorest regions to 21% in the highest wage areas, although the absolute decline of the six dirty industries is smaller in terms of the fraction of plants.

	Fraction in six dirty industries	
Wage decile	Employment	Number plants
1 Low wage	0.751	0.421
2	0.481	0.362
3	0.366	0.320
4	0.294	0.308
5	0.268	0.294
6	0.345	0.296
7	0.278	0.295
8	0.195	0.351
9	0.276	0.310
10 High wage	0.213	0.322

Table 3: DIRTY INDUSTRY CONCENTRATION: MUNICIPIOS OF BRAZIL BY DECILE OF MEDIAN WAGE OF HOUSEHOLD HEAD

Total emissions and health damage

The results so far are consistent with positive answers to all three questions posed at the end of the section which reviewed issues and evidence. The Mexican data especially, combined with the air dispersion parameters, shows that small plants are far more pollution-intensive than large ones.[11] In Brazil, small plants play a much larger role in the industrial economies of poor regions. Furthermore, these poorer regions of Brazil exhibit a greater concentration of plants in the dirtiest sectors. In contrast to poor areas, Brazil's richest *municipios* are the heartland of large enterprises in relatively clean sectors.[12]

Yet, paradoxically, positive answers to these three questions do not imply that pollution damage is actually greater in poor regions. Economic development also promotes two countervailing trends. The first is an increase in the **scale** of industrial

11 See Chapter 22, where Scott shows small-scale brick producers in Zimbabwe pollute less than large-scale producers.

12 These correlations are not just an artifact of Brazil's separation into poor and rich regions. Regressions for the *municipios* of Rio de Janeiro and São Paulo reveal similar relationships between dirty-sector share, industry scale and development.

activity, which may lead to greater pollution even if production shifts toward larger plants in cleaner sectors. The second is **urbanisation**, which increases the size of populations exposed to industrial pollution. A larger exposed population will suffer greater aggregate health damage from pollution, even if industrial emissions remain constant. In this section, these two effects are examined in turn.

Total emissions by wage decile

Total industrial emissions are estimated here in three steps:

- First, total industrial employment by sector and plant size class in each *municipio* is obtained by summing across individual plants in the Brazilian data set.
- Second, total employment within each sector-plant-size category is multiplied by the relevant pollution intensity (emissions per employee) computed from Mexico's SNIFF database.
- Finally, emissions are aggregated across Brazil's *municipios* by wage decile.

Table 4 records the resulting estimated total emissions of particulates by wage decile, both in total and by plant size.[13] Table 4 presents a striking contrast to the information in Tables 1–3, showing that most of Brazil's industrial emissions are in relatively affluent *municipios*, and most of the emissions in all wage deciles come from large plants.

	Total particulate emissions per capita			
Wage decile	Small	Medium	Large	Total
1 Low wage	18	16	415	449
2	49	60	1009	1118
3	119	135	808	1063
4	159	177	795	1131
5	161	237	1272	1669
6	162	271	1838	2270
7	153	255	2314	2722
8	42	101	2445	2587
9	89	244	2854	3187
10 High wage	28	88	1678	1793

Table 4: INDUSTRIAL EMISSIONS BY PLANT SIZE AND WAGE DECILE

13 The data is presented per capita, to control for slight variations in population within each wage decile.

Ambient quality and health damage

Estimating the impact of total emissions on human health (represented here by expected mortality) requires four additional steps:

- First, the World Bank's dispersion model (mentioned in the section on scale and pollution intensity) is used to estimate the impact of industrial emissions on air quality in each *municipio*. The model incorporates the effects of total emissions, the relative impact of plant size, and the size of the area over which the emissions are dispersed.
- The second step is conversion of the concentration increments (by plant size) into changes in the probability of mortality for *municipio* populations. For this, the particulate 'dose–response' function, developed by Ostro (1994) from a number of prior studies, is deployed. In particular, a 0.1 $\mu g/m^3$ reduction in concentration induces a fall of 0.067 per 100,000 in the mortality rate. It should be noted that this assumes the mortality effects of suspended particulates are confined to the *municipio* in which they are emitted.
- These estimated concentration increases are then multiplied by *municipio* population to obtain expected mortality from the emissions of small, medium and large plants.
- Finally, the expected number of deaths are aggregated across *municipios* to obtain expected mortality increments by wage decile.

The results are portrayed in Table 5.

	Expected deaths per 10 million people			
	Plant size			Total
Wage decile	Small	Medium	Large	
1 Low wage	0.07	0.00	0.18	0.26
2	0.35	0.04	0.63	1.02
3	1.03	0.08	0.54	1.66
4	2.63	0.29	1.70	4.61
5	2.13	0.31	1.72	4.17
6	3.72	0.42	3.69	7.83
7	4.87	0.70	8.71	14.28
8	1.98	0.57	118.77	121.33
9	5.90	1.05	15.01	21.96
10 High wage	5.95	1.03	40.86	47.83
Overall	2.80	0.44	21.05	24.29

Table 5: EXPECTED DEATHS FROM INDUSTRIAL PARTICULATE EMISSIONS IN BRAZIL

Do the poor suffer from more pollution in Brazil? In fact, the results strongly suggest the converse. Expected deaths per capita from particulate emissions are much lower in the bottom six wage deciles than in the top four. The estimated mortality rate per capita is by far the highest in the third-highest wage decile, rising monotonically up to this point, although estimated mortality is also very high even in the highest wage decile.

Overall, the great majority of these deaths are attributable to emissions from large plants.[14] However, the relative contribution of small and large plants does vary across wage deciles, as may be seen in Figure 1. In particular, the relative importance of small plants is larger in the lower wage deciles. Medium-sized plants prove relatively harmless in each wage decile, reflecting the fact that they are less emission-intensive than small plants, but also that they play a relatively minor role in employment.

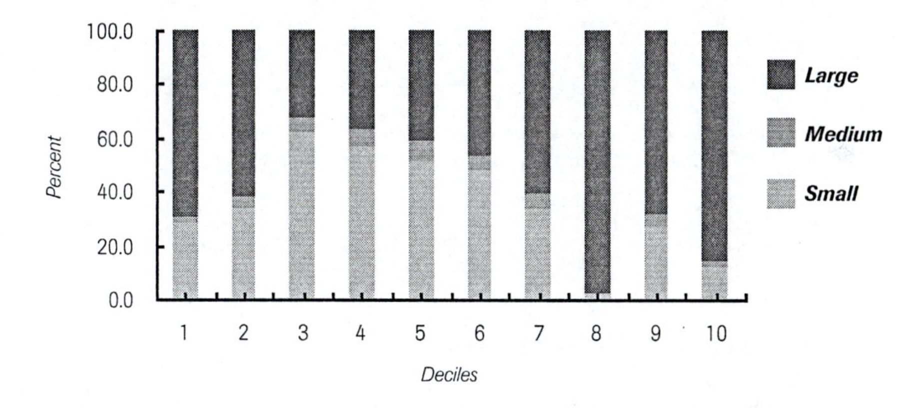

Figure 1: PERCENTAGE OF EXPECTED DEATHS ACCORDING TO PLANT SIZE BY WAGE DECILE

This data suggests that industrial air pollution in Brazil is a problem for relatively affluent areas, not poor ones. In large part this is a reflection of the concentration of industry in the more affluent *municipios*. Thus, as Table 6 shows, industrial employment per capita rises monotonically over the bottom nine wage deciles, from 2.6 employees per thousand people in the lowest wage decile to almost 66 per thousand in the ninth wage decile.

14 It should be emphasised that this study focuses on particulate air pollution from the manufacturing industry alone. Large power plants are a major source of particulate pollution, and including them would obviously increase estimates of deaths attributable to large plants. Motor vehicles are another major source of particulate pollution. For an estimate of their contribution to overall mortality in Brazilian municipalities, see von Amsberg 1998.

Wage decile	Industrial employment per 1,000 population
1 Low wage	2.57
2	8.25
3	12.11
4	17.64
5	28.56
6	33.98
7	52.18
8	61.70
9	65.59
10 High wage	48.82

Table 6: INDUSTRIAL EMPLOYMENT PER CAPITA BY WAGE DECILE

Summary and conclusions

In this chapter, we have used new data from Mexico and Brazil to analyse relationships linking income levels, the size distribution of manufacturing plants, and exposure to industrial pollution. Prior work in this field has generally been limited to water pollution and to medium-sized plants, owing to lack of data. In contrast, the present study examines air pollution and emissions of airborne suspended particulates in particular, and encompasses small plants (1–20 employees) as well as medium (21–100 employees) and large plants (more than 100 employees).

The data for Mexico indicates that emissions per employee decline with plant size in the aggregate and also within most three-digit sectors on the ISIC. With the exception of a few sectors (such as basic foods, petroleum products and industrial chemicals) the small plants are the more pollution-intensive per employee (and presumably per unit of output).

Some 73% of Brazil's *municipios* (for which data exists) report having some manufacturing within their boundaries. Among these industrial *municipios*, the share of the standard six dirty industries rises with the median wage of household heads. This same data also shows that the share of the smaller plants (no matter whether measured by the fraction of employment or proportion of plants) is greater among the low-wage *municipios*. Conversely, large plants dominate the high-wage *municipios*.

At first glance, these findings seem to imply that industrial air pollution has a 'regressive' bias and that most of the damage is inflicted by small factories. However, the introduction of scaling variables reverses this conclusion: almost all the pollution damage is in the highest income areas, and most of it is generated by large plants. The reversal occurs because production volume, plant scale and population density are highly correlated with income levels of *municipios* in Brazil and these effects are not offset by reasonable allowances for higher smokestacks among larger plants.[15]

Do the poor suffer from more pollution? Our results for Brazil strongly suggest the converse. Expected mortality from industrial air pollution is highest in the upper-income *municipios*. Furthermore, the great majority of these projected deaths are attributable to emissions from large plants, despite the greater emission intensity of smaller plants and the prevalence of smaller plants in lower-income areas.

15 It is worth noting that our conclusion for air pollution should be doubly valid for industrial water pollution, since there is no 'stack height' differential for large and small water polluters. Mixing dynamics are basically the same, so the impact of emissions on water quality is simply a function of volume relative to the local absorptive capacity of the medium. By implication, large plants should account for an even greater share of water pollution in Brazil's high-income areas.

24

Integration of environmental management systems and cleaner production

An Indonesian case study

Liana Bratasida

In recent years, the management of our living environment worldwide has emerged as an issue of major concern. Environmental degradation is now perceived as a serious threat in most industrialised countries. Meanwhile, the developing world recognises too that pollution and uncontrolled degradation of natural resources have become an impediment to sustainable economic development.

Our natural environmental assets provide three main types of essential support to humanity. First, the environment is a basic source of raw materials for the support of economic activities. Second, it has become the 'sink' into which is poured the waste products of human society. And, third, the environment continues to provide the elements of our essential life-support system. These three elements need to be in balance.

In line with the rapid industrial growth experienced in Indonesia's economic development, where industrial outputs have been the backbone of national earnings, a profound shift has come to pass. A great utilisation of natural resources has increased industrial outputs, but, on the other hand, has also resulted in tremendous waste generation.

The success of this economic development has resulted in significant environmental impacts leading to pollution and environmental degradation. On the one hand, small and medium-sized enterprises (SMEs) are considered to be a critical element in Indonesia's development, since they contribute a significant amount to gross domestic product (GDP). On the other hand, their lack of environmental awareness leads to environmental problems. Due to the locations and the social structures of SMEs, most of the victims of this pollution and environmental degradation are the surrounding population, i.e. the SME workers themselves and low-income communities. Ultimately, if these environmental issues are not managed properly, they will become a threat to Indonesian development.

Indonesian products are highly competitive due to their low cost. But, since environmental awareness has emerged globally, especially in developed countries, market trends have shifted to the point where cost is not the single most important element in international market competition. Therefore, to compete in international markets, Indonesian industries must consider and manage their environmental impacts.

SMEs in Indonesia

SMEs play a major role in the national development of Indonesia. They accomplish a wide distribution of income generation and provide employment, business opportunities and economic development. In Indonesia, SMEs are defined as all industries with total assets, not including lands and buildings, of not more than five billion rupiah (approximately US$800,000).

In 1995, the Agency for Development of Small-Scale Industry (BAPIK) announced that there were 2,157,138 enterprises, consisting of small-scale multifarious industries, small-scale agricultural and forestry-based industries.

Small-scale metal machinery and chemical industries and small-scale service industries employed 8,255,747 labourers, with total production worth approximately 29.9 trillion rupiah. Between January 1997 and October 1997 total exports reached US$2.2 billion (Indonesian Ministry of Industry and Trade 1997). Export earnings of SMEs are dominated by leather and craft industries (wood, rattan, silver handicraft, etc.).

BAPIK achievements during its sixth Five Year Development Plan (1993–98) (Indonesian Centre for Statistics Bureau 1998) include:

- 170,330 newly established enterprises
- Employment of 1,180,138 workers
- Export earnings of US$2,177 million
- 370 enterprises certified to one of the quality management standards in the ISO 9000 series
- Subcontracting work gained for 1,645 enterprises

Indonesia has had some success with a programme called Bapak Angkat, in which large companies assist and support smaller enterprises both financially and through the provision of expertise. This programme was initiated in 1992 with the commitment of chief executive officers from many of Indonesia's biggest corporations. The programme also requires SMEs to co-operate between themselves, particularly those in the same sector and geographical area.

Impact of the current economic crisis

During the past two years, Indonesia has confronted tremendous economic turmoil, with the rupiah slumping sharply against the US dollar. The present Indonesian economic and monetary situation has dramatically shown that economies are interdependent. The market demands transparency and assurance in many aspects, including the environment.

Contending with economic crisis, Indonesian industries have had to prioritise for economic survival. Efforts have been made to discourage industries not to close down production during this time.

There has been some speculation that the economic crisis would result in a decreased level of industrial output and, consequently, a decreased flow of pollutants to the environment. However, recent research among facilities required to report their waste-water discharges has concluded that the opposite is in fact occurring. While it is true that industrial output has decreased, facilities are, in many cases, abandoning end-of-pipe waste-water treatment equipment in an effort to reduce costs. The result has been an overall *increase* in environmental pollution.

SMEs suffer the most during an economic crisis. The prices of raw and other materials increase, resulting in rising product prices. Since the domestic market cannot be relied on as before, the target of industrial output has been focused primarily on international markets.

From the environmental point of view, the major obstacle to implementing environmental management programmes, especially in SMEs, is cost. If a firm looks only at the cost of implementing an environmental management system (EMS) or the cost of reaching a certain environmental standard without considering the benefits that can be gained, then cost becomes a major obstacle.

To survive during this monetary crisis, SMEs need to review their business priorities and values and focus on a more comprehensive and integrated economic/environmental management approach. They need to know how to increase product competitiveness internationally through the implementation of proper EMSs. For this reason, the Indonesian government has taken the initiative to act as a facilitator for industries in their efforts to tackle environmental issues and competition in the international market.

Institutional arrangements and national schemes for environmental management

BAPEDAL (the Environmental Impact Management Agency) was established in 1990. It is a government agency responsible for the development of environmental management in a harmonious, co-ordinated and balanced manner to support the implementation of sustainable development. To achieve this goal, several strategic environmental programmes have been developed in the areas of compliance, institutional capacity and societal partnerships.

Since 1993, BAPEDAL has worked towards the anticipated globalisation of economic and environmental concerns by shifting away from the old 'command-and-control' approach towards a balance of regulation, market-based instruments and voluntary initiatives. In 1993, BAPEDAL introduced the cleaner production (CP) concept in Indonesia.[1] Since its inception, significant progress has been achieved in key areas, including raising awareness, training, technical assistance, dissemination of information and incentive development. A National Commitment on CP was announced in 1995, followed by a Cleaner Production Action Plan in 1996.

The voluntary CP programme is well accepted by industry and is becoming increasingly popular, especially in comparison with the more traditional approach of environmental protection such as pollution control and waste management. CP refers to actions taken to influence potential causes of environmentally adverse effects, thereby averting them. The concept of CP is related closely to the 'precautionary principle'. This emerged in the early 1980s and advocates the reduction of hazardous or toxic materials into the environment, especially where there is reason to believe that harmful effects are likely to occur.

The following achievements have brought about acceptance of the CP programme by interested stakeholders:

- The development of technical guidelines on CP for specific industries, such as textiles, electroplating, tapioca, leather tanning, pulp and paper, palm oil, rubber, hotel and gold mining.
- The implementation of CP demonstration projects for the leather tanning, pulp and paper, textiles, hotel, sugar refining and tapioca industries.
- The establishment of counselling groups for specific industry sectors to facilitate information exchange and research on CP methods and techniques, such as the Counselling Group on Textile and Leather Tanning Industries.
- The development of the CP Award, of which a demonstration project in the textile industry was successfully carried out.
- The development of a CP database to disseminate information on CP both nationally and internationally.

As the international EMS standard in the ISO 14000 series—ISO 14001—came into acceptance in 1996, the government of Indonesia has kept up with this new standard as a means of promoting Indonesian industries' competitiveness in the world market.

In 1997, BAPEDAL carried out demonstration projects on the integration of CP into EMS. As such, BAPEDAL has played a significant role in introducing and preparing

1 See Hobbs's discussion on cleaner production in Chapter 11.

voluntary environmental management strategies and tools—such as CP, eco-labelling, economic incentives and ISO 14001—to various stakeholders in Indonesia. Compliance with these measures has no legal impact, only marketing advantages.

BAPEDAL works closely with the Indonesian Standardisation Board to co-ordinate standardisation activities in Indonesia, especially in the field of environmental management. Updated information is frequently disseminated to all stakeholders, particularly the Indonesian business community through seminars, workshops and training.

Implication of environmental issues for SMEs

Environmental requirements in Indonesia are becoming demanding, especially with the finalisation of ISO 14001. Unfortunately, only multinational and powerful businesses are proactive and able to keep up with international requirements, such as ISO 14001 and ISO 14020—the international standard on eco-labelling.

SMEs in Indonesia, as elsewhere, tend not to actively adopt international requirements such as ISO 14001. There are a variety of reasons for this, the main one being that SME owner-managers' main interest is profit. When the Indonesian economic climate was still stable and the domestic market could be relied on, SMEs did not encounter many business difficulties, although they still neglected environmental considerations. After the crisis, when success in the international market became the only solution for business survival, SMEs in Indonesia have been obliged to consider environmental requirements.

Indonesian companies have adopted ISO 14001 and ISO 14010 (*Guidelines for Environmental Auditing: General Principles*) voluntarily. The main application of these standards is expected to occur in commercial transactions between corporations. Therefore, standards in the ISO 14000 series can have implications for trade, particularly for Indonesian exporters which find it difficult or costly to comply with ISO standards.[2] Compliance with the standards for an individual company will, in general, be very costly. SMEs are put off by the high costs of setting up EMSs. The government must, therefore, play an active role in encouraging them to set up these and other environmental programmes.

As the majority of companies in Indonesia do not yet manage their environmental impacts within a systems approach, the standards will have a significant effect on how companies manage environmental issues. The new standards will affect the ways in which companies operate and how they conduct their affairs with suppliers, contractors, bankers and other parties. They will also affect the way in which companies manage internal procedures for production processes, use of materials, management practices and employee relationships. The adoption of standards such as ISO 14001 will also provide opportunities, having beneficial effects on the competitiveness of Indonesian producers and exporters through improving production efficiency, spurring technological change, promoting new market niches, etc.

2 See Johannson's discussion on trade barriers in Chapter 6.

Environmental management programmes for SMEs

The Indonesian government is keen to help industries that want to enter the international market by taking into account environmental issues. With the aim of reducing production costs, BAPEDAL has encouraged SMEs to implement CP. The approach is not only focused on the process, but also on the product. The goal is to improve production processes and the international marketability of the product. Process efficiency also means cost reduction. Cost efficiency is the ultimate goal to survival during the monetary crisis. The production of environmentally friendly products is a market-driven issue which, in developed countries, can affect the product's marketability.

SMEs implementing CP can improve their environmental performance and obtain economic benefits. The fundamental change for most SMEs lies in the change in attitude CP brings about—converting traditional, wasteful, production into efficient and streamlined practices.

BAPEDAL has also encouraged SMEs to implement EMSs. There are several benefits to be gained from implementing an EMS, such as environmental protection through the prevention of pollution, and compliance with existing regulations leading to better acceptance by the community.

The Indonesian government has realised the importance of supporting SMEs wanting to implement an EMS. The task of providing guidance to SMEs is increasingly crucial, for the following reasons:

- The issues need to be solved immediately.
- SMEs' problems are directly related to the low-income communities in which many of them operate.
- Pollution prevention efforts by SMEs indicate front-line management.

Environmental management within SMEs is an important part of environmental control and the Indonesian government plans to set up policies as follows:

- Reducing the pollution load of SMEs
- Providing technical assistance and incentive programmes

The objectives are as follows:

- To develop an effective preventative management system both internally (owner) and externally (government)
- To reduce pollution generated by SMEs
- To increase product competitiveness both in the domestic and international markets
- To avoid the closure of SMEs and thus save jobs as well as provide job opportunities

To meet these objectives, BAPEDAL has also encouraged SMEs to integrate CP and ISO 14001.

Case studies of EMSs and CP for SMEs in Indonesia

As part of its provision of technical assistance to industries, the government of Indonesia is conducting two case studies implementing EMS and CP for SMEs under the *Produksih and Eco-label Project* launched by BAPEDAL and the German Agency for Technical Co-operation (GTZ).

The major goal of the case studies is to establish 'model companies' in two industrial sectors (textiles and leather tanning) which will motivate other firms to follow suit by demonstrating the economic benefits of implementing an EMS and CP.

The selection of industries joining in the BAPEDAL–GTZ demonstration project was focused on various industries meeting the following criteria:

- Export-oriented
- National resources-biased
- Supporting the development of rural areas

The textile and leather tanning industries were chosen for the following reasons:

- The textile industry is one of the ten most prominent export earning industrial sectors in Indonesia. The industry has based its activity on national natural resources and human resources. During the sixth Five Year Development Plan, the textile industry contributed export earnings of US$7.4 billion and provided work for 1.15 million people. Since then, the industry has become one of the predominant sectors of activity in Indonesia. Environmental problems within the industry are mostly due to poor management of production processes and the use of old technology. In some cases, Indonesian textile environmental initiatives, such as eco-labelling and eco-friendly product schemes, have impeded exports. This appears to be a global trend.
- The production output from the leather tanning industry contributed 11,367 billion rupiah to the economy in 1995. The sector's significant environmental issues relate to the use of old technology and the utilisation of hazardous raw materials.

The major players in the framework of the project are:

- BAPEDAL and GTZ as co-ordinators
- Ministry of Industry and Trade to assist the project, encourage industries and provide local expertise by including the research institutes
- Industry as the target group
- Industrial associations and research institutions

The major objective of the project is for participating industries to improve their environmental performance *and* production efficiency. The approach is both product- and process-oriented.

Progress on CP and EMS

The major findings raised during the audits were as follows:

- There was limited information and knowledge on the international requirements for products. This information is crucial for export-oriented products (BAPEDAL–GTZ 1998).
- There was a lack of information concerning the characteristics of input materials.
- There was no systematic documentation on compliance tests.
- There were insufficient laboratory facilities for conducting specific tests required by developed countries, such as the pentachlorophenol (PCP) test.

The medium-scale textile mill implemented the following CP recommendations:

- Collection of environmental characteristics
- A standard operating procedure (SOP) has been set up for some CP aspects to assure a definite job description.
- Collection of related legislation
- Chemical packages have been re-used.
- Residual printing paste has been re-used to a high extent.
- Use of automatic valves to avoid excessive water use

Motivation to implement EMS and CP

It was observed that the participating industries are interested in EMS and CP application because of:

- Better company image through better recognition
- Increased production efficiency
- Higher product quality
- Realisation of cost reduction options
- Transfer of know-how results in an increase of knowledge.

Furthermore, the fulfilment of market and customer requirements leads to 'a cleaner product', while the present economic situation in Asia will help to accelerate the willingness of the companies to adopt the approach.

Obstacles

The barriers to adopting EMS and CP in SMEs are:

- Limited knowledge and lack of experience of EMS and CP

- Limited access to appropriate information and data
- Limited availability of sufficient funding for EMS (in the form of staff resources) and CP installations (for technical investment)
- Lack of economic incentives such as loans, subsidies, etc.
- Economic constraints lead companies to buy cheaper raw materials which result in a lower quality product
- Old machinery leads to vast amounts of pollution generation
- Human attitudes that do not reflect good practice in the production process

Integration of CP and EMS

During implementation, several overlapping areas were identified between CP and EMS. These 'grey areas' appear especially on the managerial items (see Table 1).

Lessons learned

The key to successful CP and EMS integration is commitment from top management. In some instances, governmental support is still necessary. Industries need government to act as a facilitator rather than an enforcer. When initiatives are in place, industries feel free to discuss environmental issues with other stakeholders, which results in stronger partnerships. Industries that are committed to improved environmental performance have better relationships with their customers, the local community and suppliers.

To maintain continual improvement, BAPEDAL has taken an initiative to actively assist industry on their environmental performance. BAPEDAL facilitates contact with related stakeholders. To stimulate CP and EMS initiatives, BAPEDAL has established and promoted incentive tools, such as the CP Award, a soft loan programme, information services, training and technical assistance. With this variety of incentives, industries are free to communicate and discuss their environmental problems and planned action programmes with BAPEDAL.

Prospects

There are several ideas that underpin the integration of EMS and CP; they are as follows:

- EMS can be used as a tool to ensure a continuous improvement of CP requirements and to integrate CP-related activities into all relevant business operations and strategies.
- EMS provides the structure, i.e. the system by which all activities in a production centre having a significant impact on the environment can be controlled. Therefore, managerial aspects or non-technical aspects of CP activity can be incorporated into the EMS structure.
- CP and EMS require commitment of top management. An EMS requires an environmental policy which refers to three important aspects: commitment

No.		CP options	EMS follow-up
1.	Environmental aspects	▪ MSDS on dyestuffs, auxiliaries and chemical should be available. ▪ Proof of no PCP and aryl amine content ▪ Information on MTO should be available	▪ Environmental aspect should be defined.
2.	Objective and targets	▪ Documentation should be completed. ▪ PCP test should be done.	▪ Set up written objectives and targets which are quantifiable and cover the existing environmentally related activities.
3.	Training, awareness and competence	▪ A need for CP training	▪ Set up training system procedure.
4.	Communication	▪ Prepare internal and external SOPs especially on environmentally related issues.	▪ Prepare internal and external environmental communication systems. ▪ Environmental issues should be communicated more frequently.
5.	Documentation	▪ Collection of MSDS, the declaration of the non-existence of aryl amines, of MAK (Maximum Working Place Concentration) Class III A1 and A2 and technical information regarding all working, process material and equipment ▪ Prepare report system for in-house quality check. ▪ Prepare a system for external test. ▪ Prepare SOP on handling of material, outgoing product, communication, documentation, etc. ▪ Prepare monitoring report system.	▪ Set up documentation control system and ensure the function and location of the documents.
6.	Operational control	▪ Prepare SOPs on operation and maintenance. ▪ Keep statistics on maintenance.	▪ Set up procedures on the significant environmental impact-related activities. ▪ Set up written technical instructions.
7.	Emergency preparedness and response	▪ Provide earplugs to staff in weaving department. ▪ Define operating responsibilities concerning emergency measures.	▪ Set up procedure to identify potential emergency situations. ▪ Define action plans.
8.	Monitoring and measurement	▪ Install monitoring and measurement system of all recommendations.	▪ Identify environmental performance parameters. ▪ Implement monitoring and measurement of environmental performance parameters. ▪ Increase the frequency of monitoring. ▪ Set up monitoring and measurements of EMS. ▪ Set up written monitoring and measurement procedures.

Note MSDS: material safety data sheets; PCP: pentachlorophenol; MTO: methyl turpentine oil

Table 1: AREAS OF OVERLAP IN CP AND EMS IMPLEMENTATION

to legal compliance; commitment to prevention of pollution; and commitment to continuous improvement.

These three items are basically aspects of CP. Therefore CP deals with the practical environmental measures to be undertaken, e.g. better operating practices, raw materials substitution, re-use/recycling of waste. Environmental measures to promote environmental performance can be included as environmental aspects under an EMS approach.

The challenge of integrating CP into EMS relies on:

- The extraction of managerial aspects in CP, leaving CP focused only on the technical aspects and EMS focused on the structure and managerial aspects as well as other non-technical matters.
- Technical aspects of CP cannot be separated from process steps within industry. Therefore, consideration of the integration includes three aspects: EMS aspects, CP aspects and process steps, which are closely bound together.

Below is the proposed integration of CP and EMS:

- The CP aspects have been defined in the framework of demonstration projects, BAPEDAL's CP programme, etc.
- The EMS elements are specified in ISO 14001.
- Those CP aspects defined as the key components of a CP programme/approach are integrated into EMS elements, where meaningful, to ensure a continuous consideration of respective requirements and compliance.

Conclusion

Environmental management should be carried out in an integrated and holistic manner so as to achieve the desired goal as experienced in the demonstration project. During the implementation phase, several environmental management tools should be utilised. Nevertheless, adoption of an EMS should be a priority. This would then be followed by pollution prevention through the application of CP principles and cost–benefit analysis which quantifies the benefits gained by the company and society.

25
Implementation of ISO 14001 in small and medium-sized enterprises

The Japanese experience

Keikou Terui

After the publication of ISO 14001, the number of registrations of ISO 14001 increased significantly in Japan. This indicates that both Japanese industry and society are highly aware of environmental issues, more than likely as a result of the lessons learned from severe environmental pollution and the energy shock experienced in the past. At present, SMEs represent a small percentage of the total registrations, but more can be expected to seek registration in the future. In this chapter, I will explain the present situation of, and reasons for, ISO 14001 registration among SMEs in Japan, as well as the issues faced by them when seeking registration and the support programmes available. Finally, I will present examples of ISO 14001 implementation in SMEs in Japan.

ISO 14001 registration in Japan

The number of organisations that have obtained ISO 14001 registration[1] in Japan is high. As of the end of November 1998, a total of 1,392 entities were registered (AIST 1998). Figure 1 indicates the increasing trend in the number of registrations in Japan. Japanese companies started to obtain environmental management system (EMS) registration in February 1995 based on BS 7750. After ISO 14001 and the corresponding

1 Registration is the same as certification.

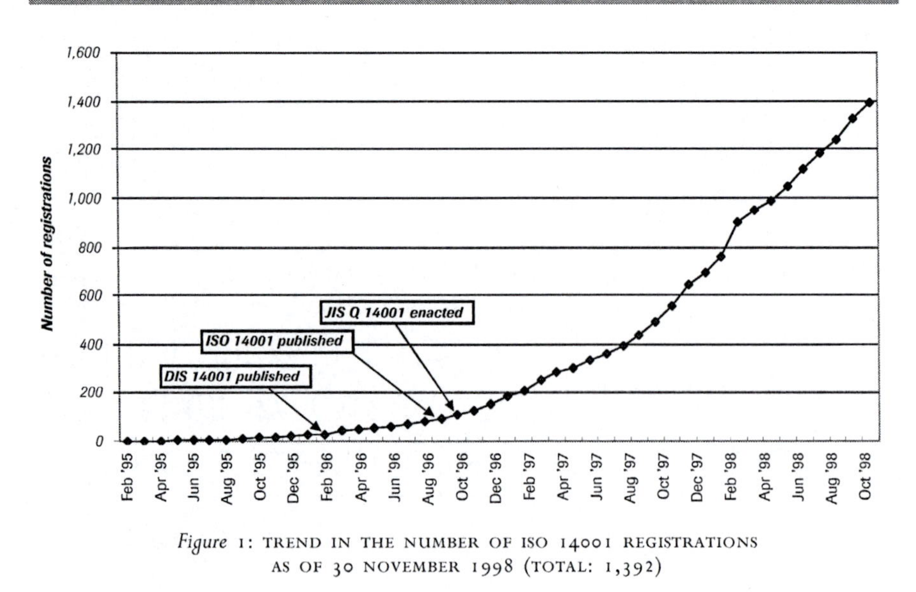

Figure 1: TREND IN THE NUMBER OF ISO 14001 REGISTRATIONS AS OF 30 NOVEMBER 1998 (TOTAL: 1,392)

Source: AIST 1998

Japanese Industrial Standard (JIS) (JIS Q 14001) were published, the number of registrations increased substantially. By 1998, the number of registrations was growing at a rate of 60–70 new registrations per month (AIST 1998).

Figure 2 shows the industrial categories of registered companies (AIST 1998). Nearly 65% of registrations were awarded in the electrical machinery and general machinery sectors. As many of the companies in these categories operate at the international level, they may consider that taking up environmental issues at the company level would positively affect their trade. The percentage of registrations in the chemical industry, which was thought to have a high awareness of environmental issues, may seem unexpectedly low. In actual fact, however, nearly 100 firms, which account for almost 75% of the total sales that occur in the Japanese chemical industry, have already taken up the industry's Responsible Care programme, the safety and environmental management system that was proposed in the 1980s (Japanese Responsible Care Council 1998). These firms are expected to steadily move towards EMS registration given that Responsible Care serves as a base for ISO 14001 registration. It should also be noted that industries in all sectors are showing interest in ISO 14001 registration. By the end of 1998, registrations had been sought by retailers, distributors, trading companies and businesses in the service sector such as healthcare, insurance and waste treatment. This is in contrast with registrations relating to the ISO 9000 series of standards which are embraced mainly by manufacturing industry. Moreover, the more recent cases of EMS registration have included administrative organisations and local government bodies, which is rather remarkable. Although only three organisations in this field have been registered so far, more are expected to be registered in the near future.

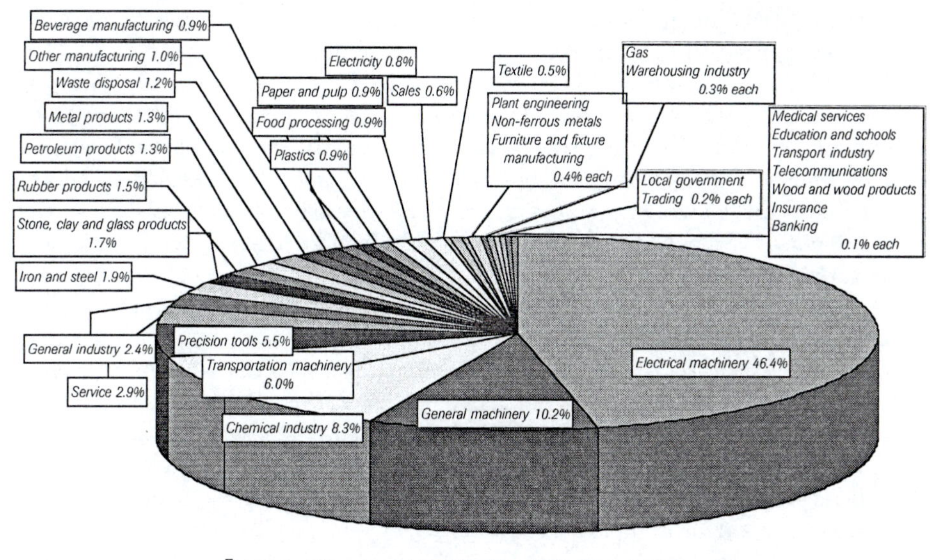

Figure 2: ISO 14001 REGISTERED BODIES IN JAPAN AS OF 30 NOVEMBER 1998 (TOTAL: 1,392)

Source: AIST 1998

Why ISO 14001 registration?

If we look at the number of registrations obtained per country, Japan has the highest compared to the UK, the Netherlands or Germany (where there are many EU Eco-Management and Audit Scheme [EMAS] registrations). Possible reasons for this are outlined below.

The first reason for the high number of ISO 14001 registrations is, given Japan's past experiences, due to the high level of interest in environmental issues in Japan. The high rate of economic growth that the country enjoyed in the 1960s brought with it horrific levels of pollution. Japan has since introduced very strict regulations aimed at reducing pollution in the industrial sector, as well as regulations concerning the release of toxic chemicals. Pollution prevention technologies, such as equipment to remove sulphur and nitrates from emissions, have been developed and were rapidly adopted (Industrial Pollution Control Association of Japan 1990). We have, therefore, been able to reduce SO_X (sulphuric oxides) concentrations in the atmosphere to one-eighth (Environment Agency 1997) of their previously recorded levels. The two oil shocks that rocked Japan in the 1970s hit especially hard because of the country's very poor natural resources base. Great efforts in energy saving and substantive changes in industrial and social structures were required in order to ride out the crisis. The intensive use of energy in major industries has dropped substantially and the energy consumption rate of electrical appliances has been greatly reduced. Among the developed countries, Japan now has the lowest energy intensity per gross domestic product (GDP) (Energy Conservation Centre 1996). These experiences have made the Japanese particularly sensitive to energy and environmental issues.

The second reason is that Japanese interest in global environmental issues was heightened by their efforts toward the various activities relating to the 1992 Earth Summit, where the deliberation on ISO 14000 series was triggered. The Japan Federation of Economic Organisations, known as Keidanren, drafted the Keidanren Global Environment Charter in 1991 and actively participated in ISO discussions on the development of the ISO 14000 series of standards. In order to accelerate the industrial sector's response to global environmental problems, the Ministry of International Trade and Industry (MITI) requested 87 major industrial groups in 1992 to establish a voluntary plan that gives consideration to environmental issues in their business activities. In an effort to solve global environmental problems, the Japanese government enacted the Basic Environmental Law, which was based on the Rio Declaration adopted at the Earth Summit. The Basic Environmental Law was used as a basis for establishing the Basic Environmental Plan, the contents of which include recommendations for companies to implement EMSs. National and local governments were also asked to support their implementation.

The third reason can be found in the widespread implementation of the ISO 9000 series of quality management system standards in Japan. As of September 1998, more than 7,700 registrations to ISO 9000 standards have been awarded and the rate of new registrations is expected to exceed 1,000 per year (Japan Accreditation Board for Conformity Assessment 1998). Due to the general acceptance of quality standards, Japanese companies had little difficulty incorporating ISO 14001 into their existing management systems.

The fourth reason is that EMS registration compliments the needs of the company. In today's society, tackling environmental issues has become an important item on the business management agenda. By opting for ISO 14001 implementation, rather than simply complying with regulations, a company can create a business structure for solving global environmental problems, for clarifying and streamlining its business management system, and for enhancing company credibility which builds confidence among stakeholders and interested parties. In short, ISO 14001 implementation can provide companies with a management tool that allows them to meet their own needs as well as societal demands for corporate responsibility. Obtaining EMS registration is considered an effective means of demonstrating their consideration of environmental issues to interested parties such as business partners, local communities, non-governmental organisations (NGOs) and governmental authorities. Many companies also believe that environmental considerations will also lead to competitive advantage in the marketplace.

ISO 14001 registration is still in its infancy and thus it is too early to draw a firm conclusion on the benefits of registration. It seems, however, that registration brings about positive benefits such as improved motivation of employees, cost reductions, improved company image and business advantages. It should also be noted that such benefits are viewed from the perspective of merits resulting not from registration of ISO 14001 *per se*, but from the establishment of an EMS and related activities within the company.

Issues on seeking registration of ISO 14001 in SMEs[2]

ISO 14001 registration of SMEs accounts for a very small percentage of the total registrations in Japan. SMEs face many obstacles when seeking registration. The Small and Medium Enterprise Agency (SMEA) of MITI conducted a survey in 1997 on issues faced by SMEs seeking ISO 14001 registration and compared these to those faced by larger companies (MITI 1998). As shown in Figure 3, the ratio of SMEs who are not interested in seeking, or do not plan to seek, registration is very high in contrast to that of larger companies where many are seeking registration. Nevertheless, the total number of SMEs far exceeds that of large companies and a quarter of SMEs responding to the survey are considering registration to some extent. An increasing number of SMEs are expected to gain ISO 14001 certification.

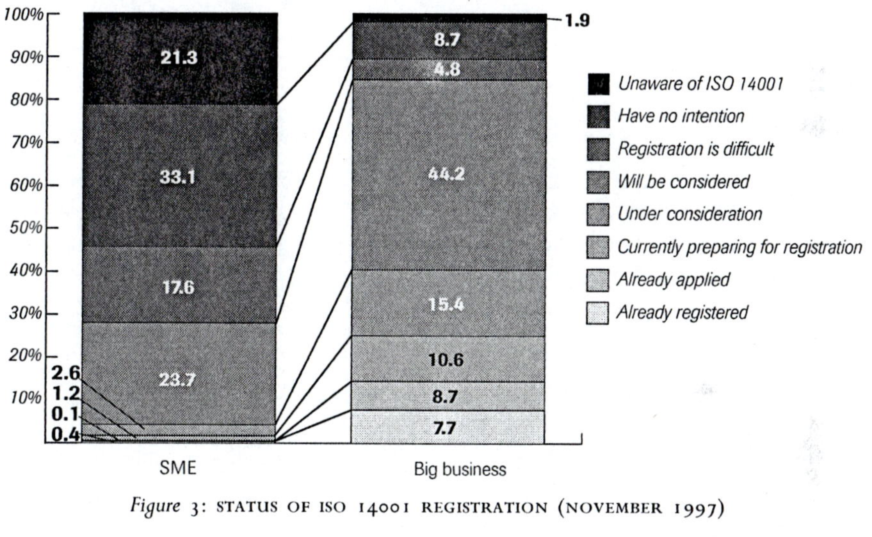

Figure 3: STATUS OF ISO 14001 REGISTRATION (NOVEMBER 1997)

Source: SMEA 1998

As for the issues in seeking registration, Figure 4 indicates that revision of the company's management system is the greatest challenge for both SMEs and larger companies (MITI 1998). SMEs have more difficulties in ensuring and developing the necessary human resources, and in obtaining information on registration procedures, than larger companies. SMEs are also faced with the high cost of EMS implementation and registration.[3] According to the result of a survey shown in Figure 5, the major

---

2 SMEs are defined in the Basic Law of Small and Medium Enterprises in Japan as one of the following: enterprises involved in retail or service sector capitalised at less than ¥10 million or with fewer than 50 employees; enterprises involved in wholesale capitalised at less than ¥30 million or with fewer than 100 employees; or enterprises involved in manufacturing, mining, transport or other business capitalised at less than ¥100 million or with fewer than 300 employees.

3 See Hillary's discussion in Chapter 10 for the barriers facing SMEs adopting EMAS.

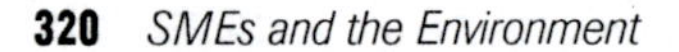

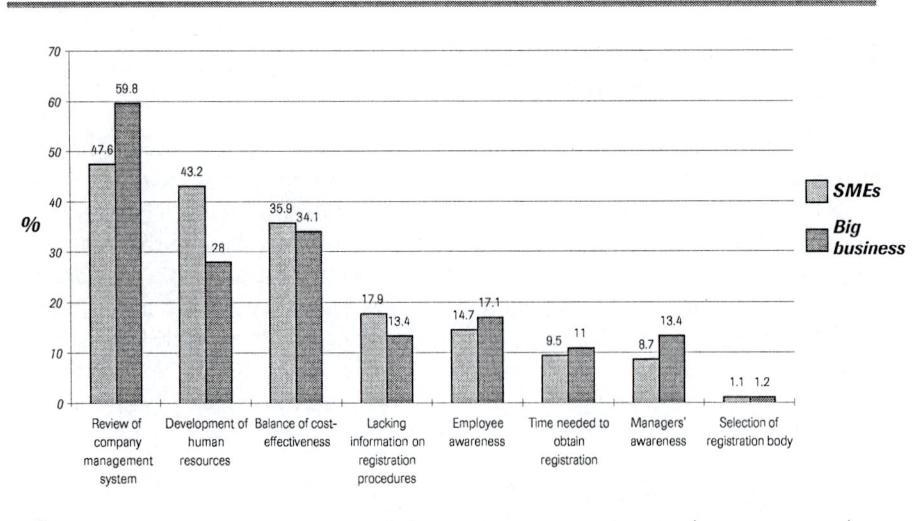

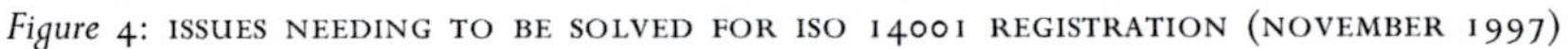

Figure 4: ISSUES NEEDING TO BE SOLVED FOR ISO 14001 REGISTRATION (NOVEMBER 1997)

Source: SMEA 1998

reasons that SMEs seek registration are to ensure the incorporation of an environmental policy into the company's activities and to improve its credibility. Other positive reasons include changing the attitudes of employees, improving quality, and compliance with international rules (MITI 1998).

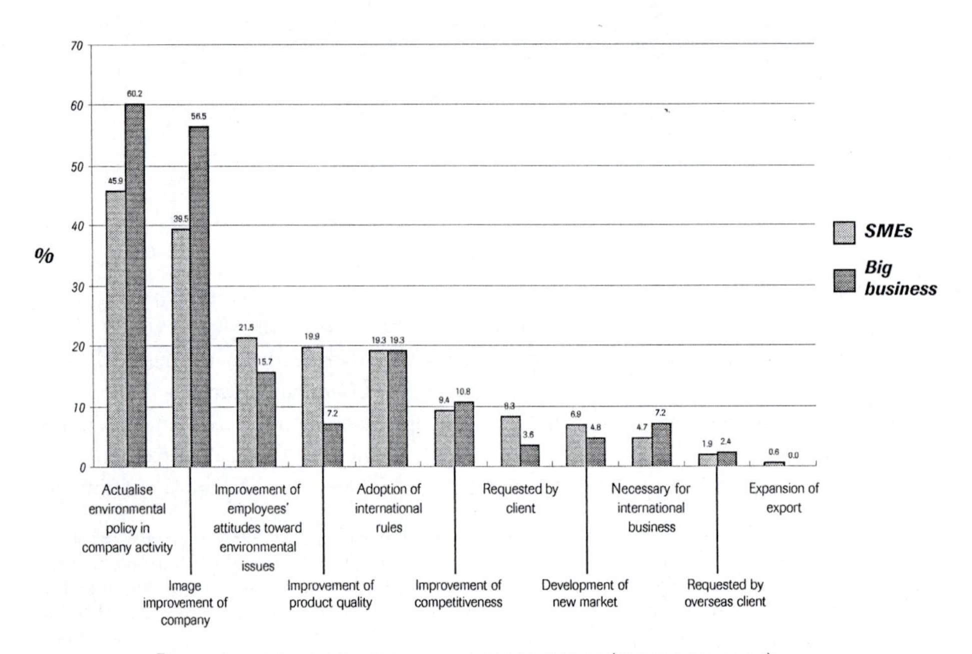

Figure 5: REASONS FOR ISO 14001 REGISTRATION (NOVEMBER 1997)

Source: SMEA 1998

In addition, Figure 6 shows that EMS registration is being utilised more and more by larger companies as a criterion for choosing business partners (MITI 1998). Registration will, therefore, become an increasing necessity for SMEs. Furthermore, given the rapid acceptance of EMSs in larger companies, an increasing number of SMEs, particularly subsidiaries and companies associated with major businesses, are expected to become registered.

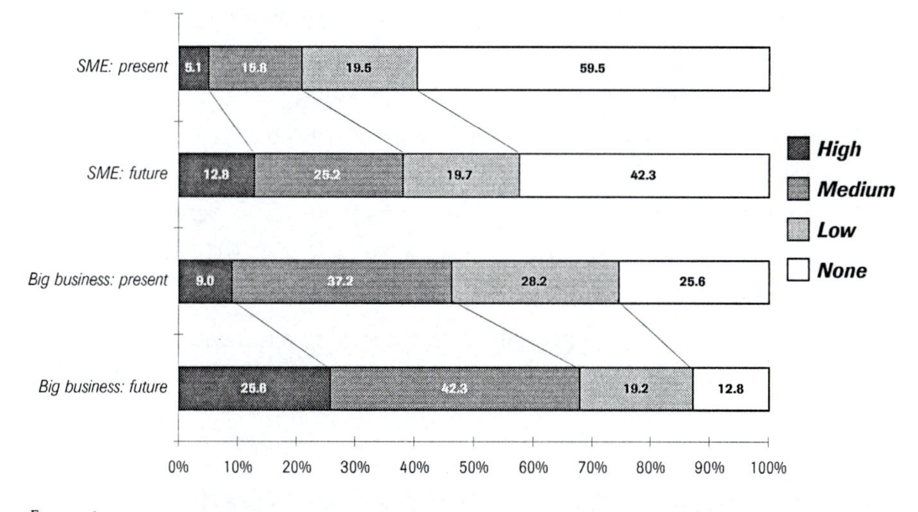

Figure 6: DESIRE TO USE ISO 14001 REGISTRATION AS A SELECTION CRITERION FOR BUSINESS PARTNERS: CURRENT (NOVEMBER 1997) AND IN THE FUTURE

Source: SMEA 1998

Support programmes for ISO 14001 registration

Programmes by national and local governments

As mentioned earlier, SMEs face many obstacles when seeking ISO 14001 registration, such as lack of human resources, information and capital. Japan enacted the Basic Environmental Plan which contains recommendations for organisations to implement EMSs and requests that national and local governments support those organisations that do so. MITI and many local governments have taken measures aimed at helping SMEs, such as providing information, organising seminars, subsidising the cost of hiring an adviser, and providing financing for the installation of environmental facilities.

Registration is not limited just to private companies, but has extended to the local government sector. Three governmental offices, Shiroi Town in Chiba Prefecture, Joetsu City in Niigata Prefecture and the Industrial Technology Centre of Shiga Prefecture, have been registered since the end 1998 and more than 20 local governments are currently preparing for registration. Since local governments in Japan have an important role to play in the enforcement of environmental regulations, the move towards seeking EMS registration might have a positive influence on their respective communities. Furthermore, one local government has relaxed the requirement in the environmental regulation concerning the publication of reports and has also eased administrative procedures relating to on-site inspections for EMS-registered organisations under certain conditions. The reasons for implementing ISO 14001 given by these local governmental bodies are:

- To raise awareness of environmental issues among businesses and in the local community
- To support local organisations seeking registration
- To influence the procurement of public projects that should give consideration to the environment
- To provide more effective and transparent administrative services
- To encourage timely assessment and revision of policies that are linked to administrative reforms

By actively addressing environmental issues, the local governmental bodies can use the resulting environmental policy as a model in order to raise awareness of environmental issues in the community.

Programmes initiated by the Japan Small Business Corporation and other organisations

The Japan Small Business Corporation is a special public corporation established for the implementation of national policies and measures related to smaller businesses under the auspices of the SMEA. It provides a variety of services for smaller businesses, such as management guidance, financing, the operation of mutual aid associations, training, and information provision. The Environmental and Safety Diffusion Office has been providing smaller businesses with information pertinent to the establishment of an environmental management and auditing system since 1996. The Office holds seminars, distributes brochures and provides telephone consultations.

In 1998, the Corporation published a collection of case studies on EMS implementation in SMEs to provide smaller businesses with help and information on setting up their own EMS (Small Business Corporation 1998). The book provides concrete information of a case-specific nature, such as 'What procedures my company followed to implement EMS', 'How my company allocated personnel, materials and money for the establishment of EMS', 'What kind of problems my company had in establishing EMS', and 'What benefits my company enjoyed from EMS'.

Five model companies were selected to implement EMSs under the guidance and training of competent consultants for six months. These companies were recommended by trade associations in order to: encourage trade associations and organisations to promote EMSs; achieve widespread implementation of EMSs among various business categories; and to give priority to business categories that have a big impact on the environment. The five companies chosen are listed in Table 1. One of them is expected to be registered to ISO 14001 in 1999.

Business category	Type of business	Capital amount (¥ million)	Number of employees
Textile dyeing and printing	Integrated processing of scarves and handkerchiefs	1	85
General industrial machinery and device manufacturing	Manufacturing of powder and fluid disposal machinery	40	145
General machinery and appliance manufacturing	Manufacturing of industrial cleaning machinery	20	293
Electrical machinery and appliance manufacturing	Manufacturing of thermal printers, etc.	25	297
Waste disposal	Waste disposal (industrial and household waste)	14*	230

* This company is not classified as an SME in the service sector under the definition provided in the Basic Law of Small and Medium Enterprise in Japan. It was chosen as a model company, however, because the Japan Small Business Corporation wanted to include the type of waste disposal business that has a significant environmental impact and is covered by small business corporations in Japan.

Table 1: LIST OF MODEL COMPANIES

Examples of implementing ISO 14001 in SMEs

Case A

The company, a plating factory for electronic devices, is located in Fukui Prefecture and has 150 employees. Its reasons for seeking registration to ISO 14001 were to strengthen the management system and improve the company's image. As a result of registration, the company achieved reductions in waste generation and made cost savings by more efficient use of resources. Employees have also gained more confidence in the company's activities. By making an effort to simplify its document system, the volume of documents in the company has decreased and it has gained know-how in document control. In addition, by advertising its ISO 14001 registration, the company has received many inquiries from larger firms and obtained new customers. The major reasons for the success are:

- Clarification of the future policy by the company's president
- Improved employee awareness by strong top-down leadership

Making the size of the system commensurate with the enterprise and linking the environmental plan to its business plan are other reasons for success.

Case B

The second case involved the registration of 14 companies by the Japan Management Association Quality Assurance (QA) Registration Centre as one organisation implementing one EMS. This co-operative association is located in an industrial park on an island in Tokyo Bay where 18 associations are established, consisting of 11 electroplaters, a plating material manufacturer, a plating equipment supplier and a waste-water treatment centre. The total number of employees is about 400, with each site having from 10–70 employees. The site covers a total of two hectares. The Association was planned just after the first oil shock and was initiated by the local government in order to reduce the use of industrial water to one-tenth of the previous level and to recycle chromate solution by treating it at the joint waste-water facility. On the 20th anniversary of its establishment, the Association decided to seek registration to ISO 14001 in order to reinforce the management system and establish its Association-wide 'plan–do–check–act' cycle. A secretariat was established and two members were appointed from each company and made responsible for developing an EMS. After the approval of the plan, it took more than one year for the group to be registered.

The establishment of the EMS led to the development and implementation of rules concerning the notification of accidents. This has decreased the number of incidents of chromate waste-water contamination and has enhanced the recycling rate. The working environment was improved significantly by implementing the five 'Ss'.[4] Furthermore, the management systems of companies within the Association has been enhanced.

Conclusion

The number of registrations of ISO 14001 in SMEs is much lower than in larger companies at present. However, since society's concern about environmental issues keeps growing, the number of ISO 14001 registrations is set to increase in future. Recently, some larger companies have declared that they intend to ask their suppliers to seek registration to ISO 14001. This means that responses to environmental issues has become a significant item on the business agenda, and one that cannot exclude SMEs. Both central and local governments are expected to strengthen support programmes for promoting EMSs in SMEs. The number of registrations of ISO 14001 in SMEs may not increase as rapidly as among larger companies, but it will grow steadily. However, SMEs that are not related to international trade may seek simpler and less expensive EMSs than ISO 14001.

4 *seiri*: removal of the unnecessary; *seiton*: arrangement of the necessary; *seiso*: cleaning; *seiketsu*: hygiene; *shitsuke*: discipline.

26
Helping small and medium-sized enterprises improve environmental management

Lessons from proactive small and micro firms*

Jason Palmer

This chapter is based on a careful selection of organisations employing up to 50 people, operating in a range of different sectors, that are proactive on environmental management. In total, seven such firms are presented, all from the UK: three retail firms, one chemical company, a house-building partnership, a small independent building society, and a charity whose primary concern is helping people recover from mental illness.

Following the EU recommendation, these companies are referred to as 'small' because they have fewer than 50 employees. 'Micro' is used to refer to firms that employ from one to nine, while 'medium-sized' is the label used for firms with between 50 and 249 employees. In all three cases it is assumed that the firm is independent and is not part of a larger group.

To put the choice of organisations for this chapter into context, it is helpful to distinguish between organisations at different levels of progress on environmental management (see Fig. 1).

* The research work underpinning this paper was funded by the Engineering and Physical Sciences Research Council (EPRSC). Dr Palmer is grateful to Rita van der Vorst, Sebastian Macmillan and Chris France for their advice during the research, and especially to Ian Cooper for his continuing guidance and commenting on a draft of this paper. Dr Palmer is also grateful to all of the case study companies for making time available for interviews.

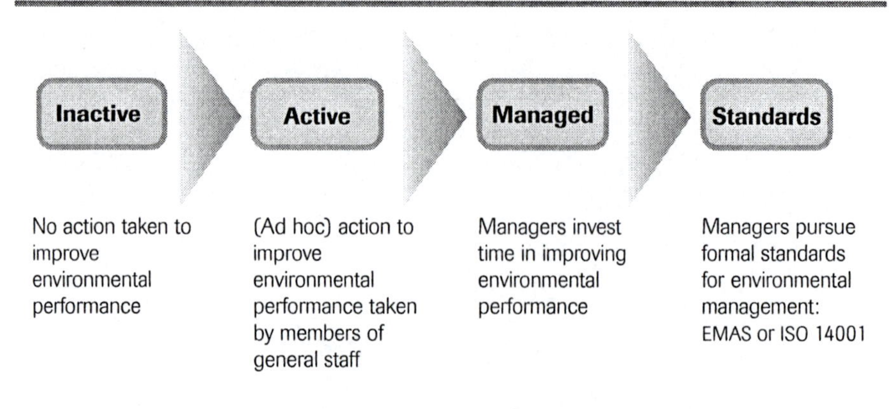

Figure 1: ORGANISATIONS AT DIFFERENT LEVELS OF PROGRESS ON ENVIRONMENTAL MANAGEMENT

Organisations do not necessarily progress from 'inactive', through 'active' and 'managed' to 'standards'. Indeed, a majority of small and medium-sized enterprises (SMEs) that address environmental performance stop at the active stage. Moreover, passing from 'active' to 'managed' or 'standards' does not automatically raise environmental performance because many management actions are restricted to the office and changes on paper alone, with no direct effect on environmental impacts (Palmer and van der Vorst 1996).

All of the case study firms presented here have identified the environment as a management issue, and each has devoted resources to improving environmental performance.

Objectives and research method

The case studies were carried out with four main objectives:

1. To understand what prompted these proactive (or 'managed') organisations to act
2. To investigate how they approach environmental management—are they working mainly informally or do they tend to apply more formalised approaches, including management systems?
3. To examine the outcomes of their action—whether they have improved environmental performance; if so, how, and what other benefits or disadvantages have resulted from their actions
4. To see what use they make of information provided by government agencies about environmental management

It proved to be extremely difficult to identify genuinely proactive SMEs: very few were cited in the literature, and a wide network of contacts working in this field was unable to offer more than ten in total. For organisations that agreed to be scrutinised, formal interviews were conducted with at least three employees: one from top management, typically the managing director, and two or more others with different areas of responsibility.

The data from these interviews was entered into comparison tables—to help analyse responses and identify trends. The intention was to look for patterns and consistency and, where appropriate, for contrasts between the organisations. This chapter summarises the seven case studies, analyses this data, and draws conclusions from this (admittedly small) selection of proactive SMEs.

The case study organisations

Company profiles and management styles

An overview of the organisations selected as case studies is shown in Table 1. The table shows what sectors of industry they operate in, says how large they are, and describes their styles of management.

There is a huge range in scale between these organisations—especially in financial terms. While Castle Project has 31 employees and a turnover of £130,000, Hampshire Chemical's 50 employees generate £30 million of sales. Almost all of the organisations examined were found to exhibit one or more unusual characteristics that differentiated them from the other case studies, and from SMEs at large.

Company profiles

Cambridge Daily Bread is a co-operative retailer selling wholefoods. It employs 16 people in total, although only eight of them work full-time. The firm seeks to inform customers and staff about environmental aspects of the products on offer, and stocks a range of fairly traded goods. It has conducted a preliminary review of the environmental effects of its activities, and used this to prioritise areas for action. The organisational structure is unusual in the extreme: all staff members participate in decision-making, they all have equal ownership of the business, and they all receive equal pay. It is also a Christian organisation.

Hampshire Chemical is recognised as one of the most proactive small companies in the chemicals industry. It has dramatically cut energy and water use through environmental management, and saved money as a result. It has established an environmental management system, and been certified to ISO 14001. The firm has succeeded in achieving very rapid growth: since 1982 turnover has increased an average of 12% per annum, and the number of employees has more than tripled. It is also unusual in having identified lifetime learning as its number one strategic priority—when the majority of SMEs do not engage in formal training and learning at all (Stanworth and Gray 1991). But the company still operates in a part of industry that is hungry for natural resources: it is energy-intensive and uses large volumes of water. It also deals in cyanide chemicals that present a significant potential threat to the environment.

Name of organisation	Cambridge Daily Bread Co-operative	Hampshire Chemical	Castle Project	Out of this World	Shared Earth	Westwind Oak Buildings	Ecology Building Society
Industrial sector	Retailing	Chemicals	Handicrafts and recycling	Retailing	Retailing and wholesale	Construction	Financial services
Part of larger group?	No	Yes, a US holding company*	Yes, a national organisation*	No	No	No	No
Number of employees	16	50	31	35	45	7	12
Turnover in 1996	£440,000	£30 million	£130,000	£890,000	£1.7 million	£150,000	£1.3 million
Management style	Informal	Mostly formal	Informal	Mostly formal	Mixed formal and informal	Mixed formal and informal	Informal
Decision-making style	Participation: consensus-based	Participation: team-based	Top-down with consultation	Top-down with consultation	Top-down with consultation	Top-down	Top-down with some consultation
Communi-cation style	Face-to-face	Mixed written and face-to-face	Face-to-face	Mixed written and face-to-face	Mixed written and face-to-face	Face-to-face	Mainly face-to-face
Level of progress (see Fig. 1)	Active	Standards	Active	Managed	Managed	Active	Managed

* Although Hampshire Chemical and Castle Project are owned by larger organisations and do not therefore meet the standard definition of SMEs perfectly, they operate with sufficient autonomy to resemble very closely a true SME.

Table 1: OVERVIEW OF CASE STUDY BACKGROUNDS AND THEIR MANAGEMENT STYLES

Castle Project is a charity set up primarily to offer opportunities for people recovering from mental illness to work, and to be rehabilitated. More recently it has started to collect and recycle materials: paper, aluminium foil, and various other waste-streams generated by local businesses. These new activities have been established as a complement to the existing manufacturing and packing activities, rather than attempting to improve the environmental performance of existing operations. Castle Project shows how even firms with very limited access to cash resources can adopt environmental management, combining environmental concern with its long-standing action on social issues.

Out of this World is a retailing co-operative comprising three stores and a head office. Stores sell a range of wholefood products and have explicit environmental, social and ethical purchasing criteria. This is a step that, to date, few retailers of any size have taken. The organisation has very ambitious aspirations: ultimately it aims to develop nationwide coverage with a branch in every major retailing centre. However, at the time of interview, it was trading at a loss—a position that is untenable in the long term.

Shared Earth is a retail and wholesaling company with five stores around the UK. It sells a range of fairly traded, non-food products: recycled stationery, crafts, books and gifts. Like Out of this World, the company acts on a broader range of issues than the environment alone. It is one of a tiny minority of SMEs that has attempted to communicate its environmental and ethical stance to stakeholders. In 1996 it published a set of combined social, environmental and financial accounts, independently audited by the New Economics Foundation: a very unusual step for a large organisation, and completely unheard of for a small one. The managing director accepts that continuous improvement of the company's operations is necessary, but he has chosen not to implement a formal management system or to register for one of the standards for environmental management.

Westwind Oak Buildings is a partnership in the construction industry specialising in oak-frame house building. It uses untreated oak timber and advises clients to complete their buildings using materials and techniques with low environmental impacts. It also offers guidance on energy efficiency. It is the only micro enterprise among these case studies, with less than ten employees. Yet, in spite of its small size, the partnership has addressed a surprisingly wide range of environmental issues: from avoiding the use of chemicals to minimising employee transport. However, the partnership is unable to affect the location of the homes it builds—consequently most of its work is new building on greenfield sites and, as such, is less than sustainable.

The **Ecological Building Society** lends to people who want to buy properties for renovation, and to other environment-oriented building projects. It also encourages borrowers to take environmental considerations into account in the design of their properties: particularly energy efficiency and the use of non-hazardous materials. Despite employing just 12 people, it has a corporate structure taken straight from the copy-book of a public company: a board with sub-committees, an annual general meeting, executive and non-executive directors, and a published annual report and accounts. But the management style is predominantly informal, and employees reportedly feel they are part of a family.

Management style

Even organisations that allocate management time to environmental considerations, and thereby establish themselves as being 'managed', can select a variety of different approaches to environmental management. They may choose to use formal techniques, such as those proposed in the standards for environmental management, and which include:

- Target setting
- Defined procedures
- Delegating responsibility to a named individual

Alternatively, they may elect to work less formally. In this case, although senior managers invest time in improving environmental performance, they may do so without applying these (bureaucratic) management techniques. Among the case study SMEs reported here, the four smaller organisations operate more informally: they have few or no formal systems or procedures, and have not delegated responsibility for environmental management to an individual, Westwind being an extreme example of this. Conversely, the largest firm—Hampshire Chemical—has set targets and uses defined procedures.

There appears to be a link between an organisation's size and the extent to which management has been formalised. This finding is consistent with other work in this field (e.g. KPMG 1997a; Palmer 1996a).

There also appears to be a link between the use of written communication and size. So, although communication within all the organisations was dominated by face-to-face meetings, the larger ones (Hampshire Chemical, Out of this World and Shared Earth) also used memoranda and other written forms of communication. The use of written communication may be attributable not only to larger numbers of employees, but also to a dispersed site structure: Out of this World and Shared Earth have little alternative to circulating some information between their stores on paper.

However, a strong message from all seven case studies was how effective internal communication is. There was remarkable consistency between what different interviewees said about their firm, and many of them—even members of general staff—said how good communication was. This could have been linked to the widespread use of consultation or participation. All but Westwind invited some form of feedback from general staff on their decisions.

There is no discernible connection between either communication mechanisms or formality of management and the dominant style of decision-making. A majority of the case studies (five out of seven) take decisions on a top-down basis. Typically, decisions were taken or proposed by the managing director or equivalent, often in consultation with other senior managers. This consultation extended more broadly to include general staff for important decisions in three of the firms.

Prompts for, and barriers to, environmental management

Viewed logically, the decision to take up, or reject, environmental management depends on the balance of prompts to barriers (shown in Fig. 2).[1] Only if the potential benefits are seen to outweigh the costs will an organisation of any size choose to act.

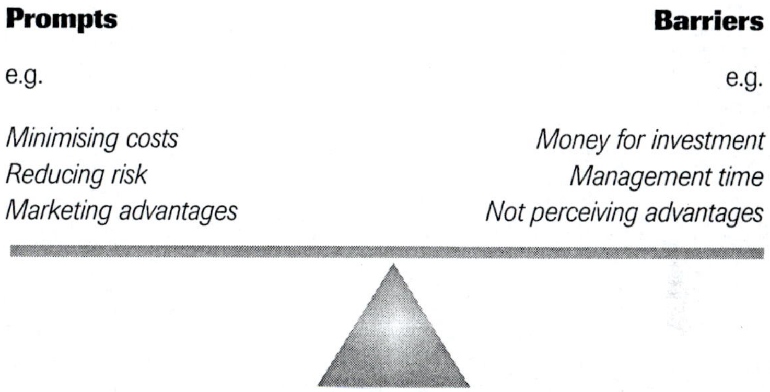

Prompts	**Barriers**
e.g.	e.g.
Minimising costs	*Money for investment*
Reducing risk	*Management time*
Marketing advantages	*Not perceiving advantages*

Figure 2: BALANCING PROMPTS AGAINST BARRIERS IN THE DECISION TO IMPLEMENT ENVIRONMENTAL MANAGEMENT

Consequently, the case studies presented here have focused in particular on the prompts and barriers to action (see Table 2). The research also sought to examine:

- How the firms engage with the outside world
- Whether they devote resources to communicating their action
- Whether they have adopted environmental management standards
- Whether they have used any of the sources of information sponsored by central government agencies

Prompts for environmental management

In every case, one or more individuals in a position of power (the managing director or equivalent) have been the driving force for change, usually with support from general staff. The main prompt for taking up environmental management was top managers' personal commitment for five of the seven case studies. However, the commitment was not always to the environment alone. Just one of the seven organisations—Westwind—expressed a commitment to the environment *per se*. For the other six, it was actually part of a broader commitment. Thus, for Daily Bread it was a reflection

1 See Chapter 10.

Name of organisation	Cambridge Daily Bread Co-operative	Hampshire Chemical	Castle Project	Out of this World	Shared Earth	Westwind Oak Buildings	Ecology Building Society
Main prompt for action	Spiritual/beliefs	Scope for cost cutting	Diversification	Personal commit-ment/ethics	Personal commitment	Personal commitment	Commitment of founders
Primary driver for change	Manager with employees' support	Managing director with employees' support	Manager's action	Managing director with members' support	Managing director with no employee resistance	Partners, with employees' support	Founders, backed by members and new employees
Primary barrier to change	Limited finances	Limited management time	Limited resources (time and money)	Limited finances	Limited resources (time and money)	Limited control over projects	Limited resources (time and money)
Actively engaging stakeholders?	No	No	No	Yes: member magazine, PC for information.	Yes: publish social and environmental accounts	No	Yes: publish member newsletter and briefings
Using standards for environmental management?	No	Yes, ISO 14001	No	No	No	No	May possibly pursue in future
Use made of government information schemes?	No	Yes, use ETBPP and local Business Link	No	No	No	Once: found scheme 'useless'	Yes, regularly use BRE information

Table 2: PROMPTS AND BARRIERS TO ACTION, AND DEALING WITH EXTERNAL STAKEHOLDERS

of members' Christianity and caring for the world and the people living in it, whereas the Ecology Building Society was equally concerned about 'self-reliance': providing the freedom for borrowers either to do their own building work, or become self-employed, or grow their own food. Out of this World and Shared Earth had an even broader set of concerns. They both felt strongly about sustainable development principles: global equity, fair trade, and participation, in addition to good environmental performance (Palmer *et al.* 1996). So, in these four organisations, the environment was actually a sub-section of a wider commitment to doing business in a particular way. Indeed, the environment was not even the *principal* concern for two of these organisations—Shared Earth and Daily Bread both put a greater emphasis on fair-trading policies than they do on environmental initiatives.

This means that, faced with a choice between importing products from distant countries and so helping to alleviate global inequity or purchasing from local manufacturers that would reduce the environmental impact of transporting goods, both firms opt for importing.

Many of those who promote environmental management (e.g. Gilbert 1993; W.S. Atkins *et al.* 1994; ISCID 1997) and central government agencies (e.g. ETBPP 1997; DoE 1995) hold that there are clear-cut financial incentives for making improvements. My own surveys have also found commercial considerations to be cited by SMEs as the most important reasons for acting on the environmental agenda (Palmer 1995; 1996b). Yet commercial incentives for action were dominant in only two of the case study organisations. And only one of them—Hampshire Chemical—had been moved to act by the potential for cost reduction.

Barriers to environmental management

Constraints on (management) time and financial resources were advanced as the most important obstacles to making more progress. Only two organisations—Daily Bread and Out of this World—did not report that limits on time represented an important bar to progress. For them, money was the limiting factor. Curiously, these two organisations are the only co-operatives among the case studies. A possible interpretation is that this organisational structure (where, traditionally, profit generation comes second to other objectives) leads inevitably to lower income, which cuts managers' room for manoeuvre on environmental issues. It may also extend the amount of time employees are able to invest in non-core activities.

Again, just two of the case studies—Hampshire Chemical and Westwind—did not mention pressure on finances as an impediment to greater action. Hampshire Chemical is unusual among SMEs in having a very high turnover per employee *and* being very profitable. While Westwind is not in this kind of privileged position—the firm operates in a very narrow market sector where product cost is rarely the primary consideration in purchase decisions—more weight is put on the choice of materials and the aesthetic appeal of their houses.

Communicating environmental performance externally

Three of the case study firms invest resources in communicating their action outside, to external stakeholders (for example, through a published environmental report). This fairly high proportion may be surprising: one would expect only a minority of such small organisations to have sufficient resources, and sufficient incentive, to make public their environmental work. Perhaps a higher proportion of 'managed' small and micro firms take their message outside than is the case for other SMEs.

Use of standards for environmental management

Just one of the case study companies—Hampshire Chemical—has established management systems meeting the requirements of any environmental management system standard: it has been certified to ISO 14001. Hampshire Chemical's wish to follow the standards route, taken together with the less formally managed case studies' rejection of standards, could suggest that formal management is a prerequisite for adopting standards (although the 'sample' size here is too small to offer any more than an indication).

On the other hand, Hampshire's core business could result in very hazardous environmental impacts, and its scale of operations generates sufficient resources to invest in formal documented systems. Ultimately, these characteristics may have been more significant in the company's decisions to adopt standards for environmental management than its style of management. The potential for hazardous emissions constitutes a prompt for proving (to employees, customers, neighbours and financial institutions) that the company has sound environmental management practices. And one way for it to prove the sophistication of its environmental controls is to get third-party approval—through recognised standards.

Yet most SMEs are not like Hampshire Chemical. They do not risk damaging human health through their environmental impacts in the short term. Few of them face strictly policed legislation. There are just 319,000 SME manufacturers in the UK, compared to about 1,800,000 service-sector SMEs (DTI 1998): a ratio of 1 to 5.6. And even among the manufacturers, only a minority has potential for large-scale environmental impacts.

These points could help to explain why the majority of SMEs are reluctant to adopt the standards for environmental management. For the other, less formally managed firms presented here either lack the resources to implement standards or feel that documented management systems are not appropriate for their situation. In essence, their low potential for hazard does not warrant the level of sophistication offered by standards and/or their scale and profitability mean that they are unable to resource the standards' model of environmental management.

Central government schemes

Only three of the case study organisations—Hampshire Chemical, the Ecology Building Society and Westwind—claim to have used any of the central government services aimed at providing information about environmental management to SMEs (see Palmer and France 1997 for a description of the major schemes). Thus, Hampshire Chemical has reportedly made extensive use of the services offered by the Environmental Technology Best Practice Programme (ETBPP), particularly the help-line. It has also benefited from a grant to help pay consultancy fees—delivered through a Training and Enterprise Council (TEC)—and attended workshops aimed at SMEs intending to implement ISO 14001, provided by its local Business Link. So Hampshire Chemical feels very positive about the assistance provided by central government agencies. The Ecology, too, is very pleased with what is offered by the Building Research Establishment (BRE), reportedly using BRE technical documents regularly. However, Westwind's reaction was less favourable. The partners were disappointed by the service received; they said it was 'useless' because the BRE could not supply the information they needed. As for the other four firms, they had not made any use of government-funded schemes. Either they were unaware of what assistance is available—in which case the schemes' methods of promotion may be ineffective—or they do not find what is being offered sufficiently attractive to invest their time in seeking out and/or using it.[2]

This could be explained by these case studies not needing any specific information in order to change. Alternatively, they may have been able to find any information that was required for the change from private sources, such as books and articles, or consultancy help.

Costs arising from environmental management

Given that time and money were cited as the main barriers to environmental management, what did the case studies say about the actual costs of their actions? They were evenly split as to whether money for investment, or management time, was the primary cost of improving environmental management (see Table 3). However, Hampshire Chemical and Castle Project referred to the investment of management time as a secondary cost of their action, making this the most commonly cited cost overall.

None of these case study organisations had attempted to quantify the total costs of their work on environmental management in terms of both time and money: only two of them had made any effort to generate these figures. The first, Shared Earth, measured the financial cost of action, but not how much management time has been absorbed by environmental initiatives, apart from knowing that a single year's environmental accounts took the managing director three months to prepare. The

2 See Bichard's discussion of the shortcomings of assistance to SMEs in Chapter 20.

Name of organisation	Cambridge Daily Bread Co-operative	Hampshire Chemical	Castle Project	Out of this World	Shared Earth	Westwind Oak Buildings	Ecology Building Society
Costs	Management time, pressure on prices	Cash for investment and management time	Cash for investment and management time	Cash for investment	Management time	Some lost sales	Management time
How much time?	Not measured	Not measured	Not measured	Not measured	Three months for accounts alone	Not measured	Not measured
How much money?	Not measured	(£75,000 on all training), plus unmeasured investments	Not measured	Not measured	£4,250	Not measured	Not measured

Table 3: THE COSTS OF IMPROVED ENVIRONMENTAL MANAGEMENT

second, Hampshire Chemical, tracked expenditure on training—claimed to be £1,500 per employee per annum, i.e. £75,000 in total—but this aggregate figure combines environmental training with all other training subjects. Other environmental management overheads, including research and development, investment in technical equipment and management time, while recognised as costs, were not measured.

Overall, the case studies displayed a very limited ability to measure, and then report to outsiders, the total cost of their work on environmental management. This is indicative of the low level of quantified information that most of them have about themselves.

Achievements arising from environmental management

Table 4 summarises the environmental and commercial achievements attributable to the organisations' actions, and compares these achievements with the objectives that had been defined, as cited by interviewees.

The most common outcomes of environmental management are:

- Reducing the amount of waste generated
- Re-using or recycling waste
- Reducing the impact of purchases

Clearly, re-using or recycling waste is linked to reductions in waste volumes, and six of the seven case studies have done both. Benefits in terms of reduced water consumption, and particularly air pollution, are less common. Not one of them has cut air pollution, and only Hampshire Chemical has pruned its use of water. Of course, the ability to reduce air pollution or effluent is dependent on generating these types of pollution in the first place—which not all firms do. So it is not surprising that there are so few reported successes in these areas.

The key point to draw from the upper part of Table 4 (describing environmental objectives and benefits) is that there is no clear trend or pattern of outcomes to come from environmental management in these organisations.[3] They are a heterogeneous set of firms, who have reported widely varying consequences of adopting environmental management.

Four of them said that they won custom as a result of their environmental work, while another one reported that sales have *probably* increased as a consequence. Shared Earth, which claimed to have increased sales, surveyed its shop customers and found that 41% cited the environment or fair trade as their main reasons for shopping in the stores. The Ecology believed that all of its savers, and some of its borrowers, chose it above other building societies because of its environmental stance. But none of the case study organisations quantified the addition to turnover that they could put down to their environmental (or ethical) achievements. Perhaps this is not possible: like the

3 See Hillary's review of benefits and disbenefits from EMS implementation in Chapter 10.

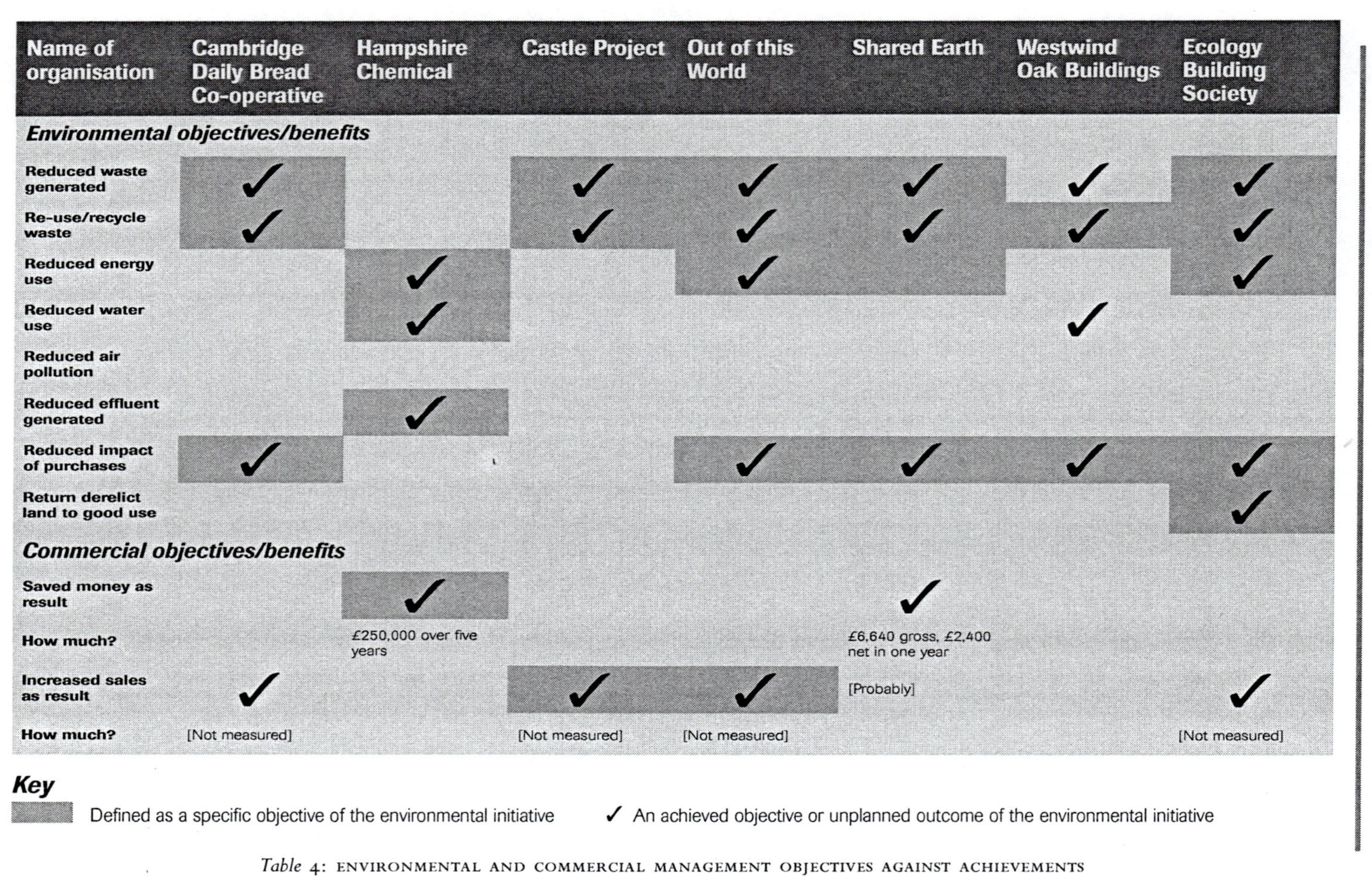

Name of organisation	Cambridge Daily Bread Co-operative	Hampshire Chemical	Castle Project	Out of this World	Shared Earth	Westwind Oak Buildings	Ecology Building Society
Environmental objectives/benefits							
Reduced waste generated	■ ✓		■ ✓	■ ✓	■ ✓	✓	■ ✓
Re-use/recycle waste	■ ✓		■ ✓	■ ✓	■ ✓	■ ✓	■ ✓
Reduced energy use		■ ✓		■ ✓	■		■ ✓
Reduced water use		■ ✓				✓	
Reduced air pollution							
Reduced effluent generated		■ ✓					
Reduced impact of purchases	■ ✓			■ ✓	■ ✓	■ ✓	■ ✓
Return derelict land to good use							■ ✓
Commercial objectives/benefits							
Saved money as result		■ ✓			✓		
How much?		£250,000 over five years			£6,640 gross, £2,400 net in one year		
Increased sales as result	✓		■ ✓	■ ✓	[Probably]		✓
How much?	[Not measured]		[Not measured]	[Not measured]			[Not measured]

Key

■ Defined as a specific objective of the environmental initiative

✓ An achieved objective or unplanned outcome of the environmental initiative

Table 4: ENVIRONMENTAL AND COMMERCIAL MANAGEMENT OBJECTIVES AGAINST ACHIEVEMENTS

largely intangible benefits of 'goodwill' or 'brand value', as entered in some organisations' balance sheets, the long-term effects of a good environmental profile are inordinately difficult to gauge.

Only two of the case study organisations have actually saved money as a result of adopting environmental management—Shared Earth and Hampshire Chemical. Further, both have actually quantified the amount saved. Shared Earth saved just 0.13% of the company's turnover in 1995, while Hampshire Chemical saved an estimated 0.16% of turnover over five years. Based on the case study findings, the tangible financial returns on environmental management for most organisations are minimal. The economic payback alone does not appear to warrant the investment of management time in better environmental management. If this is also true for organisations outside this sample, it may also help explain why, to date, so few SMEs have made the environment a management issue.

Overall, there is a good match between objectives of the various environmental initiatives and their outcomes. However, there are some inconsistencies. One is where improvement has been identified as an objective, but there has not, as yet, been any improvement—for example, Shared Earth has identified improved energy efficiency in stores as an objective, but it has not been able to reduce energy consumption. Another is where an improvement was incidental to the defined objectives—thus, in the case of Daily Bread, the environmental and ethical trading initiatives were not undertaken in order to increase sales, but more custom was said to be an unplanned outcome of their actions. Westwind unintentionally reduced waste as a result of its preference for re-using and recycling building materials. Similarly, for Shared Earth, financial savings were not a particular objective of its waste minimisation and re-use initiatives, but overall (after deducting the total costs of environmental actions) it saved £2,390.

The lower part of Table 4 (on commercial objectives and benefits) shows just how little quantified information about themselves these firms possess.

Conclusions

These case study organisations show very wide variations in how they are organised and what environmental objectives they have achieved. Yet there are some shared characteristics and certain correlations, first between styles of management and second between outcomes of environmental management.

Prompts and barriers to action

The prevailing orthodoxy, crystallised in the literature and commonly cited by government agencies, is that firms improve environmental management because of commercial incentives. But that is not the message to emerge from these case studies. Here, the key prompt for management action was personal commitment in five of the seven organisations: from one or more individuals at senior management level. And this personal commitment was not always to better *environmental* performance alone. The Ecology Building Society combined environmental with some social concerns. In Out of this World and Shared Earth the commitment was part of a broader orientation to the principles of sustainable development, majoring on ethical trading. In Daily Bread, the prompt was broader still: Christianity and caring for the world and the people in it. In these organisations, the environment was one item in a basket of non-commercial concerns, which resulted in it receiving management attention. Only in Westwind was commitment to the environment alone.

For Hampshire Chemical and Castle Project, commercial considerations were in the foreground during the decision to implement environmental management. The former was stimulated into acting by the potential for cost reductions, while the latter sought to diversify by offering new environmental services.

A majority of these organisations has reasons for acting that are different from those cited in the books and articles about environmental management. This underlines the difference between taking a proactive, 'managed' stance—making the environment a management concern—and action that is initiated by general staff, without investment of senior managers' time. Numerous surveys have suggested that cost cutting, risk reduction, or marketing potential, are the main prompts for firms to act. Conversely, in these proactive SMEs, top managers' personal commitment to good environmental practices has been the main prompt for management action in five of the firms (and was an important factor for one of the remaining two). It appears that, while commercial incentives may be necessary to persuade a smaller organisation to take some action, they are only rarely sufficient to put the environment on the boardroom agenda (or equivalent). That is, such commercial pressures may be sufficient to provoke ad hoc action, mainly by general staff, but are probably not enough to persuade top management to spend time working out complicated management systems.

Literature sources and survey findings are more consistent with what the case studies say about the barriers to improving environmental management and performance. In almost every case, the main barrier to action is pressure on resources: either money, time, or both.

Approaches to environmental management

Apart from the personal commitment that was evident in the majority of these case studies, it is difficult to identify other shared characteristics.[4] Organisational structure, decision-making processes and methods of communication all differ between the firms—although there does appear to be a tendency towards top-down decision-making, combined with some level of participation or consultation in important decisions. Further, these small and micro enterprises typically appear to have very effective corporate communication processes. As you would expect, most of them use some form of face-to-face communication, while the largest one adds written memos or other documents. The preference for face-to-face working could be simply a reflection of their small size: in small groups, meetings often prove an effective mechanism for passing information.

4 See Chapter 27.

Standards for environmental management

Only one of the case studies—Hampshire Chemical—has already implemented the standards route to improved environmental management, which is consistent with the assertion, based on survey findings, that only a small proportion of (even active) SMEs have bought into the standards (Palmer 1997). This firm is unlike the others in that it:

- Has potential for large, hazardous environmental effects
- Has resources available for non-core activities (such as environmental management), because it has a high turnover and is very profitable
- Used documented management systems before it started working on formal environmental management

Viewed logically, an organisation must have enough of a prompt—for example, in the form of significant environmental effects—to consider implementing standards. It must also generate sufficient time and money resources to invest in setting up formal systems.[5]

The commercial outcomes of environmental management

All seven case study organisations displayed very limited knowledge of how much they had invested in environmental management. Although they recognised that there had been costs in terms of either management time, or cash for investment, or both, none of them had actually measured both types of cost. This provides a clue to the level of self-knowledge that it is reasonable to expect from other SMEs. Indeed, given that the case studies are all proactive on environmental management, they are actually more likely to have sophisticated monitoring of performance generally than is the case for SMEs that have made less progress.

No financial savings were attributable to environmental management in five of these organisations. Of the remaining two companies the savings were small. This runs against the widely held claims that environmental management saves money. It may also help to explain why economic prompts are cited so rarely as a reason for taking up environmental management.

Recommendations

So what can SME managers and the agencies attempting to work with them learn from these UK case studies? Based on the findings presented here, what can they do to improve their chances of success?

5 See Whalley's discussion in Chapter 9 on the appropriateness of formal EMSs for SMEs.

People working in small and micro enterprises should note that there are alternatives to environmental management systems and published standards for environmental management. These alternatives may well be more appropriate to their needs and management cultures. Alternative approaches currently in use tend to be less formal and require fewer resources, but do not necessarily lead to lower standards of environmental performance.

SME managers should also note that it is impossible to generalise about the action they can take. Firms are so incredibly diverse, and have such different environmental impacts, that there is no single cure for all their problems. This means that, to some extent, the managers of these firms will have to find their own way—to decide for themselves what aspects of their operations show most potential for improvement, and what action they need to take.

So, too, groups trying to work with SMEs on environmental management should take on board the fact that systems and standards are not a universal cure. Smaller firms are heterogeneous, with different needs and aspirations. The low take-up of advice witnessed here may partly reflect the poor match between what is being offered (heavily dependent on management systems) and what small and micro firms want (more practical guidance that can be applied quickly and with a direct impact on environmental performance).

The agencies working with SMEs should be sceptical of the prevailing assumption that proactive SMEs are prompted to act mainly by cost savings. This suggests that the current vogue for promoting environmental management by referring to potential financial gains could be misdirected. Perhaps, by appealing to more personal, less self-centred, motives, senior managers can be wooed into improving their firms' environmental performance. These more personal motives might not be restricted to environmental considerations alone, but might also extend to the broader, social issues that feature in sustainable development.

27

The environmental champion: making a start

Westfield: an SME success story

Liz Walley

Much of what is written about small and medium-sized enterprises (SMEs) and the environment focuses on barriers to environmental progress (e.g. Welford 1994b), contributing to a powerful impression of the challenge of improving environmental performance in the SME sector. In this context it is encouraging and instructive to present a case study about a successful greening initiative which has taken place at Westfield, a small UK insurance/healthcare business employing around 60 staff. The case study has two main aims: first to illustrate what making a start on a 'greening initiative' actually means in practice—i.e. how priorities were established, the kinds of actions actually taken, how they progressed, etc., and, second, to analyse the process involved—what were the key success factors in the greening process?

This analysis focuses particularly on the role of the 'environmental champion' and the tactics and outcomes of the champion's efforts in this case. The lessons from the case study are highlighted and their general applicability examined with a view to assisting those initiating organisational greening processes elsewhere. Since the focus of this case study is on 'making a start', this chapter analyses the actions and process issues during the first phase of Westfield's greening initiatives which took place in 1996–97; progress subsequent to this—from the latter half of 1997 to December 1998—is also provided to illustrate the momentum and pace of greening following the first initiatives.

Company overview

Westfield Contributory Health Scheme was established in the UK in 1919 and provides low-cost healthcare mainly within the city of Sheffield, UK, and surrounding districts. It is established as a charitable trust (a voluntary non-profit association) and is regulated as an insurance company. The company's main office is in Sheffield with additional sales offices in Nottingham and Truro. In mid-1997 the company had 64 employees of which 56 were located at the Sheffield office. Total income—comprising mainly earned premiums—amounted to around £23 million in the 1996 financial year.

The company's main 'product' is a hospital cash plan—offered at five different contribution levels—from which members make claims for hospital, dental, optical, chiropody and other similar health costs. Hospital cash plans are looked upon by members as a means to save toward the costs associated with hospital and treatment therapies and, in this way, to provide security and peace of mind for themselves and their families. As at December 1996, some 268,000 contributors subscribed to the scheme. Claims for reimbursement, received at a rate of 1,367 per day, are normally processed within five working days. The customer profile is mainly socioeconomic groupings Cs and Ds, employed, married and, typically, in manual, skilled work or office/retail environments. Many of these contributors subscribe through a work-based scheme.

Westfield has over 3,000 affiliated companies, most of which operate payroll deductions and an increasing number elect to pay subscriptions on behalf of their employees. Other products offered by Westfield are specialist healthcare plans underwritten for specific companies (e.g. Halifax plc, Hong Kong Bank), a new physiotherapy plan specific to driving instructors, and a personal accident plan.

The hospital cash plan market is estimated to total about three million contributors. Westfield is the second-largest hospital cash plan provider in the UK with around 8% market share. The market leader in this sector is HSA Healthcare with around 24% market share, and there are approximately 30 other schemes providing similar cover to Westfield in Coventry, Leeds, Birmingham and London.

Background to environmental initiatives

The starting point for the various environmental developments initiated during the course of 1997 was an objective in the 1996 three-year Business Plan by which the company wished to move towards a paperless office and reduce the environmental impact of its activities. The Chairman and Managing Director, Graham Moore, has said that the objective should be to have a paperless office (by which is meant the transactions side of the business—processing of membership, claims forms, etc.) by 2000. Westfield has been inspired by the Co-operative Bank's initiatives, both in terms of technological developments that would enable paperless transactions and in terms of its environmental initiatives generally.

While the drive towards the paperless office has come from the Chairman, it is the Corporate Planning Manager (CPM) who has been the champion behind broadening this objective to include a wider environmental agenda. Given the nature of its business, Westfield already supports healthy living initiatives, both within the company and externally in the community in the form of substantial grants and sponsorship. The environmental initiatives have been promoted internally partly as a further dimension to this healthy living focus.

On the technology front, it is envisaged that satellite TV technology being developed currently will communicate with 'intelligent' keyboards in the home or office to enable claims forms, etc. to be processed electronically. Clearly, a company the size of Westfield does not have the resources to pioneer this kind of technological development, but hopes to be able to utilise such technology once it has been developed by bigger organisations, such as the Co-operative Bank. Westfield envisages that its bigger corporate and local government customers could have a type of automated teller machine (ATM) on site through which individual contributors/employees of that organisation could make status requests, claims and so on. This ATM-type development is very much at the conceptual stage at the moment, but Westfield is initiating discussions with various groups of customers/suppliers on better ways to communicate (this is expanded on further, below).

In order to progress its environmental objectives generally, Westfield sought the advice of the National Centre for Business and Ecology (NCBE) based in Manchester. The NCBE was established in 1996 by the Co-operative Bank in partnership with the four universities of Greater Manchester, to provide environmental expertise—in the form of wide-ranging advice, consultancy, training and research—to small and medium-sized companies wishing to make significant environmental improvements. After initial discussions, Westfield asked the NCBE to carry out an environmental review—in other words, a systematic examination of the facility's processes and procedures and a measurement of the company's impact on the environment (Welford and Gouldson 1993: 59).

An environmental review was needed in order to be able to set targets for environmental improvement and to establish baseline data on which improvement in performance could be monitored. This review was carried out in January 1997. At that time the company was developing an environmental policy and had a draft 'environmental practice' document, but no environmental management system (EMS). Other priority areas identified for the NCBE to review, in addition to the environmental review itself, were: information technology (IT) and paper use strategies required for implementation of the paperless office mandate; environmental policy development and implementation; and EMS development.

The environmental review and NCBE recommendations

The environmental review carried out by NCBE in January 1997 looked at the following areas:

- Environmental management (current internal systems and procedures, e.g. draft environment policy)
- Environmental purchasing
- Environmental legislative requirements and compliance
- Paper use and minimisation/recycling options
- Transport
- Recommendations for IT strategy for the paperless office
- Solid waste generation and minimisation/recycling options
- Energy use
- Water use and conservation
- Staff training requirements

This case study will focus only on certain aspects of this review, namely the general sections relating to environmental management, environmental purchasing and legislative compliance and the specific sections on paper, IT and solid waste.

Reproduced below are the recommendations made by NCBE to Westfield relating to specified sections of the environmental review. Also indicated is the priority rating given to each recommendation—*high*, *medium* and *low*.

Environmental management

Overall, NCBE recommended that Westfield adopt an EMS and that, as a minimum, it should establish company-wide monitoring, target-setting and review procedures for key ecological parameters. Staff training will be crucial to the successful achievement of the paperless office and to other requirements set by the policy statement.

Specific recommendations were:

- Westfield's 'environmental practice' statement should be underpinned by a systematic and effective EMS. Such a management system would ensure that the requirements are carried through (*high*).
- As an absolute minimum, systematic collection and monitoring of resource usage should be instigated to allow effective control and measurement against targets set (*high*).
- The introduction of performance tables relating to the monitoring of waste-streams and resource use: for example, the quantities of waste disposed of, waste recycled, paper used, energy used and water used. This would enforce environmental awareness and identify specific obtainable targets in order to reduce environmental effects (*high*).

Environmental purchasing

NCBE recommended that Westfield develop an environmental purchasing policy. This could be incorporated into the development of an EMS with the two being developed simultaneously (*medium*).

Paper

NCBE recommendations were:

- Westfield should reduce the total volume of paper used on a standard measure (perhaps per customer) and institute measures to achieve this in conjunction with heightened staff awareness. It should monitor paper use and provide targets, reports and incentives for staff to participate in paper reduction across all the company's activities (*high*).
- Westfield should systematically work with its suppliers to move towards using 100% post-consumer waste and totally chlorine-free paper (*medium*).
- Westfield should monitor the volume of paper recycled and should set a target for all paper to be recycled. Given that overall paper use should decrease, monitoring of paper recycled will have to be made by reference to the overall volume of paper used (*high*).
- Westfield must consider alternative, less paper-intensive, marketing strategies (*medium*).

Information technology (IT)

NCBE recommended that:

- Westfield should evaluate the use of IT-based systems to reduce the impact of its operations in other areas (including paper and transport) (*high*).
- Westfield must consider the full-life costs (including energy to run and disposal costs) in its criteria for selecting new equipment (*medium*).
- Westfield should implement a policy of buying 'environmentally sound' equipment. Purchasing policies should incorporate environmental criteria for selection of suppliers and equipment (*medium*).
- Westfield should consider re-using and/or recycling its redundant computer equipment (*medium*).

Solid waste

The recommendations were that:

- Westfield should consider introducing whole-life costing for product selection (*medium*).
- The volume and nature of all waste produced should be monitored in order to identify problem areas, set targets and suggest systematic solutions in

accordance with the priorities set out in the waste management hierarchy: reduce, re-use and recycle (*high*).

- Staff awareness of the scale of the waste problem should be raised and they should be motivated to segregate waste and provided with appropriate facilities to do so (*medium*).
- High waste-generating activities should be identified through monitoring, allowing problem areas to be isolated and solutions sought (*medium*).

Staff training

NCBE recommended that Westfield incorporates environmental issues into staff awareness and training (*high*).

Action initiated in the first phase

Given that there was no budget allocated in 1997 for following through with the recommendations in the environmental review, the CPM's criteria for determining what to do first were clearly pragmatic: namely, those initiatives that implied no cost, were straightforward to implement and would cause minimum disruption. These were readily identified and carried out or initiated and included such things as writing the environmental policy, sourcing a new paper supplier and awareness building. Similarly, it was decided to put to the bottom of the Action Plan those recommendations that are judged most difficult, expensive and/or contentious to implement, e.g. action on the use of company cars and paper use in the marketing literature. As at July 1997, the following specific initiatives had been taken.

Environmental policy

An environmental policy statement was prepared and included in the 1996 Annual Report.

Paper

A new paper supplier has been sourced for photocopying paper, the new product being more environmentally friendly than the previous type—not meeting, but moving towards, the full 100% post-consumer waste/totally chlorine-free specification.

The company has initiated steps to reduce paper usage in the medium term in a number of ways:

- Corporate customers have been advised that they will no longer automatically receive monthly statements, unless they specifically request them.
- Smaller corporate customers will be contacted by phone twice a year to enquire about stationery requirements and be asked to send back out-of-date forms.
- Individual members are asked to supply telephone numbers and are encouraged to phone in with their queries and information.

- To support the above, customer service staff have undertaken further training in telephone techniques and a member of the sales staff has transferred to customer service to encourage a more proactive approach to members.
- The company has just introduced a 'twilight shift' of customer service staff, working from 3.30 pm to 6.30 pm to enable easier phone contact with members.

More problematic is the recommendation about measuring paper usage; in order to set targets on reducing paper volumes, Westfield needs to monitor and measure existing usage. Clearly, paper needs differ according to the proportion of new and existing customers and, therefore, some kind of formula needs to be constructed that takes this into account. Similarly, if there is a scheme change this means old documentation needs to be scrapped; again, the measurement system needs to take account of this kind of change. The CPM has made enquiries with other companies on benchmarking, i.e. collecting data on comparable operations, in order to help devise a suitable measurement system.

A further problem encountered is that Westfield's output of recycled materials—cans, paper, glass—is not sufficiently large to warrant regular collections by the recycling companies. Arrangements for collection have to be made on an ad hoc basis every two to three months when volumes have accumulated sufficiently.

Information technology

A new IT system had been specified prior to the environmental review and this new system was to be implemented in the autumn of 1997. It essentially enables enhanced claims and administration facilities and a new sales tracking and prospecting system, but also includes imaging and scanning equipment which will mean claim forms, etc. can be recycled immediately after scanning. The new IT system will link all aspects of the company's operations giving faster response times and higher productivity. It will also provide e-mail to all employees. (In fact, internal communications are not perceived to be an issue by the company in terms of paper usage because, with small numbers of employees, most communication is verbal; memos are rarely used.) This up-to-date and integrated IT system will support some of the telesales-type initiatives outlined above.

Many of the recommendations in the NCBE report will apply to the next technological upgrade—which may incorporate satellite technology developments, as outlined above. Westfield can start to plan towards this vision after the successful implementation of the current IT phase. One specific idea that the company wishes to consider is the feasibility of the new part-time twilight shift ultimately working from home. Another is to start to talk to the company's partners in the medical profession—doctors, opticians, etc.—about how current and future technology could help them to communicate better: for example, by direct computer links, optical character recognition software, etc.

Staff training

Environmental awareness building and training was identified as a high priority in the review and the company has made considerable progress on this front:

- There is now a voluntary environment committee which meets in lunch hours and organises environmental initiatives within the Sheffield office.
- The CPM, together with the environment committee, organised a Green Awareness Week in July 1997. This included talks from a National Park official on what individuals can do to become more environmentally friendly, videos from a neighbouring organisation on what changes they have made in-house, having a specifically 'green' focus to the staff newsletter, articles on notice boards, special recycling collections, etc.
- New staff induction training includes information about the company's environmental policy and practices.

Subsequent developments

Although this case study concentrates on the first phase of greening initiatives, it is useful to record progress since mid-1997 to illustrate the momentum and pace of change that has resulted.

In the second half of 1997 an EMS was drafted and introduced which included appropriate checking and monitoring systems, a budget was secured for an ISO 14001 application in 1998 and the environment committee participated in some weekend conservation activities. In 1998, some environmental measures were included in the office refurbishment, the company registered for and underwent the pre-assessment exercise for ISO 14001 and, in November, it won the *Sheffield Telegraph* Green Business Award. In March 1999, the company achieved ISO 14001 certification.

Analysis of the greening process

To understand the process of greening, this analysis suggests that it is necessary to pay attention to the linkages between the context (i.e. the nature of the business, recent developments, the external environment, etc.), the process and content (or outcomes) of greening (Walley and Stubbs 1998). In other words, this approach views greening as a process in which concerned individuals (environmental champions) construct what becomes recognised as the content of green initiatives from a subtle blend of contextual factors that inspire, enable and constrain that process. Using this framework, a useful case study will highlight and show interrelations between aspects of the context, process and content of a particular initiative. This is, of course, an iterative and circular process, in that the content/outcomes of one greening initiative feed into and become part of the new context for subsequent initiatives. This context–process–content framework will be used to analyse the Westfield story and interpret significant aspects of the story generically (i.e. in terms of transferable concepts) to give it meaning and relevance to those wishing to learn general lessons from this case. These generic cues or transferable concepts are highlighted in the sections below.

Context of greening

It is important first to identify what were the significant favourable or supportive aspects of the context or business environment of Westfield that facilitated the greening initiative and, similarly, what other (possibly negative) factors were there that needed to be handled sensitively or pragmatically. From the information provided above on Westfield, the following factors are identified as significant:

- **Chair offering direction in a time of change**. The drive toward the paperless office came from the Chair of Westfield who generally gave senior management support to these change initiatives, notwithstanding the fact that it was the CPM who was the champion behind broadening out this objective to include the wider environmental agenda (see 'Process' section below).
- **Not a 'dirty' business**. Given the nature of its business, Westfield already supports healthy living initiatives, both within the company and externally in the community in the form of substantial grants and sponsorship. The environmental initiatives have thus been promoted internally partly as a further dimension to this healthy living focus.
- **Availability of affordable environmental know-how**. Westfield sought the advice and assistance of the NCBE whose specific role is to provide environmental expertise to SMEs. NCBE carried out an environmental review of the company in January 1997.
- **Departmental sub-cultures**. The environmental review report made a comprehensive range of prioritised recommendations across all areas of the company's operations. In the first six months after the review the CPM, having no specific budget available at that time, took a pragmatic and sensitive approach in initiating the most straightforward, no-cost recommendations first; similarly, she decided to put to the bottom of the Action Plan those recommendations that were judged to be most difficult, expensive and/or contentious to implement.
- **IT revamp**. Coincidentally, a new IT system, specified before the environmental review, was installed into the company in autumn 1997, which facilitated progress on other environmental objectives.

These five factors are evident from the information provided above on Westfield. Further discussions with Westfield's CPM—as part of the validation process for the write-up of this case study—revealed two further aspects of the context which were perceived to be significant, namely:

- **Umbrella quality culture**. The company has a strong quality/process improvement culture, having recently won a local quality award under the European Foundation for Quality Management's 'Business Excellence Model'. Part of that model relates to 'impact on society' and includes environmental objectives. The CPM's environmental initiatives were seen as part of her role as specified in the quality model. Also, an Investors in People initiative and a second internal goal of promoting teamwork have resulted in what the CPM describes as a culture change toward more employee consultation and involvement in the business. This was perceived as being very helpful to the environmental initiatives because activities such as chairing environmental meetings and working on external environmental projects were used as development tools.

- **Seeking a socially responsible image**. As part of its public relations strategy, the company wants to be seen as a socially responsible company.

The greening process: the role of the environmental champion

The greening process is the main focus of this analysis and for this company the environmental champion (the CPM) was at the centre of it. Significant aspects of the process are perceived and described here, in terms of transferable concepts, as follows:

- **Networking**. For Westfield's CPM, networking has produced valuable contacts with NCBE, local environmental groups (training and conservation work), the local government environmental officer, and benchmarking contacts with local companies. Such networking also helps to sustain the champion's environmental vision.[1]
- **Sense of audience** is about sending different messages to different audiences, according to their priorities, needs, etc. Although this has not been a big issue for a small company such as Westfield, an example is that cost savings resulting from environmental improvements have not been advertised to the workforce as a whole (the environment committee felt it would 'cloud the issue'), but have been communicated to the management team. A further example is that the CPM has acquired an additional title along the way—Environmental Manager. Which job title she uses depends on the audience addressed.
- **Agenda translation**. The most interesting and significant part of the greening process at Westfield has been what is termed here 'agenda translation': that is, taking what was a paperless office agenda, hijacking it and translating it into a green agenda. The term 'hijacking' has a negative connotation (see Welford 1997) but in this analysis we mean it in a positive sense and so have coined the term 'greenjacking' to mean taking another agenda and giving it a 'green spin'. Part of this greenjacking process is interpreting other benefits or gains—such as improved productivity through less paper handling—and promoting them also as environmental gains to build a feeling that progress is achievable.

1 See Chapter 15 on networking and Chapter 17 on mentoring.

Content/outcomes

The outcomes, in a specific sense, are all the environmental gains so far at Westfield: the environmental policy, EMS, reduction in paper usage, ISO 14001 registration, and so on. In a generic sense, these outcomes or content of the greening process are a commitment to systematic environmental performance monitoring and to continuous environmental improvement.

The CPM also identified other aspects of the content of Westfield's greening initiatives so far: namely, employee involvement and participation, and a broader awareness of environmental gains that could be and were being made. These outcomes contribute towards a new organisational context from which further environmental initiatives can be constructed and, interestingly, the CPM commented that new employees, having been exposed to environmental policy and procedures as part of their induction training, seem to embrace the environmental ethos faster than existing ones. It will be interesting to see in what ways the CPM can work with the new context and specifically harness particular aspects of it, to help with the next set of environment objectives. Will there be opportunities for more 'greenjacking'?

Lessons from the case study

This case study seeks to highlight the subtle process through which the new context at Westfield was constructed from the old. The CPM's role as environmental champion was at the heart of this process of networking, both internally and externally, continually maintaining an acute awareness of the agendas, interests and potential contributions of the people with whom she interacted, and translating agendas where necessary to optimise participation and contribution. In many ways this analysis supports the conclusions of the influential research by Post and Altman (1994) on successful environmental change, particularly that the effectiveness of the champion in engaging others rests heavily on expertise, top management support and a strong appreciation for the problems that every business unit faces; and also that environmental change must be visibly connected to the core values of the company if it is to take root in the minds and actions of people throughout the organisation.

The Westfield case study also provides striking evidence for the suggestion posed by Post and Altman that barriers to innovative environmental performance are reduced in organisations that link environmental change programmes to other 'umbrella' organisational change programmes, such as Total Quality Management, which offer a 'delivery system' for achieving change. In Westfield, the champion used the delivery system offered by the Business Excellence model to raise the profile of environmental performance at a time when new working practices were being developed through IT. By exploiting the disruption to routine accompanying the introduction of new systems of work, and appealing directly to the impacts on society element of the accepted Business Excellence model, she effectively managed to 'greenjack' the IT/quality agenda. In so doing, elements of a culture predisposed towards further environmental improvement were also created and an important priority now is to work on ways to sustain this emerging green company culture.

Finally, a question prompted by this case study is whether an environmental champion has to be a special kind of person? To what extent was success at Westfield attributable to the particular skills and personality of this individual and/or the generic cues and tactics identified here? Although personality issues are not addressed here, it is clear that the environmental champion route to greening requires an interpersonally skilled individual—i.e. a person possessing the kind of 'soft' skills that facilitate networking and sensitivity.

Conclusions

A striking aspect of the Westfield case study is how much progress has been made in a relatively short period of time—1996–98—and from a fairly modest start. It would seem that the key aspects of getting started in this case—and which are potentially transferable to other contexts—are as follows:

- **Starting small**. An organisation does not need an ambitious environmental objective and/or commitment before getting started. The original Business Plan objective was to 'reduce the environmental impact of the company's activities'; the actions initiated in the first phase (after the environmental review) were quite modest.
- **Working with the context**. In this case this meant such things as harnessing the paperless office agenda (and top management support of it), the healthy living focus, teamworking initiatives, etc.
- **Networking**. Internally (for awareness and to promote the message) and externally (for expertise, inspiration, benchmarking).
- **Sense of audience**. Acting pragmatically and sensitively in response to different departmental cultures and priorities, e.g. the actions initiated in the first phase.
- **Greenjacking**. Acting opportunistically to piggyback other change initiatives (in this case the IT revamp and the quality programme) and interpreting other benefits and gains as green gains.
- **Sustaining the emerging green culture**. Reinforcing connections to core values (at Westfield, the quality culture, healthy living) and publicising green gains (e.g. with the Green Business Award).

And, linked to all of the above . . .

- **A skilled environmental champion**. An individual with vision and the kind of interpersonal skills to network effectively and operate sensitively and pragmatically with different audiences.

Although there may be various and different routes to 'the greening of business', this case study highlights the role and skills of an environmental champion as being key to the process of greening. It would appear that her central role in the organisational structure facilitated and contributed significantly to the success of these actions. In SMEs, individual champions' roles are more discernible and they have more potential for significant impact.

What, then, are the implications for larger organisations? Post and Altman's research on the greening of US organisations concluded that each of the successful companies had a clearly identifiable environmental champion and that most innovative companies tried to create many champions at different levels throughout the organisation. It would seem that the success factors identified here are potentially relevant and applicable to champions at many different levels in an organisation and are recommended for consideration by those initiating organisational greening processes.

Bibliography

Abrams, S. (1998) 'An Analysis of the Drivers and Barriers for Improving Environmental Performance of Small and Medium-Sized Enterprises in the Industrial Painting Contract Sector with Wider Applications' (MSc thesis; London: Imperial College).

AIST (Agency of Industrial Science and Technology) (1998) *Survey on the Situation of ISO 14001 Registration in Japan* (Tokyo: AIST/MITI).

Andel, T. (1997) 'Information Supply Chain: Set and get your goals', *Transportation and Distribution* 3.2.

Andersen, T.V. (1998) 'Companies in Denmark with an Environmental Management System', presented at the *Article 19 Committee Meeting* (Brussels: Commission of the European Communities, 8 June 1998)

Angel, D.P., and J. Huber (1996) 'Building Sustainable Industries for Sustainable Societies', *Business Strategy and the Environment* 5.3: 127-36.

APEC (Asia Pacific Economic Co-operation) (1997) *Helping* Your *Business Grow: Guide for Small and Medium Enterprises in the APEC Region* (Canada: APEC).

APO (Asian Productivity Organisation) (1995) *ISO 14000 for Small and Medium Enterprises* (Tokyo: APO).

Ashley, S. (1993) 'Designing for the Environment', *Mechanical Engineering* March 1993: 52-5.

Australian Institute of Company Directors (1992) *Environmental Law: A Guide for Directors* (Sydney: Australian Institute of Company Directors).

Ayres, R.U., and L.W. Ayres (1996) *Industrial Ecology: Towards Closing the Materials Cycle* (Cheltenham, UK: Edward Elgar Publishing).

Baggott, R. (1989) 'Regulatory Reform in Britain: The Changing Face of Self-Regulation', *Public Administration* 67: 435-54.

Ball, S., and S. Bell (1995) *Environmental Law* (London: Blackstone Press).

Banerji, R. (1978) 'Average Size of Plants in Manufacturing and Capital Intensity', *Journal of Development Economics* 5: 155-66.

Bannock, G. (1981) *The Economics of Small Firms: Return from the Wilderness* (Oxford, UK: Basil Blackwell).

Bannock, G., and A. Peacock (1989) *Governments and Small Business* (London: Chapman).

BAPEDAL–GTZ (1998) *Cleaner Production in the Textile Industry* (Jakarta: BAPEDAL–GTZ).

Bartone, C., and L. Benavides (1997) 'Local Management of Hazardous Wastes from Small-Scale and Cottage Industries', *Waste Management and Research* 15: 3-21.

Baylis, R. (1998) *Summary of SME Responses to British Standard Institution's ES/1/-/1 Small Firms Panel Survey on Environmental Management Systems* (unpublished).

Baylis, R., L. Connell and A. Flynn (1997) 'Environmental Regulation and Management: A Preliminary Analysis of a Survey of Manufacturing Processing Companies in Industrial South Wales', *Environmental Planning Research* 14 (Cardiff, UK: Environmental Planning Research Unit, University of Wales).

BCAS (Bangladesh Centre for Advanced Studies) (1997) *Study on Textile Dyeing and Printing Industries in Bangladesh: Technical and Socio-economic Survey and Case Studies* (report prepared for ITDG; Dhaka, Bangladesh: BCAS, July 1997).

BCC (British Chambers of Commerce) (1994) *Small Firms Survey. Environment: The View of the Small Firm* (London: BCC).

BCC (British Chambers of Commerce) (1996a) *Small Firms Survey 20: Energy Efficiency* (London: BCC).

BCC (British Chambers of Commerce) (1996b) *Small Firms Survey: Businesses and their Communities* (London: BCC).

Beckerman, W. (1995) *Small is stupid: Blowing the Whistle on the Green* (London: Duckworth).

BECO Environmental Management and Consultancy and Deloitte & Touche Environmental Services (1998) *The Local Authority as a Strategic Intermediary for SMEs' Environmental Performance* (Confidential report to the European Environment Agency, unpublished; Copenhagen, June 1998).
Berger, M., and B. Guillamon (1996) 'Microenterprise Development in Latin America: A View from the Inter-American Development Bank', *Small Enterprise Development* 7.3: 4-16.
Berry, R.J. (1990) 'Environmental Knowledge, Attitudes and Action: A Code of Practice', *Science and Public Affairs* 5: 13-23.
Berry, R.J. (1993) 'An Environmental Ethic', *ECOS* 14.1: 20-26.
Bessant, J., K. Hoffman and S. Meredith (1996) *Process Innovation: A Training Package Using the 'Toolbox' Concept* (Swindon, UK: Economic and Social Science Research Council).
Biekart, J.W. (1994) *De basismetaalindustrie en het doelgroepenbeleid industrie: Analyse van proces en resultaten op weg naar* 2000 (*Basic Metal Industry and the Target Group Industry: Analysis of the Process and Results towards* 2000) (Utrecht: Stichting Natuur en Milieu).
Billatos, S.B., and N.A. Basaly (1997) *Green Technology and Design for the Environment* (Washington, DC/London: Taylor & Francis).
Biller, D., and J.D. Quintero (1995) *Policy Options to Address Informal Sector Contamination in Urban Latin America: The Case of Leather Tanneries in Bogota, Colombia* (LATEN Dissemination Note 14; Washington, DC: The World Bank).
Blackman, A., and G.J. Bannister (1996) 'Community Pressure and Clean Technologies in the Informal Sector: An Econometric Analysis of the Adoption of Propane by Traditional Brickmakers in Cd. Juarez, Mexico', *Journal of Environmental Economics and Management* 35.1: 1-21.
Blackman, A., and G.J. Bannister (1998) *Pollution Control in the Informal Sector: The Ciudad Juárez Brickmakers' Project* (Discussion Paper 98-15; Washington, DC: Resources for the Future).
Boardman, B., K. Lane, M. Hinnells, N. Banks, G. Milne, A. Goodwin and T. Fawcett (1997) *DECADE: Transforming the Cold Market* (Oxford, UK: Energy and Environment Programme, Environmental Change Unit, University of Oxford).
Bolton Report (1971) *Small Firms: Report of the Committee of Inquiry on Small Firms* (London: HMSO).
Bressers, J.Th.A., and L.A. Plettenburg (1996) 'The Netherlands', in M. Jänicke and H. Weidner (eds.), *National Environmental Policies: A Comparative Study of Capacity-Building* (Berlin: Springer Verlag).
Brown, A., and T. van der Wiele (1996) 'A Typology of Approaches to ISO Certification and TQM', *Australian Journal of Management* 21.1: 57-73.
Brugger, N., and J. Timberlake (1994) *The Cutting Edge: Small Business and Progress* (Santiago, Chile: McGraw–Hill).
BSI (British Standards Institution) (1994) *Environmental Management Systems BS 7750: 1994* (London: BSI).
BSI (British Standards Institution) (1996a) *Implementation of ISO 14001: 1996. Specification with Guidance for Use* (London: BSI).
BSI (British Standards Institution) (1996b) *Environmental Management Systems: General Guidelines on Principles, Systems and Supporting Techniques, ISO 14004: 1996* (London: BSI).
Burke, T., and J. Hill (1990) *Ethics, Environment and the Company* (London: Institute of Business Ethics).
Business in the Environment (1995) *Buying into the Environment* (London: Business in the Environment)
Business in the Environment/Coopers & Lybrand (1995) *EC Eco-Management and Audit Scheme (EMAS): Position your Business* (London: Business in the Environment).
Cairncross, F. (1994) 'The Challenge of Going Green', *Harvard Business Review* 72.4 (July/August 1994): 37-50.
Cairncross, F. (1995) *Green, Inc.* (London: Earthscan).
Calkoen, P.T., and K. ten Have (1991) *Bedrijfsinterne milieuzorgsystemen* (*Corporate Environmental Management*) (The Hague/Tilburg, Netherlands: IVA).
Caves, R., and T. Pugel (1980) *Intra-Industry Differences in Conduct and Performance: Viable Strategies in US Manufacturing Industries* (Monograph 2; New York: New York University Graduate School of Business).
CBI (Confederation of British Industry) (1994) *Environment Costs: The Effects on Competitiveness of the Environment, Health and Safety* (London: CBI).
CBI (Confederation of British Industry) (1996) *Generating Growth: An SME Policy Checklist and Agenda* (London: CBI).
CBI (Confederation of British Industry) (1998) *Worth the Risk: Improving Environmental Regulation* (London: CBI).
CCEM (Centre for Corporate Environmental Management) (1997) *Environmental Management Tools for SMEs: A Handbook* (Huddersfield, UK: CCEM).

CEC (Commission of the European Community) (1989) Council Decision 89/490/EEC of 28 July 1989 on the improvement of the business environment and the promotion of the development of enterprises, and in particular small and medium-sized enterprises in the Community (Brussels: CEC).

CEC (Commission of the European Community) (1992) *Towards Sustainability: A European Community Programme of Policy and Action in Relation to the Environment and Sustainable Development* (Com [92] 23 final, II, 27 March 1992; Brussels: CEC).

CEC (Commission of the European Community) (1993) 'Council Regulation (EEC) no 1836/93 of 29 June 1993 allowing the voluntary participation by companies in the industrial sector in a Community Eco-Management and Audit Scheme', *Official Journal of the European Communities* L168.36 (10 July 1993).

CEC (Commission of the European Community) (1996a) 'Council Directive 96/61/EC on Integrated Pollution Prevention and Control', *Official Journal of the European Communities* L257 (10 October 1996).

CEC (Commission of the European Community) (1996b) 'Council Recommendation of 3 April 1996 Concerning the Definition of Small and Medium-Sized Enterprises', *Official Journal of the European Communities* L107.39 (30 May 1996).

CEC (Commission of the European Community) (1996c) 'Telematics for Urban and Rural Areas: A Brochure Concerning Telematics for SMEs, Opportunities for Rural Areas Programme', DG XIII, *http://www.rural-europe.aeidl.be/forada/sme.html.*

CEC (Commission of the European Community) (1997a) *Lessons Learnt by SMEs in their Implementation of the EMAS Regulation* (Brussels: CEC).

CEC (Commission of the European Community) (1997b) *Euromanagement: Environment Pilot Action Position Paper on the Revision of Regulation No 1836/93 (EMAS)* (Brussels: CEC).

CEC (Commission of the European Community) (1998) *Draft Proposal for a Council Regulation Allowing the Participation by Companies in the Industrial Sector in a Community Eco-Management and Audit Scheme* (Brussels: CEC).

CEST (Centre for Exploitation of Science and Technology) (1995) *Waste Minimisation: A Route to Profit and Cleaner Production. Final Report on the Aire and Calder Project* (London: CEST).

CEST (Centre for Exploitation of Science and Technology) (1997) *Dee Catchment Waste Minimisation Project* (London: CEST).

CEST (Centre for Exploitation of Science and Technology) (1998) *The Don Rother Dearne Waste Minimisation Project* (London: CEST).

CFIB (Canadian Federation of Independent Business) (1997) *Market Development Survey Highlights* (Toronto: CFIB).

Charlesworth, K. (1998) *A Green and Pleasant Land? A Survey of Managers' Attitudes to and Experience of Environmental Management* (London: The Institute of Management).

Chemical Week (1997) 'Software lends strategic support', *Chemical Week*, 25 June 1997.

Chestermann, M. (1982) *Small Business* (London: Sweet & Maxwell, 2nd edn).

Chryssides, G.D., and J.H. Kaler (1993) *An Introduction to Business Ethics* (London: Chapman & Hall).

Clarke, P. (1972) *Small Businesses: How they survive and succeed* (Newton Abbot, UK: David & Charles).

Clements, R. (1997) 'ISO 14000: The Environmental Standard', *www.cris.com/~isogroup/14000.html* (accessed 14 August 1997).

Commoner, B. (1990) 'Can capitalists be environmentalists?', *Business and Society Review* 5: 31-35.

Coopers & Lybrand and BiE (Business in the Environment) (1995) *EC Eco-Management and Audit Scheme (EMAS): Positioning your Business* (London: BiE).

Court, P. (1996) *Encouraging the Use of Environmental Management Systems in the Small and Medium-Sized Business Sector* (Oxford, UK: Green College Centre for Environmental Policy and Understanding).

Cox, S.J., and T.R. Cox (1996) *Safety Systems and People* (Oxford, UK: Butterworth–Heinemann).

Dandridge, T.C. (1979) 'Children are not little "grown-ups": Small business needs its own organisational theory', *Journal of Small Business Management* 17.2: 53-57.

Dasgupta, N. (1998) 'Tall Blunders: Delhi's Environment and Industrial Pollution', *Down to Earth* 7.9 (30 September 1998).

Dasgupta, S., A. Mody, S. Roy and D. Wheeler (1995) 'Environmental Regulation and Development: A Cross-Country Empirical Analysis', *World Bank Policy Research Department Working Paper* 1448 (April 1995).

Dasgupta, S., M. Huq, D. Wheeler and C.H. Zhang (1996) 'Water Pollution Abatement by Chinese Industry: Cost Estimates and Policy Implications', *World Bank Policy Research Department Working Paper* 1630 (August 1996).

Dasgupta, S., M. Hettige and D. Wheeler (1998a) 'What improves environmental performance? Evidence from Mexican Industry', *World Bank Development Research Group Working Paper* 1877 (January 1998).
Dasgupta, S., R.E.B. Lucas and D. Wheeler (1998b) *Small Plants, Pollution and Poverty: New Evidence from Brazil and Mexico* (Washington, DC: The World Bank).
de Bruijn, T., and K. Lulofs (1996) *Bevordering van milieumanagement in organisaties (Supporting Environmental Management in Organisations)* (Enschede, Netherlands: Twente University Press).
Deegan, C., and B. Gordon (1996) 'A Study of the Environmental Disclosure Practices of Australian Corporations', *Accounting and Business Research* 26.3: 187-99.
Department of Workplace Relations and Small Business (1998) 'Small Business Sector in Australia', *http://www.dwrsb.gov.au/* (accessed 16 June 1998).
Desoto, H. (1989) *The Other Path: The Invisible Revolution in the Third World* (New York: HarperCollins).
Dobson, A. (1995) *Green Political Thought* (London: Routledge, 2nd edn).
Dodds, O.A. (1997) 'An Insight into the Development and Implementation of the International Environmental Management System ISO 14001', in R. Hillary (ed.), *Environmental Management Systems and Cleaner Production* (Chichester, UK: John Wiley).
Dodge, J. (1998) 'Reassessing Culture and Strategy: Environmental Improvement, Structure, Leadership and Control', in R. Welford (ed.), *Corporate Environmental Management 2: Culture and Organisations* (London: Earthscan): 104-26.
DoE (UK Department of the Environment) (1990) *This Common Inheritance* (Cmnd 1200; London: HMSO).
DoE (UK Department of the Environment) (1995) *EC Eco-Management and Audit Scheme: A Participant's Guide* (London: DoE).
Dreborg, K.H. (1996) 'Essence of Backcasting', *Futures* 28: 813-28.
Drucker, P. (1995) *Managing in a Time of Great Change* (Oxford, UK: Butterworth–Heinemann).
DTI (UK Department of Trade and Industry) Small Firms Statistics Unit (1998) *Small and Medium Sized Enterprise (SME) Statistics for the UK, 1997* (DTI Press Release 597, London: DTI, 29 July 1998).
DTI (UK Department of Trade and Industry) (1999) *Small and Medium Enterprise (SME) Statistics for the United Kingdom, 1998* (Sheffield, UK: SME Statistics Unit, DTI).
Dunlap, R.E. (1997) 'Trends in Public Opinion toward Environmental Issues 1965–1990', in P. McDonagh and A. Prothero (eds.), *Green Management: A Reader* (London: Dryden Press).
E2M (E2 Management Corporation) (1998) 'A Report on SME Interests and Needs', presented at the Sixth Meeting of ISO/TC 207, Toronto, 1998.
Eden, S. (1993) 'The Environment and Business: The Incorporation of Environmental Responsibility into Retail Ethics', in J. Holder *et al.* (eds.), *Perspectives on the Environment: Interdisciplinary Research in Action* (Aldershot, UK: Avebury).
Eden, S. (1996) *Environmental Issues and Business: Implications of a Changing Agenda* (Chichester, UK: John Wiley).
Elliot, D., D. Patton and C. Lenaghan (1996) 'UK Business and Environmental Strategy: A Survey and Analysis of East Midlands Firms' Approaches to Environmental Audit', *Greener Management International* 13 (January 1996): 30-48.
ENDS (Environmental Data Services) (1995) 'The Uphill Struggle of Greening Small Business', *ENDS Report* 250: 21-23.
ENDS (Environmental Data Services) (1996) 'Upbeat Message from Environmental Managers', *ENDS Report* 254: 6.
ENDS (Environmental Data Services) (1997) 'Trends towards Higher Fines in Waste Prosecutions', *ENDS Report* 267: 42-43.
ENDS (Environmental Data Services) (1998) 'ISO 14001 takes off worldwide', *ENDS Report* 287: 13.
Energy Conservation Centre (1996) *Japan Energy Conservation Handbook 1996–1997* (Tokyo: The Energy Conservation Centre).
Environment Agency (Japan) (1997) *White Paper on the Environment in 1996* (Tokyo: Environment Agency).
Environmental Manager (1995) 'SMEs sceptical about waste savings', *Environmental Manager* 2.8: 6.
EPA (US Environmental Protection Agency) (1996) *Environmental Leadership Program Framework Document* (305-f-96-016; Washington, DC: EPA).
Esty, D.C., and M.E. Porter (1998) 'Industrial Ecology and Competitiveness: Strategic Implications for the Firm', *Journal of Industrial Ecology* 2: 35-43.
ETBPP (Environmental Technology Best Practice Programme) (1996) *Attitudes and Barriers to Improved Environmental Performance* (Harwell, UK: ETBPP).
ETBPP (Environmental Technology Best Practice Programme) (1997) 'Cleaner Technology Boosts Profits', *Update* January 1997: 1.

ETBPP (Environmental Technology Best Practice Programme) (1998) *Attitudes and Barriers to Improved Environmental Performance 1998* (Harwell, UK: ETBPP).
Fisher, T., V. Mahajan and A. Singha (1997) *The Forgotten Sector: Non-Farm Employment and Enterprises in Rural India* (London: IT Publications).
Foster, J. (1998) *A User Perspective* (unpublished; Business in the Environment/Green Business Clubs Initiative Convention, London, 26 October 1998).
Fox, M.L. (1996) 'Integration for the Future', *Manufacturing Systems* 14.10: 98.
Fox, W. (1990) *Towards a Transpersonal Ecology: Developing New Foundations for Environmentalism* (Boston, MA: Shambhala).
Fox, W. (1996) 'A Critical Overview of Environmental Ethics', *World Futures* 46: 1-21.
Frijns, J., P. Kirai, J. Malombe and B. van Vliet (1997) *Pollution Control of Small-Scale Metal Industries in Nairobi* (Wageningen, Netherlands: Centre for Urban Environment).
Gandy, M. (1996) 'Crumbling Land: The Postmodernity Debate and the Analysis of Environmental Problems', *Progress in Human Geography* 20.1: 23-40.
Geiser, K., and M. Crul (1996) 'Greening of Small and Medium-Sized Firms: Government, Industry and NGO Initiatives', in P. Groenwegen, K. Fischer, E. Jenkins and J. Schot (eds.), *The Greening of Industry Resource Guide and Bibliography* (Washington, DC: Island Press).
Ghose, A.K. (ed.) (1997) *Mining on a Small and Medium Scale: A Global Perspective* (London: IT Publications).
Gilbert, M. (1993) *Achieving Environmental Management Standards: A Step-by-Step Guide to Meeting BS 7750* (London: Financial Times/Pitman).
Gilbert, R., and R. Harris (1984) 'Competition with Lumpy Investments', *RAND Journal of Economics* 15: 197-212.
GlobeNet (1998) 'Big Blue "Encourages" ISO 14001 Registration of Suppliers', *http://www.iso14000.net/* (accessed 16 June 1998).
GlobeScan (1998) *Sustainable Development Trends* (GlobeScan Syndicated Survey of Experts Report 1998-1; Toronto: Environics International).
Gondrand, F. (1992) *Eurospeak: A User's Guide. The Dictionary of the Single Market* (London: Nicholas Brealey Publishing).
Goodchild, E. (1998) 'The Business Benefits of EMS Approaches' (PhD thesis; Salford, UK: Salford University).
Goss, D. (1991) *Small Business and Society* (London: Routledge).
Gouldson, A., and J. Murphy (1998) *Regulatory Realities: The Implementation and Impact of Industrial Environmental Regulation* (London: Earthscan).
Gouldson, A., and J. Murphy (1998) *Regulatory Realities: The Implementation and Impact of Industrial Environmental Regulation* (London: Earthscan).
Graedel, T.E. (1998) *Streamlined Life-Cycle Assessment* (Upper Saddle River, NJ: Prentice–Hall).
Graedel, T.E., and B.R. Allenby (1995) *Industrial Ecology* (Englewood Cliffs, NJ: Prentice–Hall).
Griffin, P., D. Cunningham, J. Moriarty and B. Ryan (1995) *Factors Concerning the Uptake of Cleaner Technologies by SMEs: An Irish Perspective* (Cork, Eire: Clean Technology Centre).
Groundwork (1998) *Small Firms and the Environment* (Birmingham, UK: The Groundwork Foundation).
Grubb, M., M. Koch, A. Munson, F. Sullivan and K. Thomson (1993) *The Earth Summit Agreements: A Guide and Assessment* (London: Earthscan).
Guthrie, J., and J. Parker (1990) 'Corporate Social Disclosure Practice: A Comparative International Analysis', *Advances in Public Interest Accounting* 3: 159-76.
Haggblade, S., and C. Liedholm (1991) *Agriculture, Rural Labor Markets, and the Evolution of the Rural Non-Farm Economy* (GEMINI Working Paper 19; Bethesda, USA: GEMINI).
Haigh, N., and F. Irwin (eds.) (1990) *Integrated Pollution Control in Europe and North America* (London: The Conservation Foundation and the Institute for European Environmental Policy).
Hamza, A. (1991) *Impacts of Industrial and Small-Scale Manufacturing Wastes on Urban Environment in Developing Countries* (New York: UN Centre for Human Settlements).
Hansen, O.E., B. Sondergard and S. Meredith (1998) 'Environmental Innovations in Small and Medium-Sized Enterprises: Development of Integrated Policy Instruments Related to Innovation and Sector-Specific Conditions', paper presented at the *8th European Environmental Conference: Advances in European Environmental Policy*, London, September 1998.
Hanssen, O.J. (1997) 'Sustainable Industrial Product Systems: Integration of Life-Cycle Assessments in Product Development and Optimisation of Product Systems' (PhD thesis; Norwegian University of Science and Technology, Trondheim; Fredrikstad, Norway: Østfold Research Foundation).
Hardin, G. (1983) 'The Tragedy of the Commons', in T. O'Riordan and K.R. Turner (eds.), *An Annotated Reader in Environmental Planning and Management* (Oxford, UK: Pergamon).

Hartman, R., M. Huq and D. Wheeler (1996) 'Why paper mills clean up: Survey Evidence from Four Asian Countries', *World Bank Policy Research Department Working Paper* 1710 (December 1996).

Hass, J. (1996) 'Environmental ("Green") Management Typologies: An Evaluation, Operationalisation and Empirical Development', *Business Strategy and the Environment* 5.1: 93-107.

Heida, J.F., A.H. Hupkens, J. Janson-Rogers, S. Karreman, J. van der Kolk, H. Senders, P. Stoppelenburg and A. Vloet (1996) *Evaluatie Bedrijfsmilieuzorgsystemen (Evaluation of Corporate Environmental Management Systems)* (The Hague/Tilburg, Netherlands: KPMG/IVA).

Henderson, V.E. (1982) 'The Ethical Side of Business', *Sloan Management Review* 23: 37-47.

Hess, E., and T. Bishophric (1995) 'WasteCap: A Business-to-Business Recycling and Waste Reduction Programme', *Recycling Resources*, November 1995: 29-32.

Hettige, M., M. Huq, S. Pargal and D. Wheeler (1996) 'Determinants of Pollution Abatement in Developing Countries: Evidence from South and Southeast Asia', *World Development* 24.12 (December 1996): 1891-1904.

Hettige, M., M. Mani and D. Wheeler (1998) 'Pollution Intensity in Economic Development: Kuznets Revisited', *World Bank Development Research Group Working Paper* 1876 (February 1998).

Hillary, R. (1995) *Small Firms and the Environment: A Groundwork Status Report* (Birmingham, UK: The Groundwork Foundation).

Hillary, R. (1997a) *The Eco-Management and Audit Scheme: Analysis of the Regulation, Implementation and Support* (London: Imperial College, University of London).

Hillary, R. (1997b) *UK National Co-ordinator's Report on Euromanagement: Environment Pilot Action* (London: Imperial College, University of London).

Hillary, R. (1998) *An Assessment of the Implementation Status of Council Regulation (No 1836/93) Eco-Management and Audit Scheme (EMAS) in the European Union Member States (AIMS-EMAS)* (London: Imperial College, University of London).

Hillary, R. (1999) *Evaluation of Study Reports on the Barriers, Opportunities and Drivers for Small and Medium-Sized Enterprises in the Adoption of Environmental Management Systems* (London: Department of Trade and Industry).

Hoffman, A.J. (1993) 'The Importance of Fit between Individual Values and Organisational Culture in the Greening of Industry', *Business Strategy and the Environment* 2.4: 10-18.

Hoffman, W.M. (1991) 'Business and Environmental Ethics', *Business Ethics Quarterly* 1.2: 169-84.

Holland, L., and J. Gibbon (1997) 'SMEs in the Metal Manufacturing, Construction and Contracting Service Sectors: Environmental Awareness and Actions', *Eco-Management and Auditing* 4: 7-14.

Hollaway, J. (1993) 'Cyanide, Mercury and the Environment in Southern and Eastern Africa', paper presented to *Workshop on Mining and the Environment*, MERN, University of Bath, UK, September 1993.

Holmberg, J. (1998) 'Backcasting: A Natural Step when Making Sustainable Development Operational for Companies', *Greener Management International* 23 (Autumn 1998): 30-51.

Holmberg, J., and K.-H. Robèrt (1997) 'The Systems Conditions for Sustainability: A Tool for Strategic Planning', submitted for publication.

Holmberg, J., K.-H. Robèrt and K.-E. Eriksson (1996) 'Socio-ecological principles for a Sustainable Society: Scientific Background and Swedish Experience', in R. Costanza, S. Olman and J. Martinez-Alier (eds.), *Getting Down to Earth: Practical Applications of Ecological Economics* (International Society of Ecological Economics; Washington, DC: Island Press): 17-48.

Hooper, P.D., and D.C. Gibbs (1995) *Profiting from Environmental Protection: A Manchester Business Survey* (Manchester, UK: Manchester Metropolitan University).

Hough, J. (1982) 'Franchising: An Avenue for Entry into Small Business', in J. Stanworth, A. Westrip, D. Watkins and J. Lewis (eds.), *Perspectives on a Decade of Small Business Research: Bolton Ten Years On* (Aldershot, UK: Gower).

House of Commons Select Committee on Trade and Industry (UK) (1998) *Small and Medium-Sized Enterprises: Sixth Report* (London: HMSO).

Humphreys, N., D.P. Robin, R.E. Reidenbach and D.L. Moak (1993) 'The Ethical Decision Making Process of Small Business Owner/Managers and their Customers', *Journal of Small Business Management* 31.3: 9-22.

Hunt, C.B., and E.R. Auster (1990) 'Proactive Environmental Management: Avoiding the Toxic Trap', *Sloan Management Review*, Winter 1990: 7-18.

Hunt, J., M. Purvis and F. Drake (1997) *Business Constructions of the Environment* (Working Paper 97/12; Leeds, UK: University of Leeds).

Hutchinson, A., and C. Hutchinson (1995) 'Sustainable Regeneration of the UK's Small and Medium-Scale Enterprise Sector: Some Implications of SME Response to BS 7750', *Greener Management International* 9 (January 1995): 74-84.

Hutchinson, A., and F. Hutchinson (1997) *Environmental Business Management: Sustainable Development in the New Millennium* (London: McGraw–Hill).

Hutchinson, A., and I. Chaston (1993) 'Environmental Perceptions, Policies and Practices in the SME Sector: A Case Study', paper presented at *Business Strategy and the Environment Conference*, Bradford, UK, 23–24 September.

Hutchinson, A., and I. Chaston (1995) 'Environment Management in Devon and Cornwall's Small and Medium-Sized Enterprise Sector', *Business Strategy and the Environment* 4: 15-21.

Hutter, B.M. (1986) 'An inspector calls', *British Journal of Criminology* 26: 114-28.

ICC (International Chamber of Commerce) (1991) *Business Charter for Sustainable Development* (Paris: ICC).

ICEM (Institute for Corporate Environmental Mentoring) (1998) *Business Helping Business: White House Corporate Environmental Mentoring Conference Proceedings* (Washington, DC: National Environmental Education and Training Foundation), also available at *http://www.neetf.org/business/reports.html.*

Indonesian Centre for Statistics Bureau (1998) *Report of 1998* (Jakarta: Indonesian Centre for Statistics Bureau).

Indonesian Ministry of Industry and Trade (1997) *Small and Medium Enterprises* (Jakarta: Indonesian Ministry of Industry and Trade).

Industrial Pollution Control Association of Japan (1990) *Environmental Control Regulations in Japan* (Tokyo: Industrial Pollution Control Association).

INEM (International Network for Environmental Management) (1999) *EMAS Tool Kit for SMEs* (Hamburg: INEM).

Institute of Environmental Management Journal (1996) 'ISO 14001: The Facts', *Institute of Environmental Management Journal* 4.2: 6-15.

IoD (Institute of Directors) (1994) *Business Opinion Survey: The Environment* (London: Institute of Directors).

ISCID (Institut Supérieur de Commerce Internationale à Dunkerque) (1997) *Environmental Management Standards: Accelerator or Brake for Business?* (East Grinstead, UK: SGS Yarsley).

ISO (International Organization for Standardization) (1996a) *ISO 14001 Environmental Management Systems: Specification with Guidance for Use* (Geneva: ISO).

ISO (International Organization for Standardization) (1996b) *ISO 14004 Environmental Management System: General Guidelines on Principles, Systems and Supporting Techniques* (Geneva: ISO).

ISO (International Organization for Standardization) (1998) *The ISO Survey of ISO 9000 and ISO 14001 Certificates: Eighth Cycle* (Geneva: ISO).

ISO (International Organization for Standardization) (1999a) *ISO/FDIS 14031 Environmental Performance Evaluation: Guidelines* (Geneva: ISO).

ISO (International Organization for Standardization) (1999b) *ISO/TR 14032 Draft Technical Report Type 3: Environmental Management—Environmental Performance Evaluation. Case Studies Illustrating the Use of ISO 14031* (Geneva: ISO).

ISO/IEC (International Organization for Standardization/International Electrotechnical Commission) (1996) *ISO/IEC Guide 2:1996* (Geneva: ISO/IEC).

Jackson, T., and R. Clift (1998) 'Where's the profit in industrial ecology?', *Journal of Industrial Ecology* 2.1: 3-5.

Jacobs, M. (1991) *The Green Economy* (London: Pluto Press).

James, P. (1998) 'Environmental Performance and the Bottom Line', *Professional Management*, July 1998: 12-13.

Japan Accreditation Board for Conformity Assessment (1998) *Survey on the Situation of ISO 9000 Series Registration in Japan* (Tokyo: Japan Accreditation Board for Conformity Assessment).

Japan Responsible Care Council (1998) *Survey on the Situation of Responsible Care in Japan* (Tokyo: Japan Responsible Care Council).

Johannson, L.E. (1997) *ISO/TC 207 Fifth Meeting in Kyoto: Report on ISO 14001/14004 and their Impact on and Adoption by Canadian Small and Medium-Sized Enterprises* (Georgetown, ON, Canada: E2M, June 1997).

Johnston, N., and A. Stokes (1995) *Waste Minimisation and Clean Technology* (London: Centre for the Exploitation of Science and Technology).

Joyce, P., C. Seaman, S. Black and A. Woods (1996) 'The Social and Environmental Challenge to Small Firms: Managing the Transition to Social Responsiveness', paper presented at the *19th ISBA National Small Firms Conference*, Birmingham, UK.

Karlsson, M. (1997) 'Green Concurrent Engineering Assuring Environmental Performance in Product Development' (AFR report, 163; licentiate thesis; Lund, Sweden: International Institute for Industrial Environmental Economics, Lund University).

Kaye, B., and B. Jacobson (1996) 'Reframing Mentoring', *Training and Development*, August 1996: 44-47.

Kearney, A.T. (1997) 'Extending the Supply Chain', *Chemical Week*, 25 June 1997.

Kemp, R., *et al.* (1996) 'Environmental Training for the SME Sector', paper presented at the *Eco-Management and Audit Conference*, Leeds, UK, 2–3 July 1996.

Kent, L. (1991) *The Relationship between Small Enterprises and Environmental Degradation in the Developing World (with Emphasis on Asia)* (Development Alternatives International/US Agency for International Development, September 1991).

Keyworth, B. (1999) 'Front Line Action', *Environment Action* 18.

Kotas, T.J. (1985) *The Exergy Method of Thermal Plant Analysis* (London: Butterworths).

KPMG (1997a) *UK Environmental Reporting Survey 1996* (London: KPMG).

KPMG Environmental Consulting (1997b) *The Environmental Challenge and Small and Medium-Sized Enterprises in Europe* (The Hague: KPMG).

Leeds Development Agency (1992) *Leeds Economic Development Strategy* (Leeds, UK: Leeds Development Agency).

Leonard-Barton, D. (1995) *Wellsprings of Knowledge: Building and Sustaining the Sources of Innovation* (Boston, MA: Harvard Business School Press).

Lessem, R. (1991) 'Foreword: Green Management', in J. Davis (ed.), *Green Business. Managing for Sustainable Development* (Oxford, UK: Blackwell).

Lindfors, L.-G., K. Christiansen, L. Hoffman, Y. Virtanen, V. Juntilla, O.J. Hanssen, A. Rønning, T. Ekvall and G. Finnveden (1995) *The Nordic Guidelines on Life-Cycle Assessment* (Copenhagen: Nordic Council of Ministers).

Line, M., and T. Vogt (1996) 'Environmental Management Systems: The Challenge to SMEs', paper presented at the *Eco-Management and Audit Conference*, Leeds, UK, 2–3 July 1996.

Little, I.M.D. (1987) 'Small Manufacturing Enterprises in Developing Countries', *World Bank Economic Review* 1: 203-35.

Lucas, R. (1978) 'On the Size Distribution of Business Firms', *Bell Journal of Economics* 9: 508-23.

Lucas, R. (1996) *Pollution Levies and the Demand for Industrial Labour: Panel Estimates for China's Provinces* (Institute for Economic Development Discussion Paper; Boston, MA: Boston University).

Ludevid, M. (1995) 'El canvi global en el medi ambient: Introducciates for China's Provincesnin', *Biblioteca Universitària* 25 (Barcelona: Edicions Proa/Universitat Pompeu Fabra).

Luttropp, C. (1997) 'Design for Disassembly: Environmentally Adapted Product Development based on Prepared Disassembly and Sorting' (PhD thesis; Stockholm: Department of Machine Design, Royal Institute of Technology).

Magrab, E.B. (1997) *Integrated Product and Process Design and Development: The Product Realisation Process* (Boca Raton, FL: CRC Press).

Makower, J. (1998) 'Peer Counselling: The Promise (and Pitfalls) of Business-to-Business Mentoring', *The Green Business Letter* 1: 6-7.

Mallett, E. (1999) *Results of 1999 CFIB Survey on Internet Use among Small and Medium-Sized Firms* (Toronto: Canadian Federation of Independent Business, August 1999).

Manchester Chamber of Commerce and Industry (1998) *Survey of Member Organisations* (unpublished; Manchester, UK).

Mani, M. (1996) 'Environmental Tariffs on Polluting Imports: An Empirical Study', *Environmental and Resource Economics* 7: 391-411.

Mani, M., and D. Wheeler (1997) 'In Search of Pollution Havens? Dirty Industry in the World Economy, 1960–1995' (mimeo; World Bank Development Research Group, April 1997).

Manne, A. (1967) *Investments for Capacity Expansion: Size, Location and Time-Phasing* (Cambridge, MA: MIT Press).

McGrew, A. (1993) 'The Political Dynamics of the "New" Environmentalism', in D. Smith (ed.), *Business and the Environment* (London: Chapman).

Mead, D.C., and C. Liedholm (1998) 'The Dynamics of Micro and Small Enterprises in Developing Countries', *World Development* 26.1: 61-74.

Meredith, S., and T. Wolters (1996) *Environmental Strategic Management and Proactivity* (Occasional Paper, 2; Brighton, UK: University of Brighton Business School).

Meredith, S., and V. Biondi (1997) 'Motivating Mechanisms and Policy Options in the Diffusion of Environmental Technologies within Small and Medium-Sized Companies in the Electroplating Industry: Preliminary Findings of a European Project', paper presented at the *7th International Forum on Technology Management: Challenges for the 21st Century*, Kyoto, Japan, November 1997.

Merritt, J.Q. (1998) 'EM into SME won't go? Attitudes, Awareness and Practices in the London Borough of Croydon', *Business Strategy and the Environment* 7: 90-100.

Merseyside Innovation Centre (1998) *Merseyside Waste Minimisation Demonstration Project: Final Report to Government Office Merseyside* (ERDF 94/354 and ERDF 96/1009; Liverpool, UK, unpublished).
Mills, D. (1990) 'Capacity Expansion and the Size of Plants', *RAND Journal of Economics* 21.4: 555-66.
Mills, D., and L. Schumann (1985) 'Industry Structure with Fluctuating Demand', *American Economic Review* 75: 758-67.
Ministry of Environment, Denmark (1997) *Consolidated Environmental Protection Act* 625, 15 July 1997.
Mitchell, G.(1996) 'Problems and Fundamentals of Sustainable Development Indicators', *Sustainable Development* 4: 1-11.
MITI (Japanese Ministry of International Trade and Industry) (1998) *White Paper on Small and Medium Enterprise* (Tokyo: SMEA, MITI).
Moe, M. (1995) 'Environmental Administration in Denmark', *Environment News* 17 (Copenhagen: Danish Environmental Protection Agency).
Monczka, R.M., and J. Morgan (1997) 'What's wrong with supply chain management?', *Purchasing*, January 1997.
Morgan, J. (1997) 'Integrated Supply Chains: How to make them work', *Purchasing*, 22 May 1997.
MORI (Market Opinion Research International) (1994) *Eco-Management and Audit Scheme, Research Study* (London: Department of the Environment).
MORI (Market Opinion Research International) (1998) *Small Firms and the Environment 1998: Research Study Conducted for the Groundwork Foundation* (London: MORI).
Naess, A. (1986) 'The Deep Ecology Movement: Some Philosophical Aspects', *Philosophical Inquiry* 8.1-2: 10-31.
NALAD (National Association of Local Authorities in Denmark) (1997) *Guide for the Promotion of Cleaner Technology and Responsible Entrepreneurship* (Copenhagen: NALAD).
Nattrass, B., and M. Altomare (1999) *The Natural Step for Business: Wealth, Ecology and the Evolutionary Corporation* (Gabriola Island, Canada/Stony Creek, CT: New Society Publishers).
NCBE (National Centre for Business and Ecology) (1999a) *Business and Ecology Demonstration Project: A Small Company Initiative for the Enhancement of Environmental Performance* (Manchester, UK: NCBE).
NCBE (National Centre for Business and Ecology) (1999b) *Survey of Environmental Training Needs of SMEs* (unpublished; Manchester, UK).
Neale, A. (1997) 'Organisation Learning in Contested Environments: Lessons from Brent Spar', *Business Strategy and the Environment* 6: 93-103.
Nimtech (1999) *Environet 2000: Case Studies* (St Helens, UK: Nimtech).
O'Laoire, D., and R. Welford (1996) 'The EMS in SME', in R. Welford (ed.), *Corporate Environmental Management: Systems and Strategies* (London: Earthscan).
OECD (Organisation for Economic Co-operation and Development) (1997) *Business Development Services for SMEs: Preliminary Guidelines for Donor-Funded Interventions, Summary of the Report to the Donor Committee on Small Enterprise Development* (Paris: OECD).
Ogbonna, E., and B. Wilkinson (1996) 'Information Technology and Power in the UK Grocery Distribution Chain', *Journal of General Management* 22.2.
Oi, W. (1983) 'Heterogeneous Firms and the Organisation of Production', *Economic Inquiry* 21: 147-71.
Ostro, B. (1994) 'Estimating the Health Effects of Air Pollutants: A Method with an Application to Jakarta', *World Bank Policy Research Department Working Paper* 1301 (May 1994).
Overleggroep Chemische Industrie (Consultation Group, Chemical Industry) (1995) *Jaarrapportage 1994* (*Year Report 1994*) (The Hague: VROM).
Paehlke, R. (1995) 'Environmental Values for a Sustainable Society: The Democratic Challenge', in F. Fischer and M. Black (eds.), *Greening Environmental Policy* (London: Chapman).
Pahl, G., and W. Beitz (1996) *Engineering Design: A Systematic Approach* (London: Springer Verlag, 2nd edn).
Pallen, D. (1997) *Environmental Sourcebook for Micro-Finance Institutions* (Ottawa, Canada: Asia Branch, Canadian Agency for International Development).
Palmer, J. (1995) *How green is black? A Survey of Action and Attitudes in the Black Country Environment Initiative* (Cambridge, UK: Eclipse Research Consultants).
Palmer, J. (1996a) *'Greening' SMEs Expert Panel* (Cambridge, UK: Eclipse Research Consultants).
Palmer, J. (1996b) *Angling for Green SMEs: A Survey of Action and Attitudes among SMEs in East Anglia* (Cambridge, UK: Eclipse Research Consultants).
Palmer, J. (1997) *Environmental Management for Smaller Organisations: Executive Summary* (Cambridge, UK: Eclipse Research Consultants).
Palmer, J., and C. France (1997) 'Informing Smaller Organisations about Environmental Management: An Assessment of Government Schemes', *Journal of Environmental Planning and Management* 41.3: 355-74.

Palmer, J., and R. van der Vorst (1996) 'Are standard systems right for SMEs?', *Eco-Management and Auditing* 3.2: 91-96.
Palmer, K., W. Oates and P. Portney (1995) 'Tightening Environmental Standards: The Benefit-Cost or the No-Cost Paradigm?', *Journal of Economic Perspectives* 9.4: 119-32.
Palmer, J., I. Cooper and R. van der Vorst (1996) 'Mapping Out Fuzzy Buzzwords: Who Sits Where on Sustainable Development and Sustainability', *Sustainable Development* 5.2: 87-94.
Pargal, S., and D. Wheeler (1996a) 'Informal Regulation of Industrial Pollution in Developing Countries: Evidence from Indonesia', *Journal of Political Economy* 104.6 (December 1996): 1314-27.
Pargal, S., and D. Wheeler (1996b) 'Informal Regulation of Industrial Pollution in Developing Countries: Evidence from Indonesia', *Journal of Political Economy* 104.6 (December 1996): 1314.
Pearce, D.W., and J.J. Warford (1993) *World without End* (New York: Oxford University Press).
Pedersen, C. (1994) 'Cleaner Technology and Environmental Management at Local Authorities in Denmark', paper presented at the *1st European Roundtable on Cleaner Production Programmes*, Graz, Austria, October 1994.
Petts, J. (1998) 'Trust and Waste Management Information: Expectation versus Observation', *Journal of Risk Research* 1.4: 307-20.
Petts, J. (1999) 'The Regulator-Regulated Relationship and Environmental Protection: Perceptions in Small and Medium-Sized Enterprises', *Environment and Planning C*, in press.
Petts, J., A. Herd and M. O'hEeocha (1998a) 'Environmental Responsiveness, Individuals and Organisational Learning: SME Experience', *Journal of Environmental Planning and Management* 41.6: 711-30.
Petts, J., A. Herd, S. Gerrard and S. Horne (1998b) *Business Attitudes to Environmental Compliance* (Loughborough, UK: Centre for Hazard and Risk Management, Loughborough University).
Petts, J., A. Herd, S. Gerrard and C. Horne (1999) 'The Climate and Culture of Environmental Compliance within SMEs', *Business Strategy and the Environment* 8.1: 14-30.
Pfeffer, J., and G.R. Salancik (1978) *The External Control of Organisations: A Resource-Dependence Perspective* (New York: Harper & Row).
Poole, M., J. Coombs and K. Van Gool (1999) *The Environmental Needs of the Micro Company Sector and the Development of a Tool to Meet Those Needs* (Plymouth, UK: Payback Business Environmental Association for the Southwest).
Porter, M., and C. van der Linde (1994) 'Towards a New Conception of the Environment–Competitiveness Relationship', *Journal of Economic Perspectives* 9.4: 97-118.
Porter, M.E., and C. van der Linde (1995) 'Green and Competitive: Ending the Stalemate', *Harvard Business Review* 73.5: 120-33.
Post, J., and B. Altman (1994) 'Managing the Environmental Change Process: Barriers and Opportunities', *Journal of Organisational Change Management* 7.4: 64-81.
Purvis, M., J. Hunt and F. Drake (1999) 'From Global to Local: Expertise and the Definition of Solutions in the UK Refrigeration Industry', *Geoforum*, in press.
Quinn, J.J. (1997) 'Personal Ethics and Business Ethics: The Ethical Attitudes of Owner/Managers of Small Business', *Journal of Business Ethics* 16.2: 119-27.
Robèrt, K.-H. (1992) *The Necessary Step* (Swedish language; Falun, Sweden: Ekerlids Förlag).
Robèrt, K.-H. (1994) *The Natural Challenge* (Swedish language; Falun, Sweden: Ekerlids Förlag).
Robèrt, K.-H. (1997) 'ICA/Electrolux: A Case Report from 1992', paper presented at the *40th CIES (International Centre for Companies of the Food Trade and Industry) Annual Executive Congress*, 5–7 June 1997, Boston, MA.
Robèrt, K.-H., H. Daly, P. Hawken and J. Holmberg (1997) 'A Compass for Sustainable Development', *International Journal of Sustainable Development and World Ecology* 4: 79-92.
Robinson, D. (1998) 'SMEs and EMAS', *Environmental Policy and Procedures* 33 (special report).
Robinson, G. (1996) 'Survey of Accreditation Bodies, Accredited Certification Bodies and Certified Companies in the UK' (unpublished; London, September 1996).
Robinson, J.B. (1990) 'Future under Glass: A Recipe for People who Hate to Predict', *Futures*, October 1990.
Robison, D.H. (1988) 'Industrial Pollution Abatement: The Impact on the Balance of Trade', *Canadian Journal of Economics* 21: 702-706.
Ross, A., and J. Rowan-Robinson (1997) 'It's good to talk: Environmental Information and the Greening of Industry', *Journal of Environmental Planning and Management* 40.1: 111-24.
Rothwell, R. (1992) 'Successful Industrial Innovation: Critical Success Factors for the 1990s', *R&D Management* 22.3: 221-39.
Rothwell, R., and W. Zegveld (1982) *Innovation and the Small and Medium-Sized Firm* (London: Pinter).
Rowe, J., and D. Hollingsworth (1996) 'Improving the Environmental Performance of Small and Medium-Sized Enterprises: A Study in Avon', *Eco-Management and Auditing* 3: 97-107.

Royal Commission on Environmental Pollution (1976) *Air Pollution Control: An Integrated Approach* (5th report; London: HMSO).
Russell, I.M.J. (1993) 'Principle and Profit: A Partnership for Growth', paper presented at the *16th ISBA National Small Firms Conference*, Nottingham, UK.
Rutherfoord, R., and L.J. Spence (1998) 'Small Businesses and the Perceived Limits of Responsibility: Environmental Issues', paper presented at the *21st ISBA National Small Firms Conference*, Durham, UK.
Ryding, S.-O. (ed.) (1995) *Miljöanpassad Produktutveckling* (Stockholm: Industriförbundet, Förlags AB Industrilitteratur).
Schmidheiny, S., with the Business Council for Sustainable Development (1992) *Changing Course: A Global Perspective on Development and the Environment* (Cambridge, MA: MIT Press).
Schokkaert, E., and J. Eyckmans (1994) 'Environment', in B. Harvey (ed.), *Business Ethics: A European Approach* (London: Prentice–Hall).
Schumacher, E.F. (1973) *Small is beautiful* (London: Blond & Briggs).
Schumacher, E.F. (1989) *Small is beautiful: Economics as if People Mattered* (New York: HarperCollins).
Scott, A. (1998) 'Occupational Health and Safety in SMEs', *Small Enterprise Development* 9.3: 14-22.
Sengenberger, W., G. Loveman and M.J. Piore (eds.) (1990) *The Re-emergence of Small Enterprises: Industrial Restructuring in Industrialised Countries* (Geneva: International Institute of Labour Studies).
SETAC (Society of Environmental Toxicology and Chemistry) (1997) *Simplifying LCA: Just a Cut?* (Brussels: SETAC).
Shayler, M. (1996) 'Minimising Waste, Maximising SME Involvement: A Comparison of Two Approaches to Waste Minimisation', paper presented at the *Eco-Management and Audit Conference*, Leeds, UK, 2–3 July 1996.
Sheldon, C. (1998a) 'EMS into SME will go', *Business Standards*, April 1998: 20-21.
Sheldon, C. (1998b) 'EMS into SME won't go?' *Environmental Policy and Procedures* 34 (special report).
Sheng, P., D. Dornfeld and P. Worhach (1995) 'Integration Issues in Green Design and Manufacturing', *Manufacturing Review* 8: 95-105.
Simon, M., and A. Sweatman (1997) 'Products of a Sustainable Future', paper presented at the *International Sustainable Development Research Conference*, Manchester, UK, 7–8 April 1997 (Manchester: Design for Environment Research Group, Department of Mechanical Engineering, Manchester Metropolitan University).
Small Business Corporation (1998) *Case Study Book on the Establishment of EMS in Smaller Businesses* (Tokyo: Small Business Corporation).
Small Business Development Corporation (1999) 'About the Small Business Development Corporation', *http://www.sbdc.com.au/reference/list.htm* (accessed 30 April 1999).
Smith, A., and R. Kemp (1998) *Small Firms and the Environment 1998: A Groundwork Report* (Birmingham, UK: The Groundwork Foundation).
Smith, D. (1993) 'Towards a Paradigm Shift?', in D. Smith (ed.), *Business and the Environment: Implications of the New Environmentalism* (London: Chapman).
Smith, D. (ed.) (1993) *Business and the Environment: Implications of the New Environmentalism* (London: Chapman).
SMV-analyse (1998) *Små og Mellemstore Virksomheders Holdninger til Miljøkrav, Miljøhensyn og Grønne Afgifter* (Copenhagen: Håndværksrådet).
Springhead Trust (1995) *Small Business Survey* (Shaftesbury, UK:The Springhead Trust).
Stanworth, J., and C. Gray (1991) *Bolton Twenty Years On: The Small Firm in the 1990s* (London: Small Business Research Trust).
Starkey, R. (ed.) (1999) *Environmental Management Tools for SMEs: A Handbook* (Copenhagen: European Environment Agency).
Statistics Canada (1998) 'The Business Registrar', *Small Business Quarterly*, Spring 1998 (Industry Canada).
Stuurgroep Grafische Industrie en Verpakkingsdrukkerijen (Steering Group Printing Industry and Packaging Print Shops) (1997) *Samenvatting en Conclusies van Bedrijfstakoverleg (Evaluation Target-Group Policy)* (The Hague: VROM).
Sunga, I., and A. Marima (1998) *Insiza Mining Study: Towards the Development of an Appropriate and Sustainable Support Programme for the Small and Medium-Scale Mining Sector* (report prepared for Netherlands Development Organisation [SNV], Harare, June 1998).
Sutherland, R.J. (1991) 'Market Barriers to Energy Efficiency Investments', *The Energy Journal* 12.3: 15-34.
Suurland, J. (1994) 'Voluntary Agreements with Industry: The Case of Dutch Covenants', *European Environment* 4.4: 3-7.

Taylor, S.R. (1992) 'Green Management: The Next Competitive Weapon', *Futures*, September 1992: 669-80.

Thompson, J.K., and L. Smith (1991) 'Social Responsibility and Small Business: Suggestions for Research', *Journal of Small Business Management*, January 1991: 30-44.

Tidd, J., J. Bessant and K. Pavitt (1997) *Managing Innovation: Integrating Technological, Market and Organisational Change* (Chichester, UK: John Wiley).

Tietenberg, T. (1997) 'Disclosure Strategies for Pollution Control', paper presented at a conference on *Environmental Implications of Market-Based Policy Instruments*, Gothenburg, November 1997.

Tilley, F. (1998) 'The Gap between the Environmental Attitudes and the Environmental Behaviour of Small Firms: With an Investigation of Mechanical Engineering and Business Services in Leeds' (PhD thesis, Leeds Metropolitan University).

Tilley, F. (1999) 'The Gap between the Environmental Attitudes and the Environmental Behaviour of Small Firms', *Business Strategy and the Environment* 8.4: 238-48.

Tobey, J.A. (1990) 'The Effects of Domestic Environmental Policies on Patterns of World Trade: An Empirical Test', *Kyklos* 43.2: 191-209.

Todd, J.A. (1996) *Streamlining Environmental Life-Cycle Assessment* (New York: McGraw–Hill).

Trevino, L.K., and K.A. Nelson (1995) *Managing Business Ethics* (New York: John Wiley).

Tweede Kamer der Staten-Generaal (Second Chamber, Netherlands) (1988–89) *Bedrijfsinterne Milieuzorg* (*Memorandum on Environmental Management*) (TK 20633 no. 3).

UK Competent Body for the EU Eco-Management and Audit Scheme (1996) *Winning through Environmental Management: Case Study One* (London: Department of the Environment).

UNEP (United Nations Environment Programme) (1994) *Government Strategies and Policies for Cleaner Production* (Paris: UNEP).

UNEP (United Nations Environment Programme) (1995) *Cleaner Production Worldwide* II (Paris: UNEP).

UNEP (United Nations Environment Programme) (1999) *Cleaner Production: A Guide to Sources of Information* (Paris: UNEP).

UNEP TIE (United Nations Environment Programme Division of Technology, Industry and Economics) (1997) *Developing Better Systems for Communicating Environmental Best Practice in Business: Final Report* (Paris: UNEP TIE).

UNIDO (United Nations Industrial Development Organisation) (1995) *Study of Cleaner Production Techniques and Technologies Covering Clusters of Small-Scale Industries in Selected Areas* (Vienna: UNIDO).

University Bocconi (1997) *EMAS Pilot Projects' Third Progress Report from the European Co-ordinator of the 39 Commission DG XI Pilot Projects* (Milan: University Bocconi).

USAID (US Agency for International Development) (1997) 'EP3 Case Studies' (Washington, DC: Enviroene/USAID, *http://es.epa.gov/ep3/ep3.html*).

van Diermen, P. (1997) 'Is small beautiful? The Environmental Impact of Small-Scale Production', *Development Bulletin* 14: 28-30.

van Dijken, K., M. Frey, O. Hansen, E. Lopes, S. Meredith and P. Kalff (1998) *The Adoption of Environmental Innovation by Small and Medium-sized Enterprises: Final Report on the ENVIS Project, EU DG XII* (Zoetermeer, Netherlands: Dutch Ministries of Environment and Economic Affairs).

van Someren, T.C.R., J. van der Kolk, K. ten Have and P.T. Calkoen (1993) *Bedrijfsmilieuzorgsystemen, Tussenevaluatie* (*Evaluation of the Corporate Environmental Management Systems/Inter-evaluation*) (The Hague: KPMG/IVA).

van Weenen, H. (1999) *Design for Sustainable Development: Practical Examples of SMEs* (Dublin: European Foundation).

van Wijngaarden, J. (1995) 'Instruments for Promoting Cleaner Production, Tailored to SMEs?', paper presented at *The Second European Roundtable for Cleaner Production*, Rotterdam, Netherlands, 1–3 November 1995.

von Amsberg, J. (ed.) (1998) *Brazil. Managing Pollution Problems: The Brown Sector Agenda* (Report 16635-BR; The World Bank).

von Weizsäcker, E., A.B. Lovins and L.H. Lovins (1997) *Factor Four: Doubling Wealth, Halving Resource Use* (London: Earthscan).

VROM (Dutch Ministry of Environment) (1997) *Environmental Management: A General View* (The Hague: VROM).

Vyakarnam, S., A. Bailey, A. Myers and D. Burnett (1997) 'Toward an Understanding of Ethical Behaviour in Small Firms', *Journal of Business Ethics* 16.15: 1625-36.

W.S. Atkins, March Consulting Group and Aspects International (1994) *Project Catalyst: Report to the Project Completion Event at Manchester Airport* (London: DTI/The BOC Foundation).

Walley, E.E., and M. Stubbs (1998) 'On the Role of Environmental Champions in the Greening of Organisations: The Case of an SME', *Proceedings of the 1998 Business Strategy and the Environment Conference*, Leeds, UK, September 1998: 251-56.

Wang, H., and D. Wheeler (1996) 'Pricing Industrial Pollution in China: An Econometric Analysis of the Levy System', *World Bank Policy Research Department Working Paper* 1644 (September 1996).

WCED (World Commission on Environment and Development) (1987) *Our Common Future* (Oxford, UK: Oxford University Press).

Weitz, K.A., and A. Sharma (1998) 'Practical Life-Cycle Assessment through Streamlining', *Environmental Quality Management* 7.4 (Summer 1998): 81-87.

Welford, R. (1992) 'Linking Quality and the Environment: A Strategy for the Implementation of Environmental Management Systems', *Business Strategy and the Environment* 1.1: 1-13.

Welford, R. (1994a) 'Barriers to the Improvement of Environment: The Case of the SME Sector', in R. Welford (ed.), *Cases in Environmental Management and Business Strategy* (London: Pitman).

Welford, R. (1994b) *Cases in Environmental Management and Business Strategy* (London: Pitman).

Welford, R. (1995) *Environmental Strategy and Sustainable Development: The Corporate Challenge for the 21st Century* (London: Routledge).

Welford, R. (1997) *Hijacking Environmentalism: Corporate Responses to Sustainable Development* (London: Earthscan).

Welford, R., and A. Gouldson (1993) *Environmental Management and Business Strategy* (London: Pitman).

Welsh, J.A., and J.F. White (1981) 'A small business is not a big business', *Harvard Business Review* 59.4: 18-32.

White, A. (1996) 'Supply Chain Link-up', *Manufacturing Systems* 14.10: 94.

Woods, B. (1996) 'A Public Good, A Private Responsibility', *CERES: The FAO Review* 158 28.2 (March–April 1996): 23-27.

World Bank (China Department) (1997a) *Clear Water, Blue Skies: China's Environment in the 21st Century* (Washington, DC: World Bank).

World Bank (Environment Department) (1997b) *Pollution Prevention and Abatement Handbook* (Washington, DC: World Bank).

WWF (World Wide Fund for Nature) and NatWest Group (1997) *The Better Business Pack: Practical Management Guide* (Godalming, UK: WWF).

Zeffane, R.M., M.J. Polonsky and P. Medley (1995) 'Corporate Environmental Commitment: Developing the Operational Concept', *Business Strategy and the Environment* 3.4: 17-28.

Biographies

Erik Bichard is Director of The Co-operative Bank's National Centre for Business and Ecology. The Centre (set up in association with the four universities of Greater Manchester, UK) advises businesses on the formulation and implementation of sustainable practices. He has been academically and professionally involved in environmental assessment, pollution control and social accounting for over 20 years. He has had an extensive and varied academic career which included four higher degrees in environmental sciences, noise control and land-use planning disciplines.
e.bichard@ncbe.salford.ac.uk

Liana Bratasida is Director of Technical Development of the Environmental Impact Management Agency in Indonesia. She obtained her BSc in entomology from the Bandung Institute of Technology (ITB) in 1974, a Diploma of environmental science and technology from the Institute of Hydraulic Engineering, Delft, the Netherlands in 1979, and graduated with an MSc in environmental biology from ITB in 1987. She lectures on environmental impact assessment and audit. As a policy expert, she is a member of the Indonesian delegation on ISO TC 207 and is actively involved in the UNEP International High Level Seminar, European Roundtable and Asia Pacific Roundtable on Cleaner Production. She has been the project leader for various cleaner production and ISO 14000 projects funded by GTZ (German Government Co-operation), EMDI (Environmental Management Development in Indonesia)/CEPI (Collaborative Environmental Project in Indonesia), NORAD (the Norwegian Agency for Development Co-operation), AUSAID (Australian Aid) and UNEP (United Nations Environment Programme).
bangtek@jkt.mega.net.id

Garrette Clark has, since 1991, worked for the United Nations Environment Programme's Division of Technology, Industry and Economics (UNEP TIE) in the Sustainable Production and Consumption Unit running the information clearing house. Previously, she worked for the US Environmental Protection Agency. She has a Master's degree in public policy from the University of California at Berkeley.
garrette.clark@unep.fr

Susmita Dasgupta PhD is an economist with the Development Economics Research Group at the World Bank. During her six years in the World Bank, she has been involved in policy-related research in industry, international finance and environmental management and is currently conducting research activities in Brazil, Colombia, China, Mexico, Indonesia and Poland. Prior to joining the Bank, she was Visiting Assistant Professor of Economics at American University.
sdasgupta@worldbank.org

Theo de Bruijn PhD is a senior research associate at the Center for Clean Technology and Environmental Policy (CSTM) at the University of Twente in the Netherlands, where he teaches environmental management. His research covers areas such as industrial transformation and European environmental policy. He is the European co-ordinator for the Greening of Industry Network and is a member of the Scientific Planning Committee of the International Human Dimensions Programme on Industrial Transformation.
t.j.n.m.debruijn@cstm.utwente.nl

Charles Duff is the former Business Development Manager for Groundwork National Office. He has a BSc in Chemistry from the University of Durham, UK, and an MBA from Columbia University, New York. He worked in marketing and sales at the British Steel Corporation before becoming Director of Corporate Affairs at Norsk Hydro (UK), a post that he held for more than ten years. Since then, he has worked on secondment for the Department of the Environment and the World Business Council for Sustainable Development.

Tim Fanshawe is the SME Adviser to the Environment Agency. He is a qualified biologist and has a higher degree from Bangor in marine environmental protection. Tim has several years' experience in environmental assessment within industry and, for the last five years, has been responsible for developing environmental protection policy.
tim.fanshawe@environment-agency.gov.uk

Paul Gerrans is lecturer in finance at the School of Finance and Business Economics, Edith Cowan University, Perth, Western Australia and his research interests include environmental economics and investment evaluation. He has written papers in a number of areas including the economic valuation of water resources and their social and environmental uses, as well as the usage of investment evaluation techniques by Australian companies.
p.gerrans@cowan.edu.au

Agneta Gerstenfeld is an environmental consultant with Entropy International, based in Lancaster, UK, where she co-ordinates environmental management system implementation projects and training.
agneta@entropy-international.com

Ruth Hillary PhD is the founder of the Network for Environmental Management and Auditing (NEMA) and member of the British Standards Institution (BSI) Small Firms SME Panel. She is the series editor of the *Business and the Environment Practitioner Series* and co-editor: Europe for *Corporate Environmental Strategy*. Her PhD is *The Eco-Management and Audit Scheme: Analysis of the Regulation, Implementation and Support*, Imperial College, University of London. She acts as a consultant and has worked for the European Commission on EMAS and been project manager on many EU projects, such as the UK national co-ordinator for a Euromanagement: Environment Pilot Action on EMAS in SMEs. She is widely published and is author of *The Eco-Management and Audit Scheme: A Practical Guide* (Earthscan) and editor of *Environmental Management Systems and Cleaner Production* (John Wiley). In 1999, she completed a study for the UK Department of Trade and Industry on the opportunities and barriers facing SMEs implementing environmental management systems.
rhillary@nema.demon.co.uk

Jonathan Hobbs is Executive Director of the Business Council for Sustainable Development (North Sea Region). He was formerly the Co-ordinator of the UNEP Cleaner Production Programme, Paris (1996–99) where he was responsible for the cleaner production programme worldwide. Previous appointments include Chief Corporate Adviser (Environmental Affairs) to Eskom electricity utility, South Africa (1982–96), land-use planner in Botswana (1978–82), and Director of the Industrial Environmental Forum of Southern Africa (1989–96). He was a liaison delegate to the World Business Council for Sustainable Development (1992–97) and represented South Africa on the ISO Technical Committee 207 (1993–96). He co-ordinated business input in the negotiations over the new constitution, environmental policy, legislation and institutions during the transition from apartheid to democracy in South Africa. He is a visiting lecturer at the University of Cape Town and a member of the UK's Institute of Environmental Management Standards and Accreditation Panel. He has a BA (Hons) in land-use planning and an MSc in rural and regional resources planning from the University of Aberdeen, Scotland.
jonhobbs@bcsdnsr.demon.co.uk

John Holmberg PhD is an associate professor. He holds an MSc in engineering physics and a doctorate in physical resource theory. He has directed scientific projects on principles and indicators for sustainability, sustainable use of energy and industrial ecology. He is Scientific Director of and Strategic Adviser to The Natural Step Foundation.
frtjh@fy.chalmers.se

Jane Hunt has been working in the field of environmental knowledge for ten years. She is currently employed as a researcher at Lancaster University, UK's Centre for the Study of Environmental Change. Interests include environmental knowledge and information, adaptation to climate change, and new deliberative processes.
j.hunt@lancaster.ac.uk

Bill Hutchinson is a lecturer in management information systems at the School of Management Information Systems, Edith Cowan University, Perth, Western Australia and his research interests are environmental management and organisational systems. He has written numerous papers on using systems methodologies to manage water catchments, and the problems of scoping complex problems.

Lynn Johannson is president of E2 Management Corporation (E2M), a Canadian management consulting firm specialising in environmental management. E2M helps companies manage their environmental resources as carefully as their cash flow to gain commercial advantage from improved environmental performance. Lynn advises governments and industry internationally on strategies to support the adoption of environmental management systems. She has authored over 30 articles on environmental management and sustainability, and several guides and workbooks on ISO 14001.
etwom@inforamp.net

Robert Kemp is Environmental Development Manager at London Regional Transport. He has a BSc (Hons) in Environmental Studies from the University of Hertfordshire, UK, and is currently completing an MSc in Environmental Management for Business. After graduating, Robert managed the Hertfordshire Business and Environment Association for a period before taking up a post as research assistant to Sir Crispin Tickell at Green College, Oxford.

Robert E.B. Lucas is a Professor of Economics at Boston University. His research focuses primarily on the developing economies, particularly in south and south-east Asia and in southern Africa. He has worked on a number of aspects of labour markets and human resources, including population migration, as well as environmental issues, international trade and industrial development.

Manuel Ludevid Anglada is an economist and historian. He is a visiting scholar at Cornell University and has been Professor at ESADE (a private business school in Barcelona) for 13 years. He is currently Director of the *Environmental Business Management* postgraduate course at the Pompeu Fabra University. He is founder and partner of Echevarría & Asociados, a consultancy specialising in environmental business management.
echevarria@mx3.redestb.es

Kris R.D. Lulofs PhD is a senior researcher at the Center for Clean Technology and Environmental Policy (CSTM) at the University of Twente in the Netherlands. Recent research has focused on topics such as environmental policy, implementation of European legislation, pollution prevention and corporate environmental management. He teaches environmental management within several postgraduate and undergraduate courses. He is the programme manager of the international environmental MBA programmes at CSTM.

Ulrika Lundqvist is a PhD student in the Department of Physical Resource Theory, Chalmers University of Technology, Sweden, and holds an MSc in engineering physics. She has worked in scientific projects on indicators for sustainability and in projects with the aim of including sustainability aspects in the methods of LCA and the 'ecological footprint'.

Sandra Meredith is a Research Fellow at the Centre for Research in Innovation Management (CENTRIM) at the University of Brighton, UK. Much of her work has focused on the relationship between business and the environment. Recent research includes a European-based project which examined the role of SMEs in the environmental innovation process, on which a book is shortly to be published. Her current work is related to promoting competitiveness within SMEs through the development of agile manufacturing strategies.
s.e.meredith@bton.ac.uk

Jason Palmer EngD received his engineering doctorate in environmental technology from Brunel University in 1998. He worked as part of Eclipse Research Consultants for more than five and a half years, examining a wide range of environmental management and sustainable development issues, and spent one year conducting research work at the Martin Centre, Cambridge University, UK. Jason currently writes for the *Building Services Journal*, a magazine that promotes environmental building design.
jason_palmer@tbg.focusnet.co.uk

Charlotte Pedersen is a manager at Deloitte & Touche in Denmark, where she has worked for four years. She has an MSc in political science. Primarily, she collects, evaluates and communicates experiences about environmental management, as well as being involved in implementing environmental management systems in private and public companies. Currently, the focus is on knowledge management and intellectual capital.
charlotte.pedersen@deloitte.dk

Judith Petts PhD is Professor of Environmental Risk Management and Assistant Director of the Centre for Environmental Research and Training, the University of Birmingham, UK, working on interdisciplinary environmental research focused on risk management. She has 18 years' applied research and consultancy experience in environmental risk management, particularly in areas of perceptions and communication and is Director of the ESRC project reported in her chapter in this book.
j.i.petts@bham.ac.uk

Alan Powell has 20 years' industrial experience in commercial, sales and marketing management (with a Diploma in Marketing) in both multinationals and SMEs, and has subsequently gained a further nine years' experience in environmental affairs, specialising in business and community issues. In 1991 he founded the AVBEC green business club, and in 1992 formulated the Environmental Health Check concept for SMEs. He played a leading role in the development of the UK government's business environment association (ACBE) initiative in the early 1990s. In 1995 he developed a vocational training map for environmental management and was a member of the steering group for the development of UK national environmental management standards and National Vocational Qualifications (NVQs).
enviro_mark@yahoo.co.uk

Karl-Henrik Robèrt MD is the founder of the international NGO The Natural Step. The Natural Step bases its work on dialogue and community building around the area of resource theory and sustainable development. He is also an adjunct professor in physical resource theory. He holds an MD and has worked as a cancer scientist at the Karolinska Institute in Stockholm. He has written several books and articles on the environment and sustainability.

Hewitt Roberts is a director of Entropy International, based in Lancaster, UK, and has an MSc in environmental management and policy. He has authored and co-ordinated numerous environmental management projects at national, European and international levels, including EMS implementation, multimedia training, EMIS (environmental management information systems) development and design.

Andrew Scott is Policy Director of the international NGO, Intermediate Technology. With an educational background in economics, he specialises in the economic and environmental aspects of technology change in small enterprises.
andrews@itdg.org.uk

Ann Smith is professor in environmental management at the University of Hertfordshire, UK. She has a BSc (Hons) in Botany from the University of Adelaide, Australia, and a PhD from Birkbeck College, University of London, and has over 25 years' experience in environmental research, training and consultancy. In the past, she has been involved in a number of important projects such as the Royal Society Surface Waters Acidification Programme and a Royal Society Joint Institutional Project which investigated impacts on wildlife and habitats in the Meshchera National Park in Middle Russia. She currently directs projects that involve the provision of training and research support to business and industry in environmental management and waste minimisation, such as the East Anglian Waste Minimisation Project funded through Local Competitiveness Challenge and Environmental Assessment for SMEs funded through the European Social Fund. She has also been involved in the development and delivery of professional training courses in environmental management in the UK, Germany, Hungary and in the Former Soviet Union. Her interest in education and training

continues through a postgraduate distance learning programme, 'Environmental Management for Business', aimed at business and industry. Professor Smith co-ordinates the University's Environmental Strategy.
m.a.smith@herts.ac.uk

Richard Starkey is Senior Researcher at the Centre for Corporate Environmental Management (CCEM), University of Huddersfield, UK. He has recently edited and co-authored a handbook on environmental management tools for SMEs published by the European Environment Agency and has also co-edited the Earthscan *Reader on Business and the Environment*. He sits as a UK representative on the ISO subcommittee responsible for writing ISO 14031, the international standard on environmental performance evaluation.
sbusrs@pegasus.hud.ac.uk

Keikou Terui is Director of the Standards Conformity Division of the Agency of Industrial Science and Technology, Ministry of International Trade and Industry in charge of conformity assessment policy for management systems such as ISO 9000 and ISO 14000 series registration, product certification, and laboratory accreditation.
terui-keikou@miti.go.jp

Fiona Tilley is a Senior Research Associate in the Foresight Research Centre at Durham University Business School. Fiona completed a first degree in accountancy followed by an MSc in environmental management and continued this area of research for her doctoral thesis on corporate environmental management. Her principal research interests are corporate environmental management, small firms and qualitative research.
f.j.tilley@durham.ac.uk

Walter W. Tunnessen III is the Senior Director for Environment and Business Programmes and the Institute for Corporate Environmental Mentoring at the National Environmental Education and Training Foundation (NEETF) in Washington, DC, Prior to joining NEETF, he served as a consultant to the President's Council on Sustainable Development. He has been a project co-ordinator for the Environmental Business Association of New York State, developed and managed a state-wide environmental technology transfer and commercialisation programme in New York, and directed conservation programmes for the Sierra Club's San Diego Office. He holds an MS in environmental management and policy from the Lally School of Management at Rensselaer Polytechnic Institute, an MSEL in environmental law from Vermont Law School, and a BA in social psychology and cultural anthropology from the University of Pennsylvania.
tunnessen@neetf.org

Liz Walley joined the Department of Business Studies at Manchester Metropolitan University (MMU), UK, six years ago and teaches environmental management, business environment and management development. Prior to MMU, her 15 years' work experience was in consultancy, banking and engineering. Her research areas are environmental management and transition in Eastern Europe.
l.walley@mmu.ac.uk

Simon Whalley specialises in the development of tailored and generic responses to environmental issues for business. His particular interest has been in ensuring that smaller businesses are not disadvantaged by the growing stakeholder pressure for information on environmental risks and performance. He has represented the Federation of Small Businesses on environmental management issues at the BSI for nearly four years and is leader of the UK delegation to ISO TC 207 SC/1 environmental management systems. He is a director of Source Environmental Management Ltd and Environmental Risk Rating Ltd, acts as the UK and European 'agent' for the Wildlife and Environment Society of South Africa, writes extensively on environmental and other issues and takes an active role as a director in his family healthcare business.
swhalley@dircon.co.uk

David Wheeler PhD is a lead economist with the Development Economics Research Group at the World Bank. Since joining the World Bank, he has worked on a wide range of environmental issues, particularly in relation to industry and community enforcement. Prior to joining the Bank, he was Professor of Economics at Boston University, where his research interests included technology and human resources.

Abbreviations

ACBE	Advisory Committee on Business and the Environment
AFMA	American Furniture Manufacturers' Association
AIST	Agency of Industrial Science and Technology (Japan)
APEC	Asia Pacific Economic Co-operation
APO	Asian Productivity Organisation
ATM	automated teller machine
BAPEDAL	Environmental Impact Management Agency (Indonesia)
BAPIK	Agency for Development of Small-Scale Industry Development (Indonesia)
BATNEEC	Best Available Techniques Not Entailing Excessive Cost
BCAS	Bangladesh Centre for Advanced Studies
BCC	British Chambers of Commerce
BEDP	Business and Ecology Demonstration Project (UK)
BiC	Business in the Community
BiE	Business in the Environment
BOD	biological oxygen demand
BOE	*Butlletí Oficial de l'Estat* (Spain)
BRE	Building Research Establishment (UK)
BSI	British Standards Institution
CBI	Confederation of British Industry
CCEM	Centre for Corporate Environmental Management (UK)
CEC	Commission of the European Communities
CENTRIM	Centre for Research in Innovation Management (UK)
CFCs	chlorofluorocarbons
CFIB	Canadian Federation of Independent Business
CERES	Coalition for Environmentally Responsible Economics
CEST	Centre for Exploitation of Science and Technology
CMA	Chemical Manufacturers' Association
CNAE	Classificacao Nacional de Atividades Economicas
CO_2	carbon dioxide
COD	chemical oxygen demand
COSHH	Control of Substances Hazardous to Health
CP	cleaner production
CPM	corporate planning manager
CSTM	Centre for Clean Technology and Environmental Policy (Netherlands)
DETR	Department of the Environment, Transport and the Regions (UK)
DG III	European Commission Directorate General III: Industry
DG XI	European Commission Directorate General XI: Environment, Nuclear Safety and Civil Protection
DG XII	European Commission Directorate General XII: Science, Research and Development
DG XXIII	European Commission Directorate General XIII: Enterprise Policy, Distributive Trades, Tourism and Co-operatives

DO	dissolved oxygen
DoE	Department of the Environment (UK), now DETR
DSS	decision-support system
DTI	Department of Trade and Industry (UK)
EA	Environment Agency (UK)
EBRD	European Bank for Reconstruction and Development
EC	European Commission
ECI	environmental condition indicator
EDI	electronic data interchange
EEA	European Environment Agency
ELM	Erhvervslivets Ledelsesforum Miljøfremme (The Business Society Management Forum for Environmental Improvements, Denmark)
EMAS	EC Eco-Management and Audit Scheme
EMS	environmental management system
EMIS	environmental management information system
ENDS	Environmental Data Services (UK)
EP3	Environmental Pollution Prevention Project
EPA	Environmental Protection Agency
EPE	environmental performance evaluation
EPRSC	Engineering and Physical Sciences Research Council
ERDF	European Regional Development Fund
ERP	Enterprise Resource Planning
ESRC	Economic and Social Research Council (UK)
ETBPP	Environmental Technology Best Practice Programme (UK)
ETSU	Environmental Technology Support Unit
EU	European Union
FDIS	Final Draft International Standard
FSB	Federation of Small Businesses (UK)
FTAA	Free Trade Area for the Americas
GAAP	generally accepted accounting principles
GDP	gross domestic product
GEC	Global Environmental Change
GEMI	Global Environmental Management Initiative
GMLC	Greater Manchester, Lancashire and Cheshire (UK)
GTZ	German Agency for Technical Co-operation
HCFCs	hydrofluorocarbons
HSE	health and safety executive
IBGE	Instituto Brasileiro de Geografia e Estatistica
ICC	International Chamber of Commerce
ICEM	Institute For Corporate Environmental Mentoring (USA)
ICPIC	International Cleaner Production Information Clearinghouse
IEC	International Electrotechnical Commission
INE	Instituto Nacional de Ecologia (Mexico)
INEM	International Network for Environmental Management
IoD	Institute of Directors
IPC	Integrated Pollution Control
IPP	integrated product policy
IPPC	Integrated Pollution Prevention and Control
ISCID	Institut Supérieur de Commerce Internationale à Dunkerque
ISIC	International Standard Industrial Classification
ISO	International Organization for Standardization
IT	information technology
ITDG	Intermediate Technology Development Group
JIS	Japanese Industrial Standard
JIT	Just-in-Time
JRC	John Roberts Company (USA)
KEDKE	Central Union of Local Authorities of Greece
LACE	Local Authorities helping Companies implementing EMAS
LCA	life-cycle assessment
LIFE	Community Financial Instrument for the Environment

MAK	Maximum Working Place Concentration
MD	managing director
MIT	Massachusetts Institute for Technology
MITI	Ministry of International Trade and Industry (Japan)
MORI	Market Opinion Research International
MPI	management performance indicators
MRO	maintenance, repair and operations
MSDS	material safety data sheets
MTO	methyl turpentine oil
NAFTA	North American Free Trade Agreement
NALAD	National Association of Local Authorities in Denmark
NCBE	National Centre for Business and Ecology (UK)
NCPC	National Cleaner Production Centre
NEETF	National Environmental Education and Training Foundation (USA)
NEMA	Network for Environmental Management and Auditing
NEPP	National Environmental Policy Plan (Netherlands)
NGO	non-governmental organisation
NIPR	New Ideas in Pollution Regulation
NUTEK	National Board for Industrial and Technical Development (Sweden)
OECD	Organisation for Economic Co-operation and Development
OEM	original equipment manufacturer
OMB	owner-managed business
OPI	operational performance indicator
OSHA	Occupational Safety and Health Administration (USA)
PCP	pentachlorophenol
PPGM	Perussahaan Pelindung Getah (Malaysia)
QA	quality assurance
R&D	research and development
RDA	Regional Development Agency (UK)
SBA	Small Business Administration (US)
SCEEMAS	Small Company Environmental and Energy Management Assistance Scheme (UK)
SCORE	Service Corp Of Retired Executives
SCCPPP	Santa Clara County Pollution Prevention Program (USA)
SETAC	Society of Environmental Toxicology and Chemistry
SHE	safety, health and environment
SEMARNAP	Secrataria del Medio Ambiente, Recursos Naturales y Pesca (Mexico)
SBA	Small Business Administration (USA)
SOP	standard operating procedure
SME	small and medium-sized enterprise
SMEA	Small and Medium Enterprise Agency (Japan)
SNIFF	Sistema Nacional de Informacion de Fuentes Fijas (Mexico)
SO_2	sulphur dioxide
SO_x	sulphuric oxides
TEC	Training and Enterprise Council (UK)
TNC	transnational corporation
TOC	total organic compound
TQM	Total Quality Management
TRI	Toxic Release Inventory
UKAS	United Kingdom Accreditation Service
UMIST	University of Manchester Institute of Science and Technology, UK
UNEP	United Nations Environment Programme
UNEP TIE	United Nations Environment Programme's Division of Technology, Industry and Economics
UNIDO	United Nations Industrial Development Organisation
USAID	United States Agency for International Development
VOC	volatile organic compound
VPP	Voluntary Protection Program (USA)
VPPPA	Voluntary Protection Programs Participants' Association (USA)
VROM	Ministry of Environment (Netherlands)

WBCSD	World Business Council for Sustainable Development
WBRD	World Bank for Reconstruction and Development
WCED	World Commission on Environment and Development
WFEO	World Federation of Engineering Organisations
WWF	World Wide Fund for Nature

Index